ADVANCES IN CHEMISTRY SERIES **231**

Emulsions
Fundamentals and Applications in the Petroleum Industry

Laurier L. Schramm, EDITOR
Petroleum Recovery Institute

Based on a short course developed by the Editor entitled
"Petroleum Emulsions and Applied Emulsion Technology"
and sponsored by the Petroleum Recovery Institute, Calgary.

American Chemical Society, Washington, DC 1992

Library of Congress Cataloging-in-Publication Data

Emulsions: fundamentals and applications in the petroleum industry / Laurier L. Schramm, editor.

    p cm.—(Advances in Chemistry Series : 231.

    "Based on a short course developed by the editor entitled Petroleum Emulsions and Applied Emulsion Technology and sponsored by the Petroleum Recovery Institute, Calgary, Alberta, Canada, December, 1990."

    Includes bibliographical references and index.

    ISBN 0–8412–2006–9

    1. Emusions—Congresses. 2. Petroleum—Congresses. I. Schramm, Laurier Lincoln. II. Petroleum Recovery Institute.   III. Series.

QD1.A355 no. 231
[TP156.E6]
540 s—dc20
[665.5]

91–29680
CIP

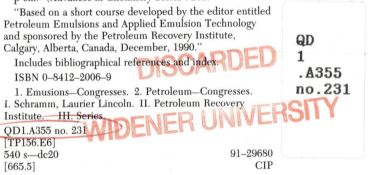

The paper used in this publication meets the minimum requirements of American National Standard for Information Sciences—Permanence of Paper for Printed Library Materials, ANSI Z39.48–1984. ∞

FOREWORD

The ADVANCES IN CHEMISTRY SERIES was founded in 1949 by the American Chemical Society as an outlet for symposia and collections of data in special areas of topical interest that could not be accommodated in the Society's journals. It provides a medium for symposia that would otherwise be fragmented because their papers would be distributed among several journals or not published at all. Papers are reviewed critically according to ACS editorial standards and receive the careful attention and processing characteristic of ACS publications. Volumes in the ADVANCES IN CHEMISTRY SERIES maintain the integrity of the symposia on which they are based; however, verbatim reproductions of previously published papers are not accepted. Papers may include reports of research as well as reviews, because symposia may embrace both types of presentation.

ABOUT THE EDITOR

LAURIER L. SCHRAMM is a senior staff research scientist and group leader for chemical improved oil recovery at the Petroleum Recovery Institute in Calgary, Alberta, Canada. He is also an adjunct associate professor of chemistry at the University of Calgary, where he lectures in applied colloid and interface chemistry. He received his B.Sc. (Hons.) in chemistry from Carleton University in 1976 and a Ph.D. in physical and colloid chemistry in 1980 from Dalhousie University, where he studied as a Killam Scholar. From 1980 to 1988 he held research positions with Syncrude Canada Ltd. in its Edmonton Research Centre.

His research interests have included many aspects of colloid and interface science applied to the petroleum industry, including research into mechanisms of new processes for the enhanced recovery of light to heavy crude oils, such as foam, polymer, and surfactant flooding, and research into fundamental and applied aspects of the hot-water flotation process for recovering bitumen from oil sands. These have involved investigations into the rheology of emulsions, suspensions, hydrocarbons, and oil foams; the electrokinetic properties of dispersed solids, oil, and gases in aqueous solutions; dynamic surface and interfacial tensions and phase attachments; and the reactions and interactions of surfactants in solution. Dr. Schramm is a member of the Chemical Institute of Canada (including serving on the Local Section Executive), the American Chemical Society, and the International Association of Colloid and Interface Scientists. He has published some 30 scientific papers and has 10 patents. This is his first book project.

CONTENTS

GLOSSARY AND INDEXES

PREFACE

Emulsions can be found in almost every part of the petroleum production and recovery process: in reservoirs, produced at wellheads, in many parts of the refining process, and in transportation pipelines. In each case the presence and nature of emulsions can determine both the economic and technical successes of the industrial process concerned. This book is intended to provide an introduction to the nature, occurrence, handling, formation, and breaking of petroleum emulsions. The primary focus is on the applications of the principles and includes attention to practical emulsion problems.

Books available up to now are either principally theoretical (such as the colloid chemistry texts), or they focus on one of the following: emulsions in general (like P. Becher's classic book), emulsions in nonpetroleum areas (as in the food industry), or narrow and highly specialized areas of petroleum emulsions (such as microemulsions). This coverage leaves an obvious gap: the petroleum emulsion area, which has an immense practical importance and, being very diverse, contains a wealth of problems of more fundamental interest.

To address this lack of an introduction to the field of petroleum emulsions, an intensive short course was sponsored by the Petroleum Recovery Institute entitled "Petroleum Emulsions and Applied Emulsion Technology" (first held December 5–6, 1990, in Calgary). This volume, after peer review and extensive revisions, has evolved out of the manual prepared for that short course. A wide range of authors' expertise and experiences have been brought together to yield the first emulsion book that focuses on the occurrence of emulsions in the petroleum industry. This broad range of authors' expertise has allowed for a variety of emulsion problems to be highlighted. These problems serve to emphasize the different methodologies that have been successfully applied to their solution. Having a distinct Canadian perspective, the coverage of types of oil is broadened rather than narrowed. Thus, examples range from emulsions containing light crude oil through heavy-oil emulsions and extend to bituminous emulsions.

This book is aimed at scientists and engineers who may encounter petroleum emulsions, whether in process design, petroleum production, or in the research and development fields. It does not assume a knowledge of colloid chemistry, the initial emphasis being placed on a review of the basic concepts important to understanding emulsions. As such, it is hoped that the book will be of interest to senior undergraduate and graduate students in science and engineering as well because topics such as this are not normally part of university curricula.

Although the aim of the book is to provide an introduction to the field, it does so in a very applications-oriented manner. Thus, the focus of the book is practical rather than theoretical. In a systematic progression, beginning with the fundamental principles of petroleum emulsions, the reader is soon introduced to characterization techniques and flow properties, and finally to industrial practice. Chapters 1–4 present the fundamental concepts and properties involved in emulsions within the context of their occurrence in the petroleum industry. Chapter 1 sets out the basic foundation for all subsequent chapters. Selected areas of special importance are then expanded in Chapter 2 on emulsion stability, Chapter 3 on characterization techniques, and Chapter 4 on rheological properties. All of these use petroleum emulsion examples for illustration, and in most cases cover the latest useful techniques available.

Chapters 5 and 6 begin the progression into more practical emulsion considerations by describing the flow properties of emulsions in pipelines and in porous media. Armed with the necessary tools, the reader is next introduced to some petroleum industry applications of emulsions. Chapters 7 and 8 cover some important areas in which emulsification is a desirable process: in some enhanced oil recovery processes and in petroleum transportation via emulsion pipelining.

The remaining chapters address the converse, and to many, more familiar, situation in which undesirable emulsions must be broken. This treatment progresses from a focus on commercial chemical demulsifiers that may be effective to pilot- and large-scale demulsification practice. A common theme in these chapters is the use of the fundamental concepts in combination with actual commercial and pilot-scale process experiences. Overall, the book shows how to approach making desirable petroleum emulsions, transporting and handling them, and breaking them when they become undesirable*.

Acknowledgments

I express my thanks to all the authors who contributed considerable time and effort to the short course and to the chapters in this book. I am very grateful

---

* One important area of petroleum emulsions that is not addressed concerns the water-in-oil "mousse" emulsions created from oil spills at sea. This topic is covered in detail in a book by J. R. Payne and C. R. Phillips entitled *Petroleum Spills in the Marine Environment: The Chemistry and Formation of Water-in-Oil Emulsions and Tar Balls;* Lewis Publishers: Chelsea, MI, 1985.

for the consistent encouragement of Conrad Ayasse, without whose support this entire project would not have been possible. In addition, the staff of the Petroleum Recovery Institute, especially Gail Swenson, Irene Comer, and Bev Fraser, contributed greatly to the hosting of a successful short course. Throughout the preparation of this book, many valuable suggestions were made by the reviewers of individual chapters, and by the editorial staff of ACS Books, particularly Janet S. Dodd and Cheryl Shanks.

LAURIER L. SCHRAMM
Petroleum Recovery Institute
Calgary, Alberta
Canada T2L 2A6

July 1991

Petroleum Emulsions

Basic Principles

Laurier L. Schramm

Petroleum Recovery Institute, 3512 33rd Street N.W., Calgary, Alberta, Canada T2L 2A6

This chapter provides an introduction to the occurrence, properties, and importance of petroleum emulsions. From light crude oils to bitumens, spanning a wide array of bulk physical properties and stabilities, a common starting point for understanding emulsions is provided by the fundamental principles of colloid science. These principles may be applied to emulsions in different ways to achieve quite different results. A desirable emulsion that must be carefully stabilized to assist one stage of an oil production process may be undesirable in another stage and necessitate a demulsification strategy. With an emphasis on the definition of important terms, the importance of interfacial properties to emulsion making and stability is demonstrated. Demulsification is more complex than just the reverse of emulsion making, but can still be approached from an understanding of how emulsions can be stabilized.

Importance of Emulsions

If two immiscible liquids are mixed together in a container and then shaken, examination will reveal that one of the two phases has become a collection of droplets that are dispersed in the other phase; an emulsion has been formed (Figure 1). Emulsions have long been of great practical interest because of their widespread occurrence in everyday life. Some important and familiar emulsions include those occurring in foods (milk, mayonnaise, etc.), cosmetics (creams and lotions), pharmaceuticals (soluble vitamin and hormone products), and agricultural products (insecticide and herbicide emulsion formulations). In addition to their wide occurrence, emulsions have impor-

0065-2393/92/0231-0001 $13.25/0

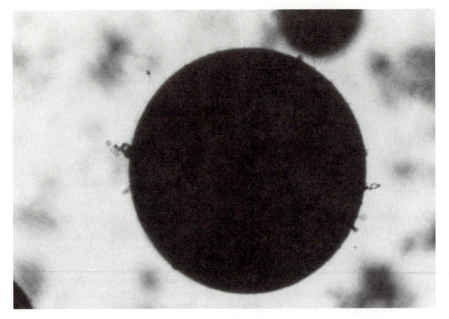

Figure 1. Photomicrograph of an emulsified droplet of a crude oil, dispersed in the aqueous solution that was used to release it from the mineral matrix in which it was originally held. An interfacial film is obvious at the surface of the droplet.

tant properties that may be desirable, for example, in a natural or formulated product, or undesirable, such as an unwanted emulsion in an industrial process. Petroleum emulsions may not be as familiar but have a similarly widespread, long-standing, and important occurrence in industry. Emulsions may be encountered at all stages in the petroleum recovery and processing industry (drilling fluid, production, process plant, and transportation emulsions). This chapter provides an introduction to the basic principles involved in the occurrence, making, and breaking of petroleum emulsions.

Crude oils consist of, at least, a range of hydrocarbons (alkanes, naphthenes, and aromatic compounds) as well as phenols, carboxylic acids, and metals. A significant fraction of sulfur and nitrogen compounds may be present as well. The carbon numbers of all these components range from 1 (methane) through 50 or more (asphaltenes). Some of these components can form films at oil surfaces, and others are surface active. It is perhaps not surprising, then, that the tendencies to form stable or unstable emulsions of different kinds vary greatly among different oils. Because of the wide range of possible compositions, crude oils can exhibit a wide range of viscosities and densities, so much so that these properties are used to distinguish light, heavy, and bituminous crude oils. One set of definitions can be compiled as follows (*1–3*):

Hydrocarbon	Viscosity Range (mPa·s), at Reservoir Temperature	Density Range (kg/m³), at 15.6 °C
Light crude oil	<10,000	<934
Heavy crude oil	<10,000	934–1000
Extra heavy crude oil	<10,000	>1000
Bitumen (tar)	>10,000	>1000

Because the viscosities correspond to ambient deposit temperatures, the variation in these properties over different temperatures is even greater than the table suggests. For example, bitumen in the Athabasca deposit of northern Alberta is chemically similar to conventional oil but has a viscosity, at reservoir temperature, of about 10^6 mPa·s (1 million times greater than that of water). During heating, as part of an oil recovery process such as hot-water flotation or in situ steam flooding, emulsions having a wide range of viscosities can be formed, particularly if they are of the water dispersed in oil type. When these different kinds of oils are emulsified, the emulsions may have viscosities that are much greater than, similar to, or much less than the viscosity of the component oil, all depending on the nature of the emulsion formed.

As shown in Table I, petroleum emulsions may be desirable or undesirable. For example, one kind of oil-well drilling fluid (or "mud") is emulsion based. Here a stable emulsion (usually oil dispersed in water) is used to lubricate the cutting bit and to carry cuttings up to the surface. This emulsion is obviously desirable, and great care goes into its proper preparation.

An emulsion may be desirable in one part of the oil production process and undesirable at the next stage. For example, in the oil fields, an in situ

Table I. Examples of Emulsions in the Petroleum Industry

Occurrence	Usual Type[a]
Undesirable Emulsions	
Well-head emulsions	W/O
Fuel oil emulsions (marine)	W/O
Oil sand flotation process, froth	W/O or O/W
Oil sand flotation process, diluted froth	O/W/O
Oil spill mousse emulsions	W/O
Tanker bilge emulsions	O/W
Desirable Emulsions	
Heavy oil pipeline emulsion	O/W
Oil sand flotation process slurry	O/W
Emulsion drilling fluid, oil-emulsion mud	O/W
Emulsion drilling fluid, oil-base mud	W/O
Asphalt emulsion	O/W
Enhanced oil recovery in situ emulsions	O/W

[a]W/O means water-in-oil; O/W means oil-in-water. *See* the section "Definition and Classification of Emulsions".

emulsion that is purposely created in a reservoir as part of an oil recovery process may change to a different, undesirable type of emulsion (water dispersed in oil) when produced at the wellhead. This emulsion may have to be broken and reformulated as a new emulsion suitable for transportation by pipeline to a refinery. Here, the new emulsion will have to be broken and water from the emulsion removed; otherwise the water would cause processing problems in the refining process.

Emulsions may contain not just oil and water, but also solid particles and even gas. In the large mining and processing operations applied to Canadian oil sands, bitumen is separated from the sand matrix in large tumblers as an emulsion of oil dispersed in water, and then further separated from the tumbler slurry by a flotation process. The product of the flotation process is bituminous froth, an emulsion that may be either water (and air) dispersed in the oil (primary flotation) or the reverse, oil (and air) dispersed in water (secondary flotation). In either case, the emulsions must be broken and the water removed before the bitumen can be upgraded to synthetic crude oil, but the presence of solid particles and film-forming components from the bitumen can make this removal step very difficult.

Some emulsions are made to reduce viscosity so that an oil can be made to flow. Emulsions of asphalt, a semisolid variety of bitumen dispersed in water, are formulated to be both less viscous than the original asphalt and stable so that they can be transported and handled. In application, the emulsion should shear thin and break to form a suitable water-repelling roadway coating material. Another example of emulsions that are formulated for lower viscosity with good stability are those made from heavy oils and intended for economic pipeline transportation over large distances. Here again the emulsions should be stable for transport but will need to be broken at the end of the pipeline.

Finally, many kinds of emulsions pose difficult problems wherever they may occur. For example, crude oil when spilled on the ocean tends to become emulsified in the form of "chocolate mousse" emulsions, so named for their color and semisolid consistency. These water-in-oil emulsions with high water content tend to be quite stable due to the strong stabilizing films that are present. Mousse emulsions increase the quantity of pollutant and are usually very much more viscous than the oil itself.

All of the petroleum emulsion applications or problems just discussed have in common the same basic principles of colloid science that govern the nature, stability, and properties of emulsions. The widespread importance of emulsions in general and scientific interest in their formation, stability, and properties have precipitated a wealth of published literature on the subject. This chapter provides an introduction and is intended to complement the other chapters in this book on petroleum emulsions. A good starting point for further basic information is one of the classic texts: Becher's *Emulsions: Theory and Practice* (4) or Sumner's *Clayton's Theory of Emulsions and*

Their Technical Treatment (5) and numerous other books on emulsions (6–11). Most good colloid chemistry texts contain introductory chapters on emulsions (12–14), and some chapters in specialist monographs (15, 16) give much more detailed treatment of advances in specific emulsion areas.

Emulsions as Colloidal Systems

Definition and Classification of Emulsions. Colloidal droplets (or particles or bubbles), as they are usually defined, have at least one dimension between about 1 and 1000 nm. Emulsions are a special kind of colloidal dispersion: one in which a liquid is dispersed in a continuous liquid phase of different composition. The dispersed phase is sometimes referred to as the internal (disperse) phase, and the continuous phase as the external phase. Emulsions also form a rather special kind of colloidal system in that the droplets often exceed the size limit of 1000 nm. In petroleum emulsions one of the liquids is aqueous, and the other is hydrocarbon and referred to as oil. Two types of emulsion are now readily distinguished in principle, depending upon which kind of liquid forms the continuous phase (Figure 2):

1. oil-in-water (O/W) for oil droplets dispersed in water
2. water-in-oil (W/O) for water droplets dispersed in oil

This kind of classification is not always appropriate. For example, O/W/O denotes a multiple emulsion containing oil droplets dispersed in aqueous droplets that are in turn dispersed in a continuous oil phase. The type of emulsion that is formed depends upon a number of factors. If the ratio of phase volumes is very large or very small, then the phase having the smaller volume is frequently the dispersed phase. If the ratio is closer to 1, then

Oil-in-water (O/W)   Water-in-oil (W/O)

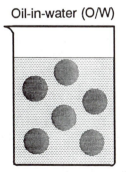   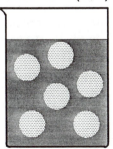

Figure 2. The two simplest kinds of emulsions. The droplet sizes have been greatly exaggerated.

other factors determine the outcome. Table I lists some simple examples of petroleum emulsion types.

Two very different broad types of colloidal dispersions have been distinguished since Graham invented the term "colloid" in 1861. Originally, colloids were subdivided into lyophobic and lyophilic colloids (if the dispersion medium is aqueous then the terms hydrophobic and hydrophilic, respectively, are used). Lyophilic colloids are formed spontaneously when the two phases are brought together, because the dispersion is thermodynamically more stable than the original separated state. The term lyophilic is less frequently used in modern practice because many of the dispersions that were once thought of as lyophilic are now recognized as single-phase systems in which large molecules are dissolved. Lyophobic colloids, which include all petroleum emulsions other than the microemulsions, are not formed spontaneously on contact of the phases because they are thermodynamically unstable compared with the separated states. These dispersions can be formed by other means, however. Most petroleum emulsions that will be encountered in practice contain oil, water, and an emulsifying agent. The emulsifier may comprise one or more of the following: simple inorganic electrolytes, surfactants, macromolecules, or finely divided solids. The emulsifying agent may be needed to reduce interfacial tension and aid in the formation of the increased interfacial area with a minimum of mechanical energy input, or it may be needed to form a protective film at the droplet surfaces that acts to prevent coalescence with other droplets. These aspects will be discussed later; the resulting emulsion may well have considerable stability as a metastable dispersion.

Most kinds of emulsions that will be encountered in practice are lyophobic, metastable emulsions. However, there remain some grey areas in which the distinction between lyophilic and lyophobic dispersions is not completely clear. A special class of aggregated surfactant molecules termed "micelles" and the microemulsions of extremely small droplet size are usually but not always considered to be lyophilic, stable, colloidal dispersions and will be discussed separately.

Stability.   A consequence of the small droplet size and presence of an interfacial film on the droplets in emulsions is that quite stable dispersions of these species can be made. That is, the suspended droplets do not settle out or float rapidly, and the droplets do not coalesce quickly. Some use of the term stability has already been made without definition.

Colloidal species can come together in very different ways. In the definition of emulsion stability, stability is considered against three different processes: creaming (sedimentation), aggregation, and coalescence. Creaming is the opposite of sedimentation and results from a density difference between the two liquid phases. In aggregation two or more droplets clump together, touching only at certain points, and with virtually no change in total surface area. Aggregation is sometimes referred to as flocculation or coagulation. In coalescence two or more droplets fuse together to form a single larger unit with a reduced total surface area.

In aggregation the species retain their identity but lose their kinetic independence because the aggregate moves as a single unit. Aggregation of droplets may lead to coalescence and the formation of larger droplets until the phases become separated. In coalescence, on the other hand, the original species lose their identity and become part of a new species. Kinetic stability can thus have different meanings. An emulsion can be kinetically stable with respect to coalescence but unstable with respect to aggregation. Or, a system could be kinetically stable with respect to aggregation but unstable with respect to sedimentation or flotation.

In summary, lyophobic emulsions are thermodynamically unstable but may be relatively stable in a kinetic sense. Stability must be understood in terms of a clearly defined process.

Microemulsions.   In some systems the addition of a fourth component, a cosurfactant, to an oil–water–surfactant system can cause the interfacial tension to drop to near-zero values, easily on the order of 10^{-3} to 10^{-4} mN/m; low interfacial tension allows spontaneous or nearly spontaneous emulsification to very small droplet sizes, ca. 10 nm or smaller. The droplets can be so small that they scatter little light; the emulsions appear to be transparent and do not break on standing or centrifuging. Unlike coarse emulsions, these microemulsions are usually thought to be thermodynamically stable. The thermodynamic stability is frequently attributed to transient negative interfacial tensions, but this hypothesis and the question of whether microemulsions are really lyophilic or lyophobic dispersions are areas of some discussion in the literature (*17*). As a practical matter, microemulsions can be formed, have some special qualities, and can have important applications.

Microemulsions can form the basis for an enhanced oil recovery (EOR) process (*18–20*). In an oil-containing reservoir, the relative oil and water saturations depend upon the distribution of pore sizes in the rock as follows. The capillary pressure (P_c), or pressure difference across an oil–water interface spanning a pore, is

$$P_c = 2\gamma \cos \theta / r_p \tag{1}$$

where γ is the oil–water interfacial tension; θ is the contact angle, which is the angle measured through the water phase at the point of oil–water–rock contact; and r_p is the pore radius. The basis for this equation is discussed further in a later section. In an idealized water-wet reservoir, the interfacial tension is fixed at some value, and the contact angle is zero. An analogy can be drawn with the rise of water in capillary tubes of differing radii. In a reservoir consisting of such capillary tubes that contain water and oil, water will be imbibed most strongly into the smallest radius pores, displacing any oil present in them, until the hydrostatic and capillary pressures in the system balance. The largest pores will retain high oil contents. Now as water is injected during a secondary recovery process, the applied water pressure increases and the larger pores will imbibe more water, displacing oil, which may be recovered at producing wells. There is a practical limit to the extent that the applied pressure can be changed by pumping water into a reservoir, however, so that after water-flooding some residual oil will still be left in the form of oil ganglia trapped in the larger pores where the viscous forces of the driving water-flood could not completely overcome the capillary forces holding the oil in place.

The ratio of viscous forces to capillary forces correlates well with residual oil saturation and is termed the capillary number (N_c). One formulation of the capillary number is

$$N_c = \eta v / \gamma \qquad (2)$$

where η and v are the viscosity and velocity, respectively, of the displacing fluid; and γ is the interfacial tension. The functional form of the correlation is illustrated in Figure 3. During water-flooding, N_c is about 10^{-6}, and at the

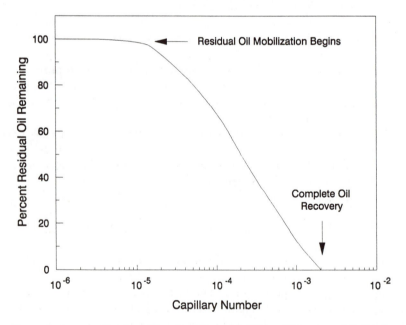

Figure 3. A generalized capillary number correlation. (Courtesy of K. Taylor, Petroleum Recovery Institute, Calgary.)

end of the water-flood the residual oil saturation is still around 45%. How could a tertiary recovery process be designed so that the remaining oil could be recovered? Lowering the residual oil saturation requires increasing the capillary number. This increase could be done by raising the viscous forces, that is, viscosity and velocity, but in practice the desired orders-of-magnitude increase will not be achieved. But, adding a suitable surfactant and cosurfactant to the water will decrease the interfacial tension from about 30 mN/m by 4 orders of magnitude and thereby increase the capillary number to about 10^{-2}.

The micelles present also help to solubilize the released oil droplets; hence, this process is sometimes referred to as micellar flooding. The emulsions can be formulated to have moderately high viscosities that help to achieve a more uniform displacement front in the reservoir; this uniform front gives improved sweep efficiency. Thus, a number of factors can be adjusted when using a microemulsion system for enhanced oil recovery. These are discussed in detail in Chapter 7.

Making Emulsions.   Much of this chapter is concerned with emulsion properties and stability, and as a practical matter chemists frequently have to contend with already-formed emulsions. Nevertheless, a few comments on how emulsions may be made are appropriate. The breaking of emulsions will be discussed later.

Emulsions of any significant stability contain oil, water, and at least one emulsifying agent. The emulsifying agent may lower interfacial tension and thereby make it easier to create small droplets. Another emulsifying agent may be needed to stabilize the small droplets so that they do not coalesce to form larger droplets, or even separate out as a bulk phase. Just a straightforward casual mixing of these components seldom, however, produces an emulsion that persists for any length of time. In the classical method of emulsion preparation, the emulsifying agent is dissolved into the phase in which it is most soluble, after which the second phase is added, and the whole mixture is vigorously agitated. The agitation is crucial to producing sufficiently small droplets, and frequently, after an initial mixing, a second mixing with very high applied mechanical shear forces is required. This latter mixing can be provided by a propeller-style mixer, but more commonly a colloid mill or ultrasound generator is employed.

A method requiring much less mechanical energy uses phase inversion (*see also* the discussion of phase inversion temperature in the section "Emulsifying Agents"). For example, if ultimately a W/O emulsion is desired, then a coarse O/W emulsion is first prepared by the addition of mechanical energy, and the oil content is progressively increased. At some volume fraction above 60–70%, the emulsion will suddenly invert and produce a W/O emulsion of much smaller water droplet sizes than were the oil droplets in the original O/W emulsion.

Physical Characteristics of Emulsions

Appearance. Not all emulsions exhibit the classical "milky" opaqueness with which they are usually associated. A tremendous range of appearances is possible, depending upon the droplet sizes and the difference in refractive indices between the phases. An emulsion can be transparent if either the refractive index of each phase is the same, or alternatively, if the dispersed phase is made up of droplets that are sufficiently small compared with the wavelength of the illuminating light. Thus an O/W microemulsion of even a crude oil in water may be transparent. If the droplets are of the order of 1-μm diameter, a dilute O/W emulsion will take on a somewhat milky-blue cast; if the droplets are very much larger, the oil phase will become quite distinguishable and apparent.

Physically the nature of the simple emulsion types can be determined by methods such as

- **Texture.** The texture of an emulsion frequently reflects that of the external phase. Thus O/W emulsions usually feel watery or creamy, and W/O emulsions feel oily or greasy. This distinction becomes less evident as the emulsion viscosity increases, so that a very viscous O/W emulsion may feel oily.

- **Mixing.** An emulsion readily mixes with a liquid that is miscible with the continuous phase. Thus, milk (O/W) can be diluted with water, and mayonnaise (W/O) can be diluted with oil. Usually, an emulsion that retains a uniform and milky appearance when greatly diluted is more stable than one that aggregates upon dilution (*15*).

- **Dyeing.** Emulsions are most readily and consistently colored by dyes soluble in the continuous phase.

- **Conductance.** O/W emulsions usually have a very high specific conductance, like that of the aqueous phase itself, but W/O emulsions have a very low specific conductance. A simple test apparatus is described in reference 15.

- **Inversion.** If an emulsion is very concentrated, it will probably invert when diluted with additional internal phase.

- **Fluorescence.** If the oil phase fluoresces, then fluorescence microscopy can be used to determine the emulsion type as long as the droplet sizes are larger than the microscope's limit of resolution (>0.5 μm).

Emulsions do not always occur in the idealized form of droplets of one phase dispersed in another. The occurrence of multiple emulsions, of the types O/W/O and W/O/W, has already been mentioned. Petroleum emulsions may also occur within another type of colloidal dispersion. For example, in a gas-flooding enhanced oil recovery process one of the ways to improve the areal sweep efficiency, that is, to maximize the amount of the reservoir contacted by injected fluids, is to inject the gas as part of a foam. However, most such foams are destabilized by contact with even small amounts of crude oil. The mechanism of destabilization appears (*21*) to involve emulsification of the oil into droplets that are small enough to permit their passage inside the foam's lamellar structure. Such emulsified oil droplets are shown in Figure 4. Once inside the foam lamellae, the oil droplets have a destabilizing effect on the foam by penetrating through and possibly spreading over the aqueous–gas interface. The limiting step, however, is apparently the emulsification and imbibition of oil into the foam.

Droplet Sizes.   As stated previously, colloidal droplets are between about 10^{-3} and 1 μm in diameter, and in practice, emulsion droplets are often larger (e.g., the fat droplets in milk). In fact, emulsion droplets usually

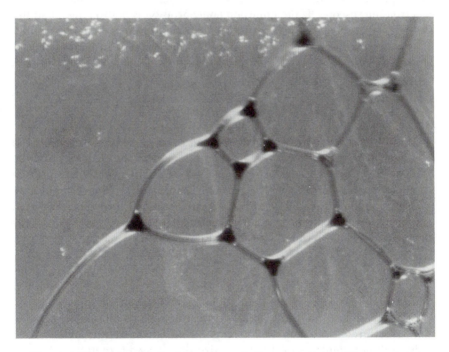

Figure 4. Photomicrograph of an enhanced oil recovery process foam containing emulsified crude-oil droplets. The droplets have traveled within the narrow lamellae to accumulate and sometimes coalesce in the plateau borders of the foam, where they are held preferentially. The presence of such emulsified oil droplets in the foam structure has a destabilizing effect on the foam.

have diameters greater than 0.2 μm and may be larger than 50 μm. Emulsion stability is not necessarily a function of droplet size, although there may be an optimum size for an individual emulsion type. Characterizing an emulsion in terms of a given droplet size is very common but generally inappropriate because there is inevitably a size distribution. The size distribution is usually represented by a histogram of sizes, or, if there are sufficient data, a distribution function.

In some emulsions, a droplet size distribution that is heavily weighted toward the smaller sizes will represent the most stable emulsion. In such cases changes in the size distribution curve with time yield a measure of the stability of the emulsions. The droplet size distribution also has an important influence on the viscosity. For electrostatically or sterically interacting droplets, emulsion viscosity will be higher when droplets are smaller. The viscosity will also be higher when the droplet sizes are relatively homogeneous, that is, when the droplet size distribution is narrow rather than wide (4).

If the droplet size is large enough, then optical microscopy can be used to determine the size and size distribution. Emulsions with somewhat smaller droplet sizes can be characterized by using cryogenic-stage scanning electron microscopy. If the emulsion concentration is not too high, and the droplets are very small, light scattering can yield droplet size information. When a beam of light enters an emulsion, some light is absorbed, some is scattered, and some is transmitted. Many dilute, fine emulsions show a noticeable turbidity given by

$$I_t / I_0 = \exp(-\tau l) \tag{3}$$

where I_t is the intensity of the transmitted beam, I_0 is the intensity of the incident beam, τ is turbidity, and l is the length of the path through the sample. From Rayleigh theory, the intensity of light scattered from each droplet depends largely on its size and shape and on the difference in refractive index between the droplet and the medium. For an emulsion, each spherical droplet scatters light having an intensity I_d at a distance x from the droplet, according to the following relationship:

$$I_d / I_0 \propto r^6 / x^2 \lambda^4 \tag{4}$$

where λ is the wavelength of the light and r is the droplet radius.

The scattering intensity is proportional to $1/\lambda^4$, so blue light ($\lambda = 450$ nm) is scattered much more than red light ($\lambda = 650$ nm). With incident white light, a dilute emulsion of 0.1–1-μm size droplets will, therefore, tend to appear blue when viewed at right angles to the incident light beam. If the droplets are smaller than 50 nm or so, the emulsion will appear to be transparent.

These approaches to determining droplet size distributions are discussed in detail in Chapter 3.

Conductivity. Conductivity can be used to distinguish O/W from W/O emulsions because the conductivity is very high when the aqueous phase is continuous and conductivity is very low when oil is the continuous phase. Of the numerous equations proposed (*4*) to describe the conductivity of emulsions (κ_E), two are cited here for illustration. If the conductivity of the dispersed phase (κ_D) is much smaller than that of the continuous phase (κ_C), $\kappa_C \gg \kappa_D$,

$$\kappa_E = \frac{8\kappa_C(2 - \phi)(1 - \phi)}{(4 + \phi)(4 - \phi)} \tag{5}$$

where ϕ is the dispersed-phase volume fraction. If, on the other hand, the conductivity of the dispersed phase (κ_D) is much greater than that of the continuous phase (κ_C), $\kappa_C \ll \kappa_D$,

$$\kappa_E = \frac{\kappa_C(1 + \phi)(2 + \phi)}{(1 - \phi)(2 - \phi)} \tag{6}$$

Further discussion of emulsion conductivity and some practical examples for emulsions flowing in pipelines are given in Chapter 5.

Rheology. *Bulk Viscosity Properties.* The rheological properties of an emulsion are very important. High viscosity may be the reason that an emulsion is troublesome, a resistance to flow that must be dealt with, or a desirable property for which an emulsion is formulated. The simplest description applies to Newtonian behavior in laminar flow. The viscosity, η, is given in terms of the shear stress, τ, and shear rate, $\dot{\gamma}$, by:

$$\tau = \eta\dot{\gamma} \tag{7}$$

where η has units of millipascal seconds. Many colloidal dispersions, including the more concentrated emulsions, do not obey the Newtonian equation. For non-Newtonian fluids, the coefficient of viscosity is not a constant, but is itself a function of the shear rate, thus:

$$\tau = \eta(\dot{\gamma})\dot{\gamma} \tag{8}$$

A convenient way to summarize the flow properties of fluids is by plotting flow curves of shear stress versus shear rate (τ versus $\dot{\gamma}$). These curves can be categorized into several rheological classifications. Emulsions are frequently pseudoplastic: As shear rate increases, viscosity decreases. This kind of flow behavior is also termed shear-thinning. An emulsion may also exhibit a yield stress, that is, the shear rate (flow) remains zero until a threshold shear stress is reached—the yield stress (τ_Y)—then pseudoplastic or Newtonian flow begins. Pseudoplastic flow that is time dependent is termed thixotropic. That is, at constant applied shear rate, viscosity decreases, and in a flow curve hysteresis occurs. Several other rheological classifications are covered in the Glossary: dilatancy, rheopexy, and rheomalaxis. Even viscosity itself is represented in many ways, as shown in Table II.

Some very useful descriptions of experimental techniques have been given by Whorlow (22) and others (23, 24). Very often measurements are made with an emulsion sample placed in the annulus between two concentric cylinders. The shear stress is calculated from the measured torque required to maintain a given rotational velocity of one cylinder with respect to the other. Knowing the geometry, the effective shear rate can be calculated from the rotational velocity. One reason for the relative lack of rheological data for emulsions, compared with that for other colloidal systems, is the difficulty associated with performing the measurements in these systems. As suggested by Figure 5, for a practical O/W emulsion, the sample may contain suspended particles in addition to the oil droplets. In attempting to conduct a measurement, a number of changes may occur in the sample

Table II. Glossary of Bulk Viscosity Terms

Term	Symbol	Explanation
Absolute viscosity	η	$\eta = \tau/\dot{\gamma}$ and can be traced to fundamental units independent of the type of instrument
Apparent viscosity	η_{APP}	$\eta_{APP} = \tau/\dot{\gamma}$ but as determined for a non-Newtonian fluid, usually by a method suitable only for Newtonian fluids
Differential viscosity	η_D	$\eta_D = d\tau/d\dot{\gamma}$
Specific increase in viscosity	η_{SP}	$\eta_{SP} = \eta_{Rel} - 1$
Intrinsic viscosity	$[\eta]$	$[\eta] = \lim_{C \to 0} \lim_{\dot{\gamma} \to 0} \eta_{SP}/C$ $[\eta] = \lim_{C \to 0} \lim_{\dot{\gamma} \to 0} (1/C) \ln \eta_{Rel}$
Reduced viscosity	η_{Red}	$\eta_{Red} = \eta_{SP}/C$
Relative viscosity	η_{Rel}	$\eta_{Rel} = \eta/\eta_0$ $\eta_0 =$ viscosity of the pure solvent or dispersion medium

chamber that make the measurements irreproducible and not representative of the original emulsion (25). These changes may include

- creaming of the droplets, causing a nonuniform distribution within the chamber, or even removal of all droplets into an upper phase away from the region in which measurements are made
- centrifugal separation of oil, water, and solid phases, making the emulsion radially inhomogeneous, and possibly breaking the emulsion
- shear-induced coalescence or finer dispersion of droplets, changing the properties of the sample
- sedimentation of the solids, causing a nonuniform distribution within the chamber, or even removal of all solids from the region in which measurements are made

It is frequently desirable to be able to describe emulsion viscosity in terms of the viscosity of the continuous phase (η_0) and the amount of emulsified material. A very large number of equations have been advanced for estimating suspension (or emulsion, etc.) viscosities. Most of these are empirical extensions of Einstein's equation for a dilute suspension of spheres:

$$\eta = \eta_0 \, (1 + 2.5\phi) \tag{9}$$

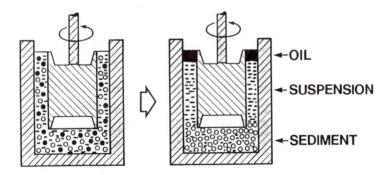

Figure 5. Some rheological measurement problems that may be encountered with practical oil-field emulsions. Initially, the O/W sample contained in the cylinders on the left may be homogeneous, containing oil droplets (•), fine particles (—), and large particles (○). After some time, the sample may become quite stratified, as shown at right.

where η_0 is the medium viscosity and ϕ is the dispersed-phase volume fraction and is <1. For example, the empirical equations typically have the form (4):

$$\eta = \eta_0(1 + \alpha_0\phi + \alpha_1\phi^2 + \alpha_2\phi^3 + \cdots) \qquad (10)$$

where the αs are empirical constants, and the empirical equations may include other terms, as in the Thomas equation,

$$\eta = \eta_0[1 + 2.5\phi + 10.5\phi^2 + 0.00273 \exp(16.6\phi)] \qquad (11)$$

These equations assume Newtonian behavior, or at least apply to the Newtonian region of a flow curve, and they usually apply if the droplets are not too large and if there are no strong electrostatic interactions. A more detailed treatment of these relationships is given in Chapter 4.

Emulsions can show varying rheological, or viscosity, behaviors. Sometimes these properties are due to the emulsifier or other agents in the emulsion. However, if the internal phase has a sufficiently high volume fraction (typically anywhere from 10 to 50%) the emulsion viscosity increases because of droplet "crowding" or structural viscosity and becomes non-Newtonian. The maximum volume fraction possible for an internal phase made up of uniform, incompressible spheres is 74%, although emulsions with an internal volume fraction of 99% have been made (15). Figure 6 shows how emulsion viscosity tends to vary with volume fraction; the drop in viscosity at $\phi = 0.74$ signifies inversion. At this point the dispersed-phase volume fraction becomes 0.26, in this example, and the lower value of ϕ is reflected by a much lower viscosity. If inversion does not occur, then the viscosity continues to increase. This condition is true for both W/O and O/W types.

A graphic and important example is furnished by the oil spill "chocolate mousse" emulsions formed when crude oil spills into seawater. These water-in-oil emulsions have high water contents that may exceed 74% and reach $\phi = 0.80$ or more without inverting. As their common name implies, these mousse emulsions not only have viscosities that are much higher than the original crude oil but can become semisolid. With increasing time after a spill, these emulsions weather (the oil becomes depleted in its lower boiling fractions), and apparently the emulsions become more stable, more solid-like, and considerably more difficult to handle and break.

Interfacial Viscosity. The foregoing discussion of rheology has dealt with the bulk viscosity properties. A closely related and very important property is the interfacial viscosity, which can be thought of as the two-dimensional equivalent of bulk viscosity, operative in the oil–water interfacial region. As droplets in an emulsion approach each other, the thinning of

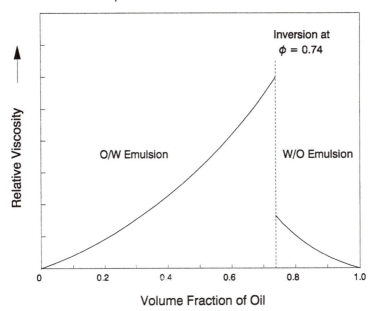

Figure 6. *The influence of volume fraction on the emulsion type and viscosity of a model emulsion. (Reproduced with permission from reference 4. Copyright 1965 Robert E. Krieger, Inc.)*

the films between the droplets, and their resistance to rupture, are thought to be of great importance to the ultimate stability of the emulsion. Thus, a high interfacial viscosity can promote emulsion stability by retarding the rate of droplet coalescence, as discussed in later sections. Further details on the principles, measurement, and applications to emulsion stability of interfacial viscosity were reviewed by Malhotra and Wasan (26).

Properties of the Interfaces

In simple two-phase colloidal systems, a thin intermediate region or boundary, known as the interface, lies between the dispersed and dispersing phases. Interfacial properties are very important because emulsified droplets have a large interfacial area, and even a modest interfacial energy per unit area can become a considerable total interfacial energy to be accommodated. For example, suppose we wish to emulsify one barrel (159 L) of oil into water. For this illustration consider the oil to be in 1 large drop that we repeatedly subdivide into drops of half the previous radius. Thus, the initial drop of $r = 33.6$ cm becomes eight drops of $r = 16.8$ cm, and so on. Figure 7 shows that the total surface or interfacial area produced increases by a factor of 2 with each cut. The initial interfacial area of 1.42 m^2 increases to 7.4×10^5

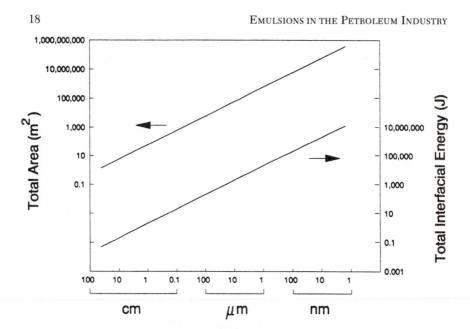

Sphere Radius

Figure 7. Total area and energy changes involved in emulsifying 1 barrel of oil into water by dispersing into progressively finer droplets.

m^2 by the time droplets of 0.64-μm radius have been produced. This increase is greater than 5 orders of magnitude! The larger interfacial area will have a significant total free energy as shown in Figure 7. If the interfacial tension is 35 mN/m, then by the time the droplet size is $r = 0.64$ μm, the total energy will have increased from 0.05×10^4 to 2.6×10^4 J. This 2.6×10^4 J of energy had to be added to the system to achieve the emulsification. If this amount of energy cannot be provided, for example, by mechanical shear, then another alternative is to use surfactant chemistry to lower the interfacial free energy, or interfacial tension.

The energy plotted in Figure 7 was obtained by multiplying the total area by the interfacial tension. Now if a small quantity of a surfactant was added to the water, possibly a few tenths of a percent, that lowered the interfacial tension to 0.35 mN/m, it would lower the amount of mechanical energy needed in the example by a factor of 100. From the area per molecule that the adsorbed emulsifying agent occupies, the minimum amount of emulsifier needed for the emulsion can also be estimated. In practice, lowering interfacial tension alone may not be sufficient to stabilize an emulsion, in which case other interfacial properties must be adjusted as well. These simple calculations do, however, show how important the interfacial properties can become when colloidal-sized species are involved, as in emulsions.

Surface and Interfacial Tensions.   Regarding the molecules in a liquid, the attractive van der Waals forces between molecules are felt equally by all molecules except those in the interfacial region. This inequality in the van der Waals forces pulls the interfacial molecules toward the interior of the liquid. The interface thus has a tendency to contract spontaneously. For this reason, droplets of liquid and bubbles of gas tend to adopt a spherical shape, because this shape reduces the surface free energy. For two immiscible liquids, a similar situation applies, except that it may not be so immediately obvious how the interface will tend to curve. There will still be an imbalance of intermolecular forces and a configuration that minimizes the interfacial free energy.

The surface free energy has units of millijoules per square meter (1 mJ/m^2 = 1 erg/cm^2), reflecting the fact that area expansion requires energy. Surface free energies are usually described in terms of contracting forces acting parallel to the surface or interface. Surface tension ($\gamma°$), or interfacial tension (γ), is the force per unit length around a surface, or the free energy required to create new surface area. Thus, the units of surface and interfacial tension are millinewtons per meter (1 mN/m = 1 dyne/cm). These units for surface and interfacial tension are numerically equal to the surface free energy. Interfacial tensions are frequently intermediate between the values of the surface tensions of the liquids involved and are smallest when the liquids are the most chemically similar (for pure liquids).

Many methods for the measurement of surface and interfacial tensions, details of the experimental techniques, and their limitations are described in several good reviews (27–29). Some methods that are used most in emulsion work are the du Nouy ring, drop weight or volume, pendant drop, and the spinning drop. The spinning drop technique is applicable to the very low interfacial tensions encountered in the enhanced oil recovery and microemulsion fields (30). In all cases, when solutions rather than pure liquids are involved, appreciable changes can take place with time at the surfaces and interfaces.

Young–Laplace Equation.   Interfacial tension causes a pressure difference to exist across a curved surface, the pressure being greater on the concave side (i.e., on the inside of a droplet). In an interface between phase A in a droplet and phase B surrounding the droplet, the phases will have pressures p_A and p_B. If the principal radii of curvature are R_1 and R_2, then

$$\Delta p = p_A - p_B = \gamma(1 / R_1 + 1 / R_2) \tag{12}$$

Equation 12 is the Young–Laplace equation (31). It shows that $p_A > p_B$; the pressure inside a droplet exceeds that outside. For spherical droplets in an emulsion,

$$\Delta p = p_A - p_B = 2\gamma / R \tag{13}$$

so that Δp varies with the radius, R. (More details are given in refs. 13 and 31.) The Young–Laplace equation forms the basis for some important methods for measuring surface and interfacial tensions, such as the pendant, sessile, and spinning drop methods and the maximum bubble pressure method (27–30). In primary oil recovery from underground reservoirs, the capillary forces described by this equation are responsible for holding back much of the oil (residual oil) in parts of the pore structure in the rock or sand. Any secondary or enhanced (tertiary) oil recovery process strategies are intended to overcome these same forces (32). In the example involving microemulsions in enhanced oil recovery (discussed under the heading "Microemulsions"), the Young–Laplace equation was used, without introduction, to demonstrate how lowering interfacial tension can facilitate emulsification and incremental oil recovery. In that example r, the pore radius, was used in place of R in the Young–Laplace equation, and the contact angle was included.

Contact Angles and Wetting. When a droplet of oil in water comes into contact with a solid surface, the oil may form a bead on the surface, or it may spread and form a film. A liquid having a strong affinity for the solid will seek to maximize its contact (interfacial area) and form a film. A liquid with much weaker affinity may form into a bead. The affinity is termed the wettability. Because there can be degrees of spreading, another quantity is needed. The contact angle, θ, in an oil–water–solid system is defined as the angle, measured through the aqueous phase, that is formed at the junction of the three phases. Whereas interfacial tension is defined for the boundary between two phases, the contact angle is defined for a three-phase junction.

If the interfacial forces acting along the perimeter of the droplet are represented by the interfacial tensions, then an equilibrium force balance can be written as

$$\gamma_{W/O} \cos \theta = \gamma_{S/O} - \gamma_{S/W} \qquad (14)$$

where the subscripts refer to water (W), oil (O), and solid (S). Equation 14 is Young's equation. The solid is completely water-wetted if $\theta = 0$ and only partially wetted otherwise. Equation 14 is frequently used to describe wetting phenomena, so two practical points are important. In theory complete nonwetting by water would mean that $\theta = 180°$, but this contact angle is not seen in practice. Also, values of $\theta < 90°$ are often considered to represent "water-wetting", and values of $\theta > 90°$ are considered to represent "non-water-wetting". This assignment is rather arbitrary because it is based on correlation with visual appearance of droplets on surfaces.

These considerations come into play in oil recovery schemes applied to reservoirs of mixed wettability or where the rock is predominantly oil-wetting. Another example is the case of the so-called Pickering emulsions,

emulsions stabilized by fine particles. A film of close-packed particles has considerable mechanical strength, which contributes to the stability of an emulsion. The most stable emulsions occur when the contact angle is close to 90°, so that the particles will collect at the interface. Combining Young's equation with the oriented wedge theory (*see later*, "Emulsifying Agents") allows some predictions to be made. If the contact angle for a fine particle at the O/W interface is $\theta < 90°$, then most of the particle will reside in the aqueous phase. In this case an O/W emulsion is indicated. Conversely, if $\theta > 90°$, then the particle will be mostly in the oil phase, and W/O is predicted.

Adsorption at Interfaces: Surface Activity. *Surfactants.* Some compounds, like short-chain fatty acids, can be partly soluble in both water and oil. This dual solubility is because such molecules are amphiphilic or amphipathic; that is, they have one part that has an affinity for the oil (the nonpolar hydrocarbon chain), and one part that has an affinity for the water (the polar group). The energetically most favorable orientation for these molecules is at the oil–water interface, so that each part of the molecule can reside in the solvent for which it has the greatest affinity (*see* Figure 8).

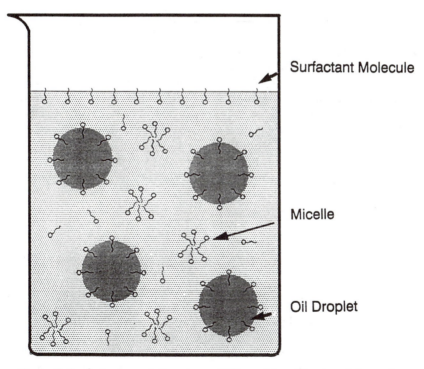

Surfactant Molecule

Micelle

Oil Droplet

Figure 8. Surfactant associations in an O/W emulsion. The size of the surfactant molecules compared to the oil droplets has been exaggerated for the purposes of illustration.

These molecules that form oriented monolayers at interfaces show surface activity and are termed surfactants. As there will be a balance between adsorption and desorption (due to thermal motions), the interfacial condition requires some time to establish. Because of this time required, surface activity should be considered a dynamic phenomenon. This fact can be seen by measuring surface tension versus time for a freshly formed surface.

A consequence of surfactant adsorption at an interface is that it provides an expanding force acting against the normal interfacial tension. If π is this expanding pressure (surface pressure), then $\gamma = \gamma_{initial} - \pi$. Thus, surfactants tend to lower interfacial tension; if a low enough value of γ is reached, emulsification can take place because only a small increase in surface free energy is required, for example, when $\pi \sim \gamma_{initial}$. If solute–solvent forces are greater than solvent–solvent forces on the other hand, then molecular migration away from the interface can occur and cause increased surface tension (e.g., NaCl [aq]).

Gibbs has thermodynamically described the lowering of surface free energy that results from surfactant adsorption. The general Gibbs adsorption equation for a binary, isothermal system containing excess electrolyte is

$$\Gamma_s = - (1 / RT) (d\gamma / d \ln C_s) \tag{15}$$

where Γ_s is the surface excess of surfactant (mol/cm^2), C_s is the solution concentration of the surfactant (M), and γ may be either surface or interfacial tension (mN/m). This equation can be applied to dilute surfactant solutions in which the surface curvature is not great and where the adsorbed film can be considered a monolayer. The packing density of surfactant in a monolayer at the interface can be calculated as follows. The surface excess in a tightly packed monolayer can be calculated from the slope of the linear portion of a plot of surface tension versus the logarithm of solution concentration (see Figure 9). From this slope, the area per adsorbed molecule (a_s) can be calculated from

$$a_s = 1 / (N_A \Gamma_s) \tag{16}$$

where N_A is Avogadro's number. Numerous examples are given by Rosen (33).

Surface Films.   Insoluble polar molecules (e.g., long-chain fatty acids) exhibit an extreme kind of adsorption at liquid surfaces. That is, they can be made to concentrate in one molecular layer at the surface. These interfacial films often provide the stabilizing influence in emulsions because they can both lower interfacial tension and increase the interfacial viscosity. Increasing interfacial viscosity provides a mechanical resistance to coalescence. Such systems also lend themselves to the study of size, shape, and

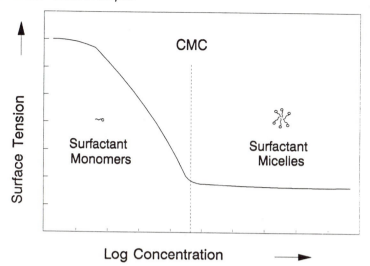

Figure 9. The association behavior of surfactants in solution, showing the critical micelle concentration (CMC).

orientation of molecules at an interface. Having an adsorbed layer lowers the surface tension (to γ) by the surface pressure $\pi = \gamma_{initial} - \gamma$, as already noted. In a surface balance (pioneered by Langmuir and others) π versus the available area of surface (A) can be determined directly. Another approach is to measure the lowered surface tension (Wilhelmy plate) and calculate π. For very low film pressures, an ideal gas law analogy,

$$\pi A = nRT = nN_A kT \qquad (17)$$

allows calculation of the effective area per molecule (A/nN_A) in the monolayer; here n is the number of moles, R is the gas constant, T is absolute temperature, and k is the Boltzmann constant.

Classification of Surfactants.   Surfactants are classified according to the nature of the polar (hydrophilic) part of the molecule, as illustrated in Table III. In-depth discussions of surfactant structure and chemistry can be found in references 33–35.

In aqueous solution, dilute concentrations of surfactant act much as normal electrolytes, but at higher concentrations very different behavior results. This behavior (illustrated in Figures 8 and 9) was explained by McBain in terms of organized aggregates called micelles in which the lipophilic parts of the surfactants associate in the interior of the aggregate and leave hydrophilic parts to face the aqueous medium. (Details are given in refs. 16 and 31.) The concentration at which micelle formation becomes

Table III. Surfactant Classifications

Class	Examples	Structures
Anionic	Na stearate	$CH_3(CH_2)_{16}COO^-Na^+$
	Na dodecyl sulfate	$CH_3(CH_2)_{11}SO_4^-Na^+$
	Na dodecyl benzene sulfonate	$CH_3(CH_2)_{11}C_6H_4SO_3^-Na^+$
Cationic	Laurylamine hydrochloride	$CH_3(CH_2)_{11}NH_3^+Cl^-$
	Cetyl trimethylammonium bromide	$CH_3(CH_2)_{15}N^+(CH_3)_3Br^-$
Nonionic	Polyoxyethylene alcohol	$C_nH_{2n+1}(OCH_2CH_2)_mOH$
	Alkylphenol ethoxylate	$C_9H_{19}-C_6H_4-(OCH_2CH_2)_nOH$
Zwitterionic	Lauramidopropyl betaine	$C_{11}H_{23}CONH(CH_2)_3N^+(CH_3)2CH_2COO^-$
	Cocoamido-2-hydroxypropyl sulfobetaine	$C_nH_{2n+1}CONH(CH_2)_3N^+(CH_3)_2CH_2CH(OH)CH_2SO_3^-$

significant is called the critical micelle concentration (CMC). The CMC is a property of the surfactant and several other factors, because micellization is opposed by thermal and electrostatic forces. A low CMC is favored by increasing the molecular mass of the lipophilic part of the molecule, lowering the temperature (usually), and adding electrolyte. Surfactant molecular weights range from a few hundred up to several thousand.

Some typical CMC values for low electrolyte concentrations at room temperature are

Surfactant Class	CMC (M)
Nonionic	10^{-5}–10^{-4}
Anionic	10^{-3}–10^{-2}
Amphoteric	10^{-3}–10^{-1}

The solubilities of micelle-forming surfactants show a strong increase above a certain temperature, termed the Krafft point (T_K). This increase in solubility is explained by the fact that the single surfactant molecules have limited solubility, whereas the micelles are very soluble. Referring to Figure 10, below the Krafft point the solubility of the surfactant is too low for micellization, and solubility alone determines the surfactant monomer concentration. As temperature increases, the solubility increases until at T_K the CMC is reached. At this temperature a relatively large amount of surfactant can be dispersed in micelles, and solubility increases greatly. Above the

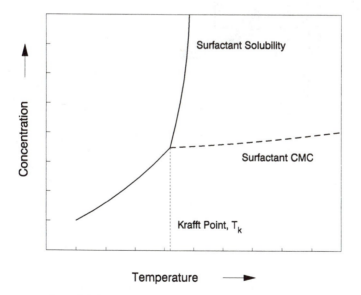

Figure 10. The solubility–micellization behavior of surfactants in solution, showing the Krafft point.

Krafft point, maximum reduction in surface or interfacial tension occurs at the CMC because now the CMC determines the surfactant monomer concentration.

Cohesion, Adhesion, and Spreading.   Two phases A and B may have an interface between them, AB. Cohesion, adhesion, and spreading can be defined for the changes shown in Figure 11 (involving always unit surface area). The work of cohesion represents the energy required to increase interfacial area by two square units. Thus the energy required to disperse oil into finer and finer droplets, to make an emulsion, increases as the interfa-

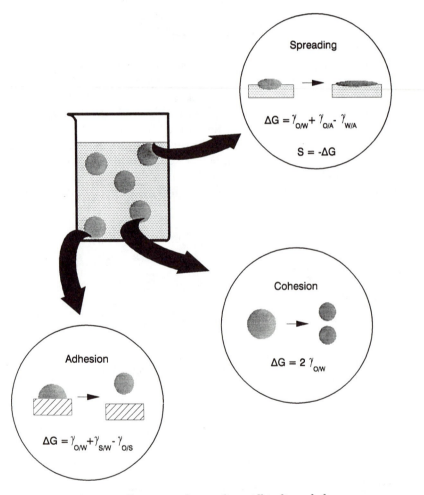

Figure 11. Cohesion, adhesion, and spreading. All indicated changes are per unit area, ΔG is the Gibbs free energy change, and the subscripts denote oil (O), water (W) and air (A).

cial tension between the droplets increases. The work of adhesion is the energy involved when two surfaces, initially in contact with each other, are separated to each contact a third phase. The work of adhesion relates to, for example, surface energy changes involved when an oil droplet initially in contact with a solid particle becomes released into water. Spreading occurs when, for example, emulsified oil droplets reach the air–water interface and spread over the surface. This spreading is a mechanism for demulsification; a negative spreading coefficient should thus tend to contribute to emulsion stability. Spreading is also a mechanism for defoaming by emulsions (36). Petroleum emulsions have been used to prevent the formation of foams, or destroy foams already generated, in various industrial processes (37).

Stability of Emulsions

Meaning of Stability.   Most emulsions are not thermodynamically stable. Rather they possess some degree of kinetic stability, and it is important to distinguish the degree of change and the time scale. As mentioned previously, coalescence and aggregation are processes in which particles, droplets, or bubbles are brought together with (coalescence) or without (aggregation) large changes in surface area. Thus, there can be different kinds of kinetic stability. This discussion of colloid stability will explore the reasons why colloidal dispersions can have different degrees of kinetic stability and how these are influenced, and can therefore be modified, by solution and surface properties. The discussion is carried further, and in more detail, in Chapter 2.

Encounters between particles in a dispersion can occur frequently because of Brownian motion, sedimentation, or stirring. The stability of the dispersion depends upon how the particles interact when these encounters happen. The main cause of repulsive forces is the electrostatic repulsion between like-charged objects. The main attractive forces are the van der Waals forces between objects.

Electrostatic Forces.   *Charged Interfaces.*   Most substances acquire a surface electric charge when brought into contact with a polar medium such as water. For emulsions, the origin of the charge can be ionization, as when surface acid functionalities ionize when oil droplets are dispersed into an aqueous solution, or the origin can be adsorption, as when surfactant ions or charged particles adsorb onto an oil droplet surface. Solid particles can have additional mechanisms of charging. One is the unequal dissolution of cations and anions that make up the crystal structure (in the salt type minerals, for example). Another is the diffusion of counterions away from the surface of a solid whose internal crystal structure carries an opposite charge because of isomorphic substitution (in clays, for example). The surface charge influences the distribution of nearby ions in the polar medium. Ions of opposite charge (counterions) are attracted to the surface, but those of like charge (co-ions) are repelled. An electric double layer, which is diffuse because of mixing caused by thermal motion, is thus formed.

In a practical petroleum emulsion situation, the degree of surface charging is more complicated. An example is the bitumen–water interface, which becomes negatively charged in alkaline aqueous solutions as a result of the ionization of surface carboxylic acid groups belonging to natural surfactants present in the bitumen. The degree of negative charging is very important to the success of bitumen recovery processes from in situ oil sands and also separation processes from surface oil sands, such as the hot-water flotation process (38–42). The degree of negative charge at the interface depends on the pH and ionic strength of the solution (38, 39) and also on the concentration of natural surfactant monomers present in the aqueous phase (40, 41). With bitumen, as with other heavy crude oils, more than one kind of surfactant may be produced (42), and the solution concentrations of the surfactants depend also on reaction conditions (temperature, etc.) and on the extent of competing reactions such as the adsorption of the surfactants onto solid (clay) particles that are present (43, 44).

Electric Double Layer.   The electric double layer (EDL) consists of the charged surface and a neutralizing excess of counterions over co-ions, distributed near the surface (*see* Figure 12). The EDL can be viewed as being composed of two layers:

- an inner layer that may include adsorbed ions
- a diffuse layer in which ions are distributed according to the influence of electrical forces and thermal motion

Gouy and Chapman proposed a simple quantitative model for the diffuse double layer, assuming, among other things, an infinite, flat, uniformly charged surface and point-charge ions. (Further details are given in refs. 14, 31, and 45.) Taking the surface potential to be ψ°, the potential ψ at a distance x from the surface is approximately

$$\psi = \psi^\circ \exp\left(-\kappa x\right) \tag{18}$$

The surface charge density is given as $\sigma^\circ = \epsilon\kappa\psi^\circ$, where ϵ is the permittivity and κ is not conductivity but a special variable defined in equation 19; thus, ψ° depends on surface charge density and the solution ionic composition (through κ). The variable $1/\kappa$ is called the double-layer thickness, and for water at 25 °C it is given by

$$\kappa = 3.288\sqrt{I} \quad (\text{nm}^{-1}) \tag{19}$$

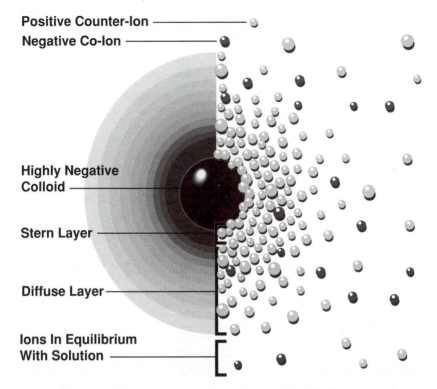

Positive Counter-Ion

Negative Co-Ion

Highly Negative Colloid

Stern Layer

Diffuse Layer

Ions In Equilibrium With Solution

Figure 12. Simplified illustrations of the electrical double layer around a negatively charged colloidal emulsion droplet. The left view shows the change in charge density around the droplet. The right view shows the distribution of ions around the charged droplet. (Courtesy of L. A. Ravina, Zeta-Meter, Inc., Long Island City, NY.)

where I is the ionic strength, given by $I = (1/2) \Sigma_i c_i z_i^2$, in which c_i are the individual ion concentrations and z_i are the respective ion-charge numbers. For a 1–1 electrolyte,

$$1/\kappa = 1 \text{ nm for } I = 10^{-1} \text{ M}$$
$$1/\kappa = 10 \text{ nm for } I = 10^{-3} \text{ M} \tag{20}$$

In fact an inner layer exists because ions are not really point charges and an ion can approach a surface only to the extent allowed by its hydration sphere. The Stern model specifically incorporates a layer of specifically adsorbed ions bounded by a plane, the Stern plane (*see* Figure 13 and refs. 14, 31, and 45). In this case the potential changes from ψ° at the surface, to $\psi(\delta)$ at the Stern plane, to $\psi = 0$ in bulk solution.

Electrokinetic Phenomena. Electrokinetic motion occurs when the mobile part of the electric double layer is sheared away from the inner layer

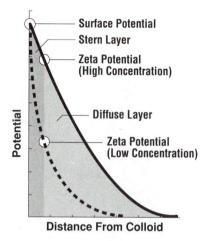

Figure 13. Simplified illustration of the surface and zeta potentials for a charged emulsion droplet dispersed in high and low electrolyte concentration aqueous solutions. (Courtesy of L. A. Ravina, Zeta-Meter, Inc., Long Island City, NY.)

(charged surface). The four types of electrokinetic measurements are electrophoresis, electro-osmosis, streaming potential, and sedimentation potential, of which electrophoresis finds the most use in industrial practice. Good descriptions of practical experimental techniques in electrophoresis and their limitations can be found in references 46–48.

In electrophoresis an electric field is applied to a sample and causes charged droplets or particles and any attached material or liquid to move toward the oppositely charged electrode. Thus the results can be interpreted only in terms of charge density (σ) or potential (ψ) at the plane of shear. The latter is also known as the zeta potential. Because the exact location of the shear plane is generally not known, the zeta potential is usually taken to be approximately equal to the potential at the Stern plane (Figure 13):

$$\zeta = \psi(\delta) \tag{21}$$

where δ is the distance from the droplet surface to the Stern plane. In microelectrophoresis the dispersed droplets are viewed under a microscope, and their electrophoretic velocity is measured at a location in the sample cell where the electric field gradient is known. This measurement must be done at carefully selected planes within the cell because the cell walls become charged as well and cause electro-osmotic flow of the bulk liquid inside the cell.

The electrophoretic mobility, μ_E, is defined as the electrophoretic velocity divided by the electric field gradient at the location where the velocity was measured. It remains then to relate the electrophoretic mobility to the zeta potential (ζ). Two simple relations can be used to calculate zeta potentials in limiting cases:

- Hückel theory. For droplets of small radius (r) with "thick" electric double layers, meaning that $\kappa r < 1$, it is assumed that Stokes' law applies and the electrical force is equated to the frictional resistance of the droplet, $\mu_E = \zeta\epsilon/(1.5\eta)$.

- Smoluchowski theory. For large droplets with "thin" electric double layers, meaning droplets for which $\kappa r > 100$, $\mu_E = \zeta\epsilon/\eta$.

These theories are discussed in more detail in references 14 and 50.

With these relations zeta potentials can be calculated for many practical systems. Within each set of limiting conditions the electrophoretic mobility is independent of particle size and shape as long as the ζ potential is constant. For intermediate values of κr, the Henry equation and many other equations apply (*46, 47, 49*).

Repulsive Forces.   In the simplest example of colloid stability, emulsion droplets would be stabilized entirely by the repulsive forces created when two charged surfaces approach each other and their electric double layers overlap. The repulsive energy, V_R, for spherical droplets is given approximately as

$$V_R = \frac{B\epsilon k^2 T^2 r \gamma^2}{z^2} \exp\left(-\kappa H\right) \tag{22}$$

where the spheres have radius r and are separated by distance H, B is a constant (3.93×10^{39} A^{-2}s^{-2}), z is the counterion charge number, and

$$\gamma = \frac{\exp\left[ze\psi(\delta)/2kT\right] - 1}{\exp\left[ze\psi(\delta)/2kT\right] + 1} \tag{23}$$

Dispersion Forces.   van der Waals postulated that neutral molecules exert forces of attraction on each other that are caused by electrical interactions between three types of dipolar configurations. The attraction results from the orientation of dipoles that may be (1) two permanent dipoles, (2) dipole–induced dipole, or (3) induced dipole–induced dipole. Induced dipole–induced dipole forces between nonpolar molecules are also called London dispersion forces. Except for quite polar materials, the London dispersion forces are the more significant of the three. For molecules the force varies inversely with the sixth power of the intermolecular distance.

For dispersed droplets (or particles, etc.) the dispersion forces can be approximated by adding the attractions between all interdroplet pairs of molecules. When added this way, the dispersion force between two droplets decays less rapidly as a function of separation distance than is the case for individual molecules. For two spheres of radius r in a vacuum, separated by distance H, the attractive energy V_A can be approximated by

$$V_A = -\frac{Ar}{12H} \tag{24}$$

for $H < 10$–20 nm and $H \ll r$. The constant A is known as the Hamaker constant and depends on the density and polarizability of atoms in the particles. Typically 10^{-20} J $< A < 10^{-19}$ J. When the particles are in a medium other than vacuum, the attraction is reduced. This reduced attraction can be accounted for by using an effective Hamaker constant

$$A = \left(\sqrt{A_2} - \sqrt{A_1} \right)^2 \tag{25}$$

where the subscripts denote the medium (1) and particles (2).

The effective Hamaker constant equation shows that the attraction between particles is weakest when the particles and medium are most chemically similar ($A_1 \sim A_2$). The Hamaker constants are usually not well known and must be approximated.

DLVO Theory.   Derjaguin and Landau, and independently Verwey and Overbeek (45), developed a quantitative theory for the stability of lyophobic colloids, now known as the DLVO theory. It was developed in an attempt to account for the observation that colloids coagulate quickly at high electrolyte concentrations, slowly at low concentrations, and with a very narrow electrolyte concentration range over which the transition from one to the other occurs. This narrow electrolyte concentration range defines the critical coagulation concentration (CCC). The DLVO theory accounts for the energy changes that take place when two droplets (or particles) approach each other, and involves estimating the energy of attraction (London–van der Waals) versus interparticle distance and the energy of repulsion (electrostatic) versus distance. These, V_A and V_R, respectively, are then added together to yield the total interaction energy V. A third important force occurs at very small separation distances where the atomic electron clouds overlap and causes a strong repulsion, called Born repulsion. The theory has been developed for several special cases, including the interaction between two spheres, and refinements are constantly being made.

The value of V_R decreases exponentially with increasing separation distance and has a range about equal to κ^{-1}, and V_A decreases inversely with increasing separation distance. Figure 14 shows a single attractive energy curve and two different repulsive energy curves, representing two very different levels of electrolyte concentration. The figure shows the total interaction energy curves that result in each case. Either the attractive van der Waals forces or the repulsive electric double-layer forces can predominate at different interdroplet distances.

With a positive potential energy maximum, a dispersion should be stable if $V \gg kT$, that is, if the energy is large compared to the thermal energy of the particles ($15kT$ is considered insurmountable). In this case colliding droplets should rebound without contact, and the emulsion should be stable to aggregation. If, on the other hand, the potential energy maximum is not very great, $V \sim kT$, then slow aggregation should occur. The height of the energy barrier depends on the surface potential, $\psi(\delta)$ and on the range of the repulsive forces, κ^{-1}. Figure 14 shows that an energy minimum can occur at larger interparticle distances. If this energy minimum is reasonably deep compared to kT, then a loose, easily reversible aggregation should occur.

Practical Guidelines. The DLVO calculations can become quite involved, requiring considerable knowledge about the systems of interest. Also, they present some problems. For example, some distortion of the spherical emulsion droplets will occur as they approach each other and begin to seriously interact; these interactions cause a flattening. Also, our view of

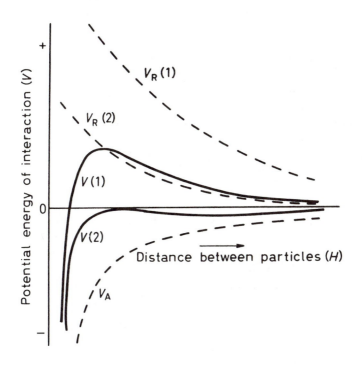

Figure 14. The effect of different repulsive potential energy curves (1 and 2) on the total interaction energy for a given attractive energy curve. (Reproduced with permission from reference 16. Copyright 1981 Butterworth Heinemann Ltd.)

the validity of the theory is changing as more becomes known about the influence of additional forces such as those due to surface hydration. The DLVO theory nevertheless forms a very useful starting point in attempting to understand complex colloidal systems such as petroleum emulsions. Empirical "rules of thumb" can be used to give a first estimate of the degree of colloidal stability that a system is likely to have if the zeta potentials of the droplets are known.

Zeta Potential Criteria.   Many types of colloids tend to adopt a negative surface charge when dispersed in aqueous solutions having ionic concentrations and pH typical of natural waters. For such systems one rule of thumb stems from observations that the colloidal particles are quite stable when the zeta potential is about –30 mV or more negative, and quite unstable because of agglomeration when the zeta potential is between +5 and –5 mV. An expanded set of guidelines, developed for particle suspensions, is given in reference 48. Such criteria are frequently used to determine optimal dosages of polyvalent metal electrolytes, such as alum, used to effect coagulation in treatment plants.

For example, as stated earlier, in order to separate bitumen from oil sands, a significant negative charge must be present at the oil–water interface. This negative charge is needed to aid in the release of the bitumen from the sand matrix, and is also needed to prevent attachment of the released and emulsified droplets to codispersed solid particles such as clays. In fact, in the flotation recovery process for bitumen the optimal processing condition correlates with maximizing the negative charge, or zeta potential, on the droplets *(41)*. On-line measurements of the emulsified bitumen droplet zeta potentials can be used to monitor and control the continuous oil recovery process, as indicated in Figure 15 *(50)*. In the Figure 15 example, the maximum (negative) emulsion droplet zeta potential achieved was about –35 mV, which is consistent with the "good stability" guideline just described.

Schulze–Hardy Rule.   The transition from stable dispersion to aggregation usually occurs over a fairly small range of electrolyte concentration. This condition makes it possible to determine aggregation concentrations, often referred to as critical coagulation concentrations (CCC). The Schulze–Hardy rule summarizes the general tendency of the CCC to vary inversely with the sixth power of the counterion charge number (for indifferent electrolyte). Further details and original references are given in references 13 and 51.

A prediction from DLVO theory can be made by deriving the conditions under which $V = 0$ and $dV/dH = 0$. The result is

$$CCC = \frac{9.75 B^2 \epsilon^3 k^5 T^5 \gamma^4}{e^2 N_A A^2 z^6} \tag{26}$$

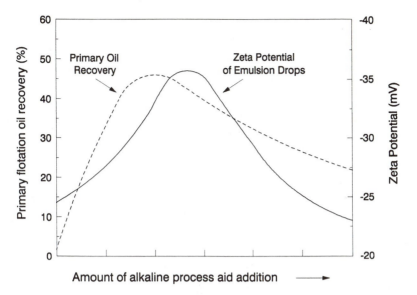

Figure 15. Plot of oil recoveries versus process aid addition level from the hot water flotation processing of an oil sand in a continuous pilot plant. Also shown is the correspondence with the zeta potentials, measured on-line, of emulsified bitumen droplets in the extraction solution. (Plotted from data in reference 50.)

showing that for high potentials ($\gamma \to 1$), the CCC varies inversely with z^6. As an illustration, for a hypothetical emulsion, equation 26 predicts a CCC of 1.18 M in solutions of sodium chloride. The critical coagulation concentrations in polyvalent metal chlorides would then decrease as follows:

Dissolved Salt	z	CCC (mol/L)
NaCl	1	1.18
CaCl$_2$	2	0.018
AlCl$_3$	3	0.0016

Broad Influence on Stability.   In general, when electrical surface charge is an important determinant of stability, it is easier to formulate a very stable O/W emulsion than a W/O emulsion because the electric double-layer thickness is much greater in water than in oil. (This condition is sometimes incorrectly stated in terms of greater charge being present on droplets in an O/W emulsion). This is not to say that W/O emulsions cannot be stabilized, however. Many reasonably stable oil-field W/O emulsions are stabilized by another mechanism: the protective action of viscoelastic, possibly rigid, films formed on the droplets by macromolecules or solid particles.

Kinetics. Thus far this chapter has mostly been concerned with an understanding of the direction in which reactions will proceed. However, from an engineering point of view, it is just as important to know the rates at which such reactions will proceed. Two principal factors determine the rate of aggregation of droplets in an emulsion: the frequency of droplet encounters and the probability that the thermal energy of the droplets is sufficient to overcome the potential energy barrier to aggregation. The rate of aggregation can be given as $-(dn/dt) = k_2 n^2$ where k_2 is the rate constant and n is the number of droplets per unit volume at time t. For $n = n_0$ at $t = 0$,

$$1/n = k_2 t + 1/n_0 \qquad (27)$$

During the process of aggregation, k_2 may not remain constant.

If the energy barrier to aggregation is removed (e.g., by adding excess electrolyte) then aggregation is diffusion controlled; only Brownian motion of independent droplets or particles is present. For a monodisperse suspension of spheres, Smoluchowski developed an equation for this "rapid coagulation"

$$n = \frac{n_0}{1 + 8\pi D r n_0 t} \qquad (28)$$

where r is the radius, and the diffusion coefficient $D = kT/(6\pi\eta r)$. Now $k_2^0 = 4kT/(3\eta)$, where k_2^0 is the rate constant for diffusion-controlled aggregation.

When there is an energy barrier to aggregation, only a fraction $1/W$ of encounters lead to attachment. The variable W is the stability ratio, $W = k_2^0/k_2$. Using W gives "slow coagulation" (hindered) times. In this case, the interaction energy and hydrodynamic viscous drag forces must be considered (16).

Finally, particles can also be brought into interaction distances by stirring or sedimentation so that the relative motions of two adjacent regions of fluid, each carrying particles, can cause particle encounters. Coagulation due to such influence is called "orthokinetic coagulation" as distinguished from the Brownian induced "perikinetic coagulation". The theory for orthokinetic coagulation is much more complicated than that for perikinetic and is not discussed here. However, shear can also cause dispersion if the energy introduced allows the interaction energy barrier to be overcome.

Stability to Coalescence. Up to this point, stability to aggregation has been considered. However, once aggregation has taken place in an emulsion, there remains the question of stability to coalescence. Usually emulsions made by mixing together two pure liquids are not very stable. To increase the stability, an additional component is usually needed, and it forms a film around the dispersed droplets to provide a barrier to both aggregation and coalescence. Although numerous agents and mechanisms are effective, the additional component is frequently a surfactant. Stability to coalescence involves the mechanical properties of the interfacial films. Mechanical properties will be considered further in the next section.

Considering stability to both aggregation and coalescence, the factors favoring emulsion stability can be summarized as follows:

- Low interfacial tension—low interfacial free energy makes it easier to maintain large interfacial area.

- Mechanically strong film—this acts as a barrier to coalescence and may be enhanced by adsorption of fine solids or of close-packed surfactant molecules.

- Electric double-layer repulsion—this repulsion acts to prevent collisions and aggregation.

- Small volume of dispersed phase—this reduces the frequency of collisions and aggregation. Higher volumes are possible (for close-packed spheres the dispersed-phase volume fraction would be 0.74), but in practice the fraction can even be higher.

- Small droplet size, if the droplets are electrostatically or sterically interacting.

- High viscosity—this slows the rates of creaming and coalescence.

An assessment of emulsion stability involves the determination of the time variation of some emulsion property such as those described in the earlier section "Physical Characteristics of Emulsions". The classical methods are well described in reference 9. Some newer approaches include the use of pulse nuclear magnetic resonance spectroscopy or differential scanning calorimetry (52).

Emulsifying Agents.   One general theory of emulsion type states that if an emulsifying agent is preferentially wetted by one of the phases, then more of the agent can be accommodated at the interface if that interface is convex toward that phase; that is, if that phase is the continuous phase. This theory works for both solids and soaps as emulsifying agents. For surfactant molecules it is referred to as Bancroft's rule; the liquid in which the surfactant is most soluble becomes the continuous phase. Very often, mixtures of emulsifying agents are more effective than single components. Some mixed emulsifiers may form a complex at the interface and thus yield low interfacial tension and a strong interfacial film.

A second general rule specifies that soaps of monovalent metal cations tend to produce O/W emulsions, but those of polyvalent metal cations will tend to produce W/O emulsions. Figure 16 illustrates the concept. In the example shown, the calcium ions each coordinate to two surfactant molecules that are aligned with their polar groups near the metal ion. This coordination forces the hydrocarbon tails into a wedgelike orientation. The hydrocarbon tails in a close-packed interfacial layer are most easily accommodated if the oil phase is the continuous phase. Thus, the oriented-wedge theory predicts that the calcium soap will produce a W/O emulsion. For the sodium soap, the charged polar groups of the surfactant tend to repel each other more strongly. This fact, together with the single cation–surfactant coordination, makes it most favorable for the polar groups to be in the continuous phase, and an O/W emulsion results.

An analogous rule to the oriented-wedge and Bancroft theories states that the liquid that preferentially wets the solid particles will tend to form the continuous phase. Thus if there is a low contact angle (measured through the water phase), then an O/W emulsion should form. Exceptions occur for each of these rules, and sometimes one will work where the others do not. They do remain useful for making initial predictions.

An empirical scale developed for categorizing single-component or mixed (usually nonionic) emulsifying agents, using this principle, is the hydrophile–lipophile balance or HLB scale. This dimensionless scale ranges from 0 to 20; a low HLB (<9) refers to a lipophilic surfactant (oil-soluble) and a high HLB (>11) to a hydrophilic (water-soluble) surfactant. In gen-

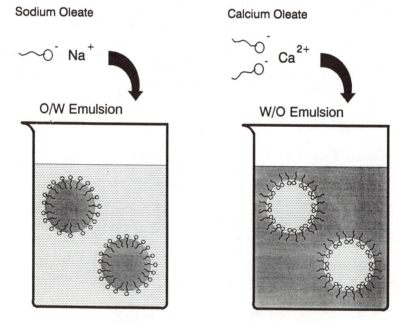

Figure 16. The oriented-wedge theory of emulsion type.

eral, W/O emulsifiers exhibit HLB values in the range of 3–8, and O/W emulsifiers have HLB values of about 8–18. Empirical tables of HLB values required to make emulsions out of various materials have been published (53). If the value is not known, then laboratory emulsification tests are required, using a series of emulsifying agents of known HLB values.

For example, to make a 15% O/W emulsion of a paraffinic mineral oil, a set of emulsion tests is performed with a series of blends of emulsifiers A (HLB 4.7) and B (HLB 14.9):

Emulsifier Blend	HLB	Emulsion Test
100% A	4.7	no emulsion
87% A + 13% B	6	no emulsion
68% A + 32% B	8	moderate emulsion
48% A + 52% B	10	stable emulsion
28% A + 72% B	12	moderate emulsion
6% A + 94% B	14	no emulsion
100% B	14.9	no emulsion

In this case an HLB of 10 is required to make the O/W mineral oil emulsion.

Various compilations and equations for determining emulsifier HLB values have been published (15, 53, 54). For example, in polyoxyethylene alcohols, $C_nH_{2n+1}(OCH_2CH_2)_mOH$, a class of nonionic surfactants, the HLB can be calculated from $HLB = E/5$, where E is the percentage by mass of ethylene oxide in the molecule. Experimentally, the unknown HLB of an emulsifier can be determined by mixing it with an emulsifier of known HLB and an oil for which the HLB required for emulsification is known. A series of tests such as just illustrated in the example can be used to determine the unknown HLB. From the previous definition of spreading it might seem that if an oil does not spread on water, then it should be emulsified in it, and vice versa. Ross et al. (55) did, in fact, find a correlation between HLB and the spreading coefficients of oil and water. They found that for a given type of emulsion, both low interfacial tension and a negative value of the appropriate spreading coefficient are necessary.

A limitation of the HLB system is that other factors are important as well. Also, the HLB is an indicator of the emulsifying characteristics of an emulsifier but not the efficiency of an emulsifier. Thus, although all emulsifiers having a high HLB will tend to promote O/W emulsions, the efficiency with which those emulsifiers act will vary considerably for any given system. For example, usually mixtures of surfactants work better than pure compounds of the same HLB.

Just as solubilities of emulsifying agents vary with temperature, so does the HLB, especially for the nonionic surfactants. A surfactant may thus stabilize O/W emulsions at low temperature but W/O emulsions at some higher temperature. The transition temperature, that at which the surfactant changes from stabilizing O/W to W/O emulsions, is known as the phase inversion temperature (PIT). At the PIT, the hydrophilic and oleophilic natures of the surfactant are essentially the same (another term for this is the HLB temperature). As a practical matter, emulsifying agents are chosen so that their PIT is far from the expected storage and use temperatures of the desired emulsions. In one method (56) an emulsifier with a PIT of about 50 °C higher than the storage–use temperature is selected. The emulsion is then prepared at the PIT where very small droplet sizes are most easily created. Next, the emulsion is rapidly cooled to the desired use temperature, at which now the coalescence rate will be slow, and a stable emulsion results.

Protective Agents and Sensitization.   The stability of a dispersion can be enhanced (protection) or reduced (sensitization) by the addition of material that adsorbs onto particle surfaces. Protective agents can act in several ways. They can increase double-layer repulsion if they have ionizable groups. The adsorbed layers can lower the effective Hamaker constant. An adsorbed film may necessitate desorption before particles can approach closely enough for van der Waals forces to cause attraction. If the adsorbed material extends out significantly from the particle surface, then an entropy decrease can accompany particle approach (steric stabilization). Finally, the adsorbed material may form such a rigid film that it poses a mechanical barrier to droplet coalescence.

Oil-field W/O emulsions may be stabilized by the presence of a protective film around the water droplets. Such a film can be formed from the asphaltene and resin fractions of the crude oil. When droplets approach each other during the process of aggregation, the rate of oil film drainage will be determined initially by the bulk oil viscosity, but within a certain distance of approach the interfacial viscosity becomes important. A high interfacial viscosity will significantly retard the final stage of film drainage and promote kinetic emulsion stability. If the films are viscoelastic, then a mechanical barrier to coalescence will be provided, yielding a high degree of emulsion stability. More detailed descriptions are given in references 26, 57, and 58.

Sensitizing agents can be protective agents that are added in smaller amounts than would be used normally. Sensitizing agents have several possible mechanisms of action. If the additive is oppositely charged to the dispersed particles, then decreased double-layer repulsion will result. In some kinds of protecting adsorption, a bilayer is formed with the outer layer having lyophilic groups exposed outwards. Addition of enough additive to form only the single layer will have lyophobic groups oriented outward with a sensitizing effect. If the additive is of long chain length, sometimes a bridging between particles occurs and induces aggregation (at higher concentrations protective action takes over because the potential bridging sites become covered).

Creaming, Inversion, and Demulsification

Creaming.   Droplets in an emulsion will have some tendency to rise or settle according to Stokes' law. An uncharged spherical droplet in a fluid will sediment if its density is greater than that of the fluid. The driving force is that of gravity; the resisting force is viscous and is approximately proportional to the droplet velocity. After a short period of time the particle reaches terminal (constant) velocity, dx/dt, when the two forces are matched. Thus

$$\frac{dx}{dt} = \frac{2r^2(\rho_2 - \rho_1)g}{9\eta} \tag{29}$$

where r is the particle radius, ρ_2 is the droplet density, ρ_1 is the external fluid density, g is the gravitational constant, and η is the bulk viscosity. If the droplet has a lower density than the external phase, then it rises instead (negative sedimentation). Emulsion droplets are not rigid spheres, so they may deform in shear flow. Also, with the presence of emulsifying agents at the interface, the droplets will not be noninteracting, as is assumed in the theory. Thus, Stokes' law will not strictly apply and may underestimate or even overestimate the real terminal velocity.

The process in which emulsion droplets rise or settle without significant coalescence is called creaming (Figure 17). This process is not emulsion breaking, but produces two separate layers of emulsion that have different droplet concentrations and are usually distinguishable from each other by color or opacity. The term comes from the familiar separation of cream from raw milk. According to Stokes' law, creaming will occur faster when there is a larger density difference and when the droplets are larger. The rate of separation can be enhanced by replacing the gravitational driving force by a centrifugal field. Centrifugal force, like gravity, is proportional to the mass, but the proportionality constant is not g but $\omega^2 x$, where ω is the angular velocity (equal to $2\pi \times$ revolutions per second) and x is the distance of the particle from the axis of rotation. The driving force for sedimentation becomes $(\rho_2 - \rho_1)\omega^2 x$. Because $\omega^2 x$ is substituted for g, one speaks of multiples of g, or "gs", in a centrifuge. The centrifugal acceleration in a centrifuge is not really a constant throughout the system but varies with x. The actual distance from top to bottom of a sedimenting column is usually small compared to the distance from the center of revolution, so the average acceleration is used. The terminal velocity then becomes

$$\frac{dx}{dt} = \frac{2r^2(\rho_2 - \rho_1)\,\omega^2 x}{9\eta} \tag{30}$$

The problem of deaerated bituminous froth produced from the hot-water flotation of bitumen from oil sands can serve as an illustration. This

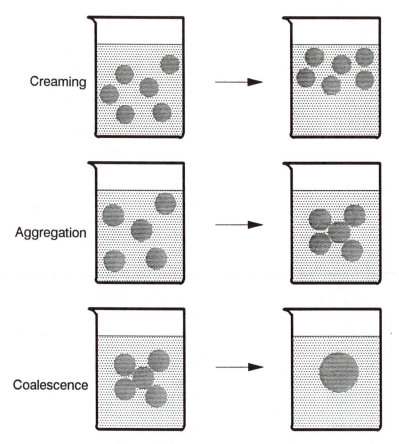

Figure 17. Creaming, aggregation, and coalescence in an O/W emulsion.

froth is a W/O emulsion from which the water must be removed prior to upgrading and refining. At process temperature (80 °C) the emulsion viscosity is similar to that of the bitumen, but the density, because of entrained solids, is higher. Taking η = 500 mPa·s (59) and ρ_1 = 1.04 g/mL, the rate of rise of 20-μm diameter water droplets under gravitational force will be very slow. According to equation 29:

$$\frac{dx}{dt} = -3.05 \times 10^{-6} \text{ cm/s}$$

$$\frac{dx}{dt} = 96 \text{ cm/year upward}$$

In a commercial oil sands plant, a centrifuge process is used to speed up the separation. The continuous centrifuges can operate at 2500 × g, the droplets having to travel 9 cm to reach the product stream. With the centrifugal force added, according to equation 30, the droplet velocity would become

$$\left(\frac{dx}{dt}\right)' = 2500 \left(\frac{dx}{dt}\right)$$

$$\left(\frac{dx}{dt}\right)' = 7.63 \times 10^{-3} \text{ cm/s outward}$$

This result is 2500 × faster than with gravity alone, but the residence time in the centrifuge would have to be about 20 min, and this length of time is not practical. To speed up the separation, naphtha is added to the level of 25%. The naphtha lowers the viscosity to about 4.5 mPa·s (59) and lowers the density of the continuous phase to 0.88 g/mL. Now the water droplets would sediment rather than rise under gravitational force, and although the emulsion density is much reduced, the absolute value of the density difference changes very little: $\Delta\rho$ = –0.07 g/mL originally, and becomes $\Delta\rho$ = +0.09 g/mL! The overall effect is to lower the viscosity by about 2 orders of magnitude. The droplet velocity now becomes $(dx/dt)''$ = 1.1 cm/s, which yields a satisfactory residence time of about 8 s.

In general, for given liquid densities, creaming will occur more slowly the greater the electrical charge on the droplets and the higher the emulsion viscosity. Although a distinct process, creaming does promote coalescence by increasing the droplet crowding and hence the probability of droplet–droplet collisions.

Inversion.   Inversion refers to the process in which an emulsion suddenly changes form, from O/W to W/O, or vice versa. This process was encountered in the example of changing emulsion viscosity given in the earlier section, "Physical Characteristics of Emulsions", subsection "Bulk Viscosity Properties" (Figure 6). The maximum volume fraction possible for an internal phase made up of uniform, incompressible spheres is 74%. Although emulsions with a higher internal volume fraction do occur, usually inversion occurs when the internal volume fraction exceeds some value reasonably close to ϕ = 0.74. Other factors have a bearing as well, of course, including the nature and concentration of emulsifiers and physical influences such as temperature or the application of mechanical shear.

The exact mechanism of inversion remains unclear, although obviously some processes of coalescence and dispersion are involved. In the region of the inversion point multiple emulsions may be encountered. The process is also not always exactly reversible. That is, hysteresis may occur if the inversion point is approached from different sides of the composition scale. Figure 18 shows the irreversible inversion of a diluted bitumen-in-water emulsion brought about by the application of shear (60).

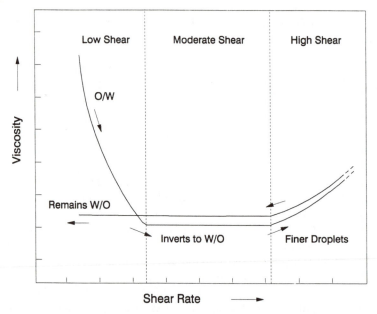

Figure 18. Example of the shear-induced inversion of an emulsion of diluted bitumen in water. (Reproduced with permission from Syncrude Canada Limited.

Demulsification (Emulsion Breaking). The stability of an emulsion is often a problem. Demulsification involves two steps. First, agglomeration or coagulation of droplets must occur. Then, the agglomerated droplets must coalesce. Only after these two steps can complete phase separation occur. Either step can be rate determining for the demulsification process. A typical W/O petroleum emulsion from a production well might contain 60–70% water. Some of this (free water) will readily settle out. The rest (bottom settlings°) requires some kind of specific emulsion treatment.

The first step in systematic emulsion breaking is to characterize the emulsion to be broken in terms of its nature (O/W or W/O), the nature of the two phases, and the sensitivity of the emulsifiers. On the basis of such an evaluation, a chemical addition could be made to neutralize the effect of the emulsifier, followed by mechanical means to complete the phase separation. For example, butter results from the creaming, breaking, and inversion of emulsified droplets in milk.

It follows directly from the previous considerations of emulsion stability that if an emulsion is stabilized by electrical repulsive forces, then demulsifi-

---

°Hence the field term "bottom settlings and water", or BS&W, used to characterize petroleum samples. (*See also* the Glossary.)

cation could be brought about by overcoming or reducing these forces. In this context the addition of electrolyte to an O/W emulsion could be used to achieve the critical coagulation concentration, in accord with the Schulze–Hardy rule. Similarly, demulsifying agents, designed to reduce emulsion stability by displacing or destroying the effectiveness of protective agents, can be applied. An example is antagonistic action, that is, addition of an O/W promoter to break a W/O emulsion (*see* the earlier section "Protective Agents and Sensitization"). These considerations are discussed in further detail in Chapter 9.

Apart from the aforementioned chemical treatments, a variety of physical methods are used in emulsion breaking. These physical methods are all designed to accelerate coagulation and coalescence. For example, oil-field W/O emulsions may be treated by some or all of settling, heating, electrical dehydration, chemical treatment, centrifugation, and filtration. The mechanical methods, such as centrifuging or filtering, rely on increasing the collision rate of droplets and applying an additional force that drives coalescence. An increase in temperature will increase thermal motions to enhance the collision rate, and also reduce viscosities (including interfacial viscosity), and thus increase the likelihood of coalescence. In the extremes, very high temperatures will cause dehydration due to evaporation, and freeze–thaw cycles will break some emulsions. Electrical methods may involve electrophoresis of oil droplets, causing them to collide, to break O/W emulsions. With W/O emulsions, the mechanism involves deformation of water droplets, because W/O emulsions are essentially nonconducting emulsions. Here the electric field causes an increase in the droplet area and disrupts the interfacial film. Increased droplet contacts increase the coalescence rate and thereby break the emulsion. More details on the application of these methods in large-scale continuous processes are given in Chapter 10.

List of Symbols

a_s	area per adsorbed molecule
A	available area of surface .sp 0.2 Hamaker constant
B	a constant $(3.93 \times 10^{39} \ A^{-2}s^{-2})$
C_s	solution concentration of the surfactant (M)
c_i	individual ion concentrations
D	diffusion coefficient
dx/dt	terminal (constant) velocity (in creaming)
e	elementary electronic charge
g	gravitational constant
H	distance separating spherical emulsion droplets
I	ionic strength

I_0	intensity of incident light beam (in light-scattering experiments)
I_d	intensity of light scattered (in light-scattering experiments)
I_t	intensity of transmitted beam of light (in light-scattering experiments)
k	Boltzmann constant
k_2	rate constant
$k_2{}^0$	rate constant for diffusion-controlled aggregation
l	length of path through sample (in light-scattering experiments)
n	number of moles; number of droplets per unit volume at time t
n_0	number of droplets per unit volume at time $t = 0$
N_A	Avogadro's number
N_c	capillary number
p, p_A, p_B	pressure, pressure in phase A, pressure in phase B
P_c	capillary pressure
r	radius of small droplets
r_p	pore radius
R	radius of spherical droplets
	also used as the gas constant
$R_1\,R_2$	principal radii of curvature
t	time
T	absolute temperature
T_K	temperature, termed the Krafft point, at which solubilities of surfactants show a strong increase due to micelle formation
v	velocity of displacing fluid
V	total interaction energy
V_A	attractive energy for spherical droplets
V_R	repulsive energy for spherical droplets
W	stability ratio
x	distance from droplet at which intensity of scattered light is taken (in light-scattering experiment)
z	counterion charge number
z_i	individual ion-charge numbers

Greek

$\alpha_0, \alpha_1, \alpha_2$	empirical constants in equations describing emulsion viscosity
γ	oil–water interfacial tension also used as a term in the repulsive energy expression for spherical droplets
$\dot\gamma$	shear rate
γ°	surface tension
Γ_s	surface excess of surfactant
δ	distance from droplet surface to Stern plane
ϵ	permittivity
ζ	zeta potential
η	viscosity

$[\eta]$	intrinsic viscosity
η_0	viscosity of continuous phase
η_{APP}	apparent viscosity
η_D	differential viscosity
η_{Red}	reduced viscosity
η_{Rel}	relative viscosity
η_{SP}	specific increase in viscosity
θ	contact angle of an oil–water interface in contact with a solid surface
$1/\kappa$	double-layer thickness
κ_C	conductivity of continuous phase
κ_D	conductivity of dispersed phase
κ_E	conductivity of emulsions
λ	wavelength of light (in light-scattering experiment)
μ_E	electrophoretic mobility
π	expanding pressure (surface pressure)
ρ_1	external fluid density
ρ_2	droplet density
σ	charge density
$\sigma°$	surface charge density
τ	turbidity
	shear stress
τ_Y	yield stress
ϕ	dispersed-phase volume fraction
ψ	potential
$\psi°$	surface potential
ω	angular velocity

Acknowledgments

I thank Eddy Isaacs (Alberta Research Council) and Karin Mannhardt (Petroleum Recovery Institute) for very helpful discussions and suggestions regarding the manuscript. Valuable suggestions offered by the external referees are also gratefully acknowledged.

References

1. Martinez, A. R. In *The Future of Heavy Crude and Tar Sands;* Meyer, R. F.; Wynn, J. C.; Olson, J. C., Eds.; UNITAR: New York, 1982; pp xvii–xviii.
2. Danyluk, M.; Galbraith, B.; Omana, R. In *The Future of Heavy Crude and Tar Sands;* Meyer, R. F.; Wynn, J. C.; Olson, J. C., Eds.; UNITAR: New York, 1982; pp 3–6.

3. Khayan, M. In *The Future of Heavy Crude and Tar Sands;* Meyer, R. F.; Wynn, J. C.; Olson, J. C., Eds.; UNITAR: New York, 1982; pp 7–11.
4. Becher, P. *Emulsions: Theory and Practice,* 2nd Ed.; ACS Monograph Series 162; American Chemical Society: Washington, DC, 1966.
5. Sumner, C. G. *Clayton's The Theory of Emulsions and Their Technical Treatment,* 5th Ed.; Blakiston: New York, 1954.
6. *Encyclopedia of Emulsion Technology;* Becher, P., Ed.; Dekker: New York, 1983; Vol. 1.
7. *Encyclopedia of Emulsion Technology;* Becher, P., Ed.; Dekker: New York, 1985; Vol. 2.
8. *Encyclopedia of Emulsion Technology;* Becher, P., Ed.; Dekker: New York, 1988; Vol. 3.
9. Lissant, K. J. *Demulsification: Industrial Applications;* Surfactant Science Series 13; Dekker: New York, 1983; Vol. 13.
10. *Micellization, Solubilization, and Microemulsions;* Mittal, K. L., Ed.; Plenum: New York, 1977; Vols. 1 and 2.
11. *Macro- and Microemulsions;* Shah, D. O., Ed.; ACS Symposium Series 272; American Chemical Society: Washington, DC, 1985.
12. Osipow, L. I. *Surface Chemistry Theory and Industrial Applications;* Reinhold: New York, 1962.
13. Ross, S.; Morrison, I. D. *Colloidal Systems and Interfaces;* Wiley: New York, 1988.
14. *Colloid Science;* Kruyt, H. R., Ed.; Elsevier: Amsterdam, Netherlands, 1952; Vol. 1.
15. Griffin, W. C. In *Kirk–Othmer Encyclopedia of Chemical Technology,* 2nd ed.; Interscience: New York, 1965; Vol. 8, pp 117–154.
16. Shaw, D. J. *Introduction to Colloid and Surface Chemistry,* 3rd ed.; Butterworth's: London, 1981.
17. *Microemulsions Theory and Practice;* Prince, L. M., Ed.; Academic Press: New York, 1977.
18. Neogi, P. In *Microemulsions: Structure and Dynamics;* Friberg, S. E.; Bothorel, P., Eds.; CRC Press: Boca Raton, FL, 1987; pp 197–212.
19. Poettmann, F. H. In *Improved Oil Recovery;* Interstate Compact Commission: Oklahoma City, OK, 1983; pp 173–250.
20. Bansal, V. K.; Shah, D. O. In *Microemulsions: Theory and Practice;* Prince, L. M., Ed.; Academic Press: New York, 1977; pp 149–173.
21. Schramm, L. L.; Novosad, J. J. *Colloids Surf.* **1990,** *46,* 21–43.
22. Whorlow, R. W. *Rheological Techniques;* Wiley: New York, 1980.
23. Van Wazer, J. R.; Lyons, J. W.; Kim, K. Y.; Colwell, R. E. *Viscosity and Flow Measurement;* Wiley: New York, 1963.
24. Fredrickson, A. G. *Principles and Applications of Rheology;* Prentice-Hall: Englewood Cliffs, NJ, 1964.
25. Schramm, L. L. *J. Can. Pet. Technol.* **1989,** *28,* 73–80.
26. Malhotra, A. K.; Wasan, D. T. In *Thin Liquid Films;* Ivanov, I. B., Ed.; Surfactant Science Series 29; Dekker: New York, 1988; pp 829–890.
27. Harkins, W. D.; Alexander, A. E. In *Physical Methods of Organic Chemistry;* Weissberger, A., Ed.; Interscience: New York, 1959; pp 757–814.
28. Padday, J. F. In *Surface and Colloid Science;* Matijevic, E., Ed.; Wiley-Interscience: New York, 1969; Vol. 1, pp 101–149.
29. Miller, C. A.; Neogi, P. *Interfacial Phenomena Equilibrium and Dynamic Effects;* Dekker: New York, 1985.

30. Cayias, J. L.; Schechter, R. S.; Wade, W. H. In *Adsorption at Interfaces;* Mittal, K. L., Ed.; ACS Symposium Series 8; American Chemical Society: Washington, DC, 1975; pp 234–247.
31. Adamson, A. W. *Physical Chemistry of Surfaces,* 4th Ed.; Wiley: New York, 1982.
32. Lake, L. W. *Enhanced Oil Recovery;* Prentice-Hall: Englewood Cliffs, NJ, 1989.
33. Rosen, M. J. *Surfactants and Interfacial Phenomena,* 2nd ed.; Wiley: New York, 1989.
34. *Industrial Applications of Surfactants;* Karsa, D. R., Ed.; Royal Society of Chemistry: London, 1987.
35. *Anionic Surfactants: Physical Chemistry of Surfactant Action;* Lucassen-Reynders, E. H., Ed.; Dekker: New York, 1981.
36. Ross, S. *J. Phys. Colloid Chem.* **1950,** *54,* 429–436.
37. Currie, C. C. In *Foams;* Bikerman, J. J., Ed.; Reinhold: New York, 1953; pp 297–329.
38. Takamura, K. *Can. J. Chem. Eng.* **1982,** *60,* 538–545.
39. Takamura, K.; Chow, R. S. *J. Can. Pet. Technol.* **1983,** *22,* 22–30.
40. Schramm, L. L.; Smith, R. G.; Stone, J. A. *AOSTRA J. Res.* **1984,** *1,* 5–14.
41. Schramm, L. L.; Smith, R. G. *Colloids Surf.* **1985,** *14,* 67–85.
42. Schramm, L. L.; Smith, R. G. *Can. J. Chem. Eng.* **1987,** *65,* 799–811.
43. Schramm, L. L.; Smith, R. G. *AOSTRA J. Res.* **1989,** *5,* 87–107.
44. Smith, R. G.; Schramm, L. L. *Fuel Proc. Technol.* **1989,** *23,* 215–231.
45. Verwey, E. J. W.; Overbeek, J. Th. G. *Theory of the Stability of Lyophobic Colloids;* Elsevier: New York, 1948.
46. Hunter, R. J. *Zeta Potential in Colloid Science;* Academic Press: New York, 1981.
47. James, A. M. In *Surface and Colloid Science;* Good, R. J.; Stromberg, R. R., Eds.; Plenum: New York, 1979; Vol. 11, pp 121–186.
48. Riddick, T. M. *Control of Stability Through Zeta Potential;* Zeta Meter: New York, 1968.
49. O'Brien; White, L. R. *J. Chem. Soc. Faraday 2* **1978,** *74,* 1607–1626.
50. Schramm, L. L.; Smith, R. G. Canadian Patent 1,265,463, February 6, 1990.
51. van Olphen, H. *An Introduction to Clay Colloid Chemistry,* 2nd Ed.; Wiley: New York, 1977.
52. Cavallo, J. L.; Chang, D. L. *Chem. Eng. Prog.* **1990;** *86,* 54–59.
53. *The HLB System;* ICI United States, Inc. (now ICI Americas, Inc.): Wilmington, DE, 1976.
54. *McCutcheon's Emulsifiers and Detergents;* MC Publishing Co.: Glen Rock, NJ, 1990; Vol. 1.
55. Ross, S.; Chen, E. S.; Becher, P.; Ranauto, H. J. *J. Phys. Chem.* **1959,** *63,* 1681.
56. Shinoda, K.; Saito, H. *J. Colloid Interface Sci.* **1969,** *30,* 258–263.
57. Cairns, R. J. R.; Grist, D. M.; Neustadter, E. L. In *Theory and Practice of Emulsion Technology;* Smith, A. L., Ed.; Academic Press: New York, 1976; pp 135–151.
58. Jones, T. J.; Neustadter, E. L.; Whittingham, K. P. *J. Can. Pet. Technol.* **1978,** *17,* 100–108.
59. Schramm, L. L.; Kwak, J. C. T. *J. Can. Pet. Technol.* **1988,** *27,* 26–35.
60. Schramm, L. L.; Hackman, L. P., unpublished results.

RECEIVED for review December 18, 1990. ACCEPTED revised manuscript April 26, 1991.

2

Practical Aspects of Emulsion Stability

E. E. Isaacs and R. S. Chow

Alberta Research Council, Oil Sands and Hydrocarbon Recovery, Edmonton, Alberta, Canada T6H 5X2

The formations of emulsions by droplet breakup mechanisms is described in relation to the rheological properties of the dispersed and continuous phases. Once emulsions are formed, their stability is largely determined by the molecular, electric double-layer, steric, and hydrodynamic forces. The application of Derjaguin, Landau, Verwey, and Overbeek (DLVO) theory to successfully predict the stability of oil-in-water emulsions is described. The stability of water-in-crude-oil emulsions can be monitored with electrokinetic sonic analysis; the change in size of the water droplets is indicated by the change in the ultrasound vibration potential signal. With this development, the dewatering characteristics of chemical demulsifiers can be assessed rapidly. For water dispersed in conventional crude oil, a combination of oil-soluble and water-soluble demulsifiers gave the best results.

Emulsion Formation

Thermodynamic Concepts. Generally, two methods are used to prepare dispersions, namely, building up particles from molecular units (nucleation and growth) and subdivision of larger bulk materials into smaller units (grinding or emulsification). The process of emulsification, that is, dispersion of liquids in liquids, is governed by the surface forces. The free energy of formation of droplets from a bulk liquid (ΔG_{form}) is illustrated in Figure 1 and is given by

$$\Delta G_{form} = \Delta A \gamma_{12} - T \, \Delta S_{conf} \tag{1}$$

0065-2393/92/0231-0051 $07.75/0
© 1992 American Chemical Society

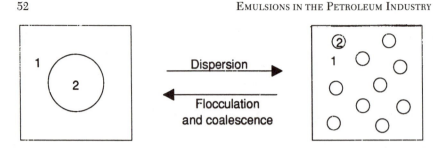

Figure 1. Schematic representation of emulsion formation and breakdown.
(Adapted from reference 1.)

where ΔA is the increase in interfacial area, γ_{12} is the interfacial tension between the two liquids, and $T\Delta S_{conf}$ is the entropy contribution resulting from the increase in configurational entropy when a large number of droplets is formed. Usually $\Delta A\gamma_{12} \gg T\Delta S_{conf}$, and hence emulsification is a nonspontaneous process. However, the energy required for emulsification is orders of magnitude larger than the thermodynamic energy ($\Delta A\gamma_{12}$) for creating a new surface. This larger energy requirement results from the additional effect of creating a curved interface with a different radius. The additional energy required can be expressed in the Young–Laplace equation,

$$\Delta P = \gamma_{12} \left[\frac{1}{r_1} + \frac{1}{r_2} \right] \qquad (2)$$

where ΔP is the Laplace pressure difference and r_1 and r_2 are the principal radii of curvature. The presence of surfactant (which lowers γ_{12}) lowers the energy required for emulsification.

Droplet Breakup. As described by the Young–Laplace equation, before creation of a new surface can take place, deformation of the dispersed phase is required. In laminar flow, this deformation is produced by viscous forces exerted by the surrounding bulk liquid, that is, $\eta_1(dv/dz)$, where η_1 is the viscosity of the continuous phase and dv/dz is the velocity gradient. The energy input to create the necessary velocity gradient can be of the order of 10^3 times the thermodynamic energy ($\Delta\gamma_{12}$). The excess energy is dissipated as heat.

The phenomenon of droplet breakup is of great importance in the preparation of emulsions. If a stream of liquid is injected with little turbulence into another liquid with which it is immiscible, the cylinder that may form is unstable, breaks down in several spots, and breaks up into droplets (Figure 2a). If the injection rate is such as to produce turbulence, the disruption is faster, and many smaller droplets are produced (Figure 2b). If in addition the liquid impinges against a surface, many smaller droplets will be formed.

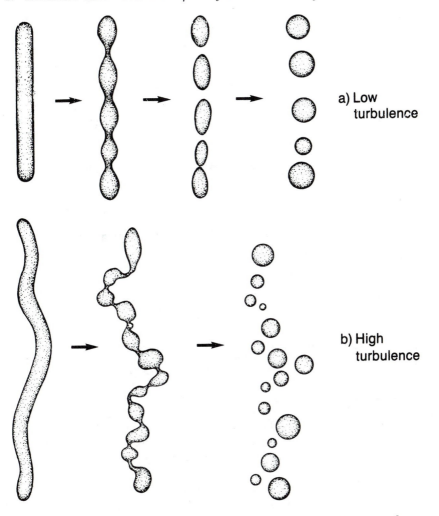

Figure 2. Schematic of the breakup of a liquid stream injected into another immiscible liquid. (Reproduced with permission from reference 2. Copyright 1983 Dekker.)

In actual situations several processes occur simultaneously. The details of any particular dispersion processes are also affected by the viscosity of each phase, the shear in the system, the interfacial energy, the pressure of solid particles, and dissolved substances. In nonuniform shear flow (e.g., tubular Poiseuille flow), for example, droplet breakup can be related to the bulk rheological properties of the dispersed and continuous phases and the critical Weber number (We_c) as shown in Figure 3 (3). The We is a dimensional group defined by

$$We = \frac{\eta_1 \dot{\gamma}_e R}{\gamma} \tag{3}$$

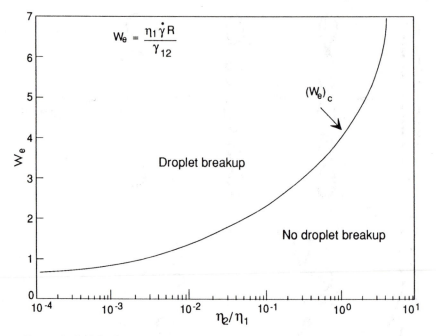

Figure 3. Droplet breakup as a function of viscosity ratio depicted schematically from the work of Chin and Han (3). The solid line represents the critical Weber number value above which droplet breakup will occur.

where $\dot{\gamma}_e$ is the rate of extension (shear rate multiplied by deformation parameter); R is the radius of the particle; and η_1 and η_2 are the viscosities of the continuous and dispersed phases, respectively. At a given η_1/η_2, lowering γ_{12} (through the use of surfactants) lowers the energy (described by the We_c) required for droplet breakup. Figure 3 shows that the greater the viscosity ratio (η_1/η_2), the easier it is to form the emulsion. This concept provides an explanation for the observations that in heavy-oil reservoirs (high η_1/η_2 ratio), water-in-oil emulsions are produced in preference to oil-in-water emulsions.

Stability in Oil-in-Water Emulsions

Stable emulsions often form during industrial processing. On the microscopic scale, the reasons that the droplets remain dispersed fall into two broad categories: (1) physical barriers to coalescence and (2) electrical repulsion between droplets. An example of a physical barrier is the presence of finely divided solids at the oil–water interface. Of primary concern, however, is the consideration of electrical forces because their influence is significant at relatively longer distances. Electrical repulsive forces arise when the double layers surrounding charged droplets overlap, and thus the "thickness" of the double layer in relation to the size of the particle is an important parameter.

Application of DLVO Theory. Some of the concepts and expressions of Derjaguin, Landau, Verwey, and Overbeek (DLVO) theory of colloid stability have been described in Chapter 1, or can be found in many different textbooks (4, 5). The application of DLVO theory to oil-in-water colloids with special reference to the stability of bitumen-in-water emulsions will be discussed here.

Theoretical Aspects. The basic concept of DLVO theory is that the stability of a colloid can be described in terms of the repulsive and attractive interactions between droplets:

$$V_{tot}(h) = V_{rep}(h) + V_{attr}(h) \tag{4}$$

where V_{tot}, V_{rep}, and V_{attr} are the total, repulsive, and attractive interactions, respectively; and h is the separation distance between the particles. Previously, when DLVO theory was applied to bitumen-in-water emulsions (6), it was assumed that the repulsive interaction originating from the overlapping double layers surrounding two droplets of radius a and could be expressed by:

$$V_{rep}(h) = \frac{64\pi a n_0 kT\gamma^2}{\kappa^2} \exp(-\kappa h) \tag{5}$$

and

$$\gamma = \tan h \left[\frac{ze\zeta}{4kT}\right] \tag{6}$$

$$\kappa^2 = \frac{e^2 \Sigma n_i z_i}{\epsilon kT} \tag{7}$$

where n_0 is the number of ions per unit volume, k is the Boltzmann constant, T is the absolute temperature, z is the valency of the ion, e is the electronic charge, ϵ is the permittivity of the continuous phase, ζ is the zeta potential of the droplet, and κ is the Debye–Hückel function that characterizes the extension of the double layer. The origin of the attractive interaction was assumed to be the van der Waals or dispersion force, which can be expressed as

$$V_{attr}(h) = \frac{-Aa}{12h}\left[1 - \frac{5.32h}{\lambda}\ln\left[1 + \frac{\lambda}{5.32h}\right]\right] \tag{8}$$

where A is the Hamaker constant that can be calculated from Lifshitz theory (7, 8) and λ is the London wavelength of roughly 100 nm (9). The Hamaker constant used in the study was 1.7×10^{-20} J (6).

Determination of the Electrophoretic Mobility. To evaluate the equation for the double-layer interaction (eq 5), the zeta potential, ζ, must be known; it is calculated from the experimentally measured electrophoretic mobility. For emulsions, the most common technique used is particle electrophoresis, which is shown schematically in Figure 4. In this technique the emulsion droplet is subjected to an electric field. If the droplet possesses interfacial charge, it will migrate with a velocity that is proportional to the magnitude of that charge. The velocity divided by the strength of the electric field is known as the electrophoretic mobility. Mobilities are generally determined as a function of electrolyte concentration or as a function of solution pH.

Many different commercial instruments can be used for particle electrophoresis. They can generally be divided into two categories: (1) those in which the velocity of the particle is determined by observing the particle through a microscope and determining the velocity by timing the movement over a grid of known dimension, and (2) those that determine the velocity by the Doppler shift of scattered radiation from the moving particle. Instruments in the second category have the advantage of measuring the velocity of a large number of particles in a short time, but they are generally more expensive.

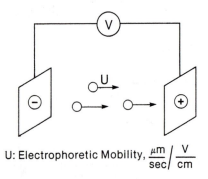

U: Electrophoretic Mobility, $\dfrac{\mu m}{sec} \Big/ \dfrac{V}{cm}$

Figure 4. Schematic diagram of particle electrophoresis.

Calculation of the Zeta Potential. The conversion of electrophoretic mobility to zeta potential is complicated somewhat by the existence of the electrophoretic relaxation effect. Figure 5 shows a schematic diagram of this effect. As we expose an emulsion droplet and its surrounding double layer to an electric field, the double layer distorts to the shape shown in the figure. This distorted double layer now creates its own electric field that

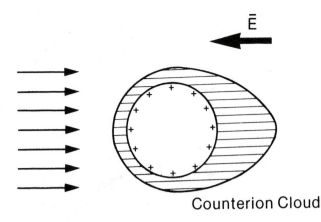

$\bar{E}$

Counterion Cloud

Figure 5. A schematic diagram of the electrophoretic relaxation effect. The distorted ion cloud around the particle generates its own electric field that opposes the motion of the particle.

opposes the motion of the particle. The magnitude of the opposition is related to the thickness of the double layer in relation to the size of the particle. This relationship can be expressed by the parameter κa, where κ is as defined for eq 7 and a is the radius of the particle.

In the limits of small κa ($\kappa a \ll 1$) and large κa ($\kappa a \gg 300$), simple equations can convert the electrophoretic mobility, U, to zeta potential. These are the Hückel and Smoluchowski equations, respectively. They may be expressed by

$$\text{Hückel equation } (\kappa a \ll 1) \qquad U = \frac{\zeta \epsilon}{1.5 \eta} \qquad (9)$$

$$\text{Smoluchowski equation } (\kappa a \gg 300) \qquad U = \frac{\zeta \epsilon}{\eta} \qquad (10)$$

where η is the viscosity of the continuous phase. The limited range of applicability of these equations, however, leaves a large area (i.e., $1 \gg \kappa a \gg 300$) where the extent of the relaxation effect would need to be accounted for. For example, in a solution containing 0.01 M univalent electrolyte, $\kappa = 3.31 \times 10^{-8}$ m^{-1}, and for 1.0-μm radius droplets, $\kappa a = 330$, a situation just barely covered by the Smoluchowski equation.

O'Brien and White (*10*), taking into account the relaxation effect, derived the relationship between electrophoretic mobility and zeta potential. The results of the theoretical calculation are shown in Figure 6. Figure 6 shows the relationship between electrophoretic mobility, U, and zeta potential for two values of κa, 114 and 285. Each value of κa has its own relation-

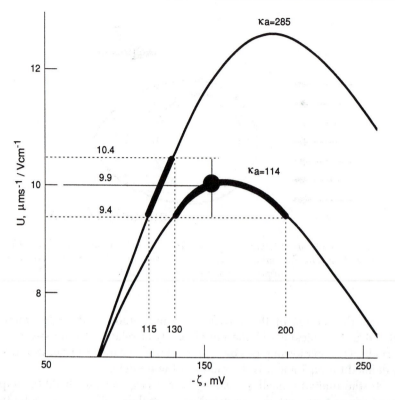

Figure 6. An example of the relationship between electrophoretic mobility and zeta potential.

ship between zeta potential and electrophoretic mobility; practically this means that knowing both electrolyte concentration and particle size is vital. In addition, the obvious maximum in mobility makes it difficult to determine a unique value of zeta potential when the measured mobility is close to the maximum value. The combination of these two factors can make it difficult to determine a zeta potential when dealing with emulsions that commonly possess a wide distribution of particle sizes. We will assume that we are dealing with an emulsion that is polydisperse and has particle that range in radius from 2.0 to 5.0 μm. The electrophoresis measurements are being conducted in a solution containing 3.0×10^{-4} M NaCl. The value for κa in these situations is 114 for a 2-μm droplet and 285 for a 5-μm droplet. If the mean electrophoretic mobility measured was 9.9 μm/s·cm/V with a variation of ±5.0%, without knowledge of the size of the droplets (almost a certainty because of the dark field illumination used in most particle electrophoresis apparatus), the zeta potential could range from −115 to −200 mV.

In reference to the example of the stability of bitumen-in-water emulsions, production samples were obtained from the Alberta Research Council's 150-cm physical simulator. These were acquired as steam injection experiments with the simulator were being conducted on Athabasca oil sands. The emulsion was diluted and electrophoretic mobilities were determined with a Rank Brothers MkII electrophoresis apparatus, in which the velocity is determined by timing the movement of the particles. Measurements were conducted as a function of electrolyte concentration in the presence of NaCl, $CaCl_2$, and $Al_2(SO_4)_3$. Mobilities were converted to zeta potentials by using the Smoluchowski equation (eq 10) and are shown in Figure 7. All curves show similar trends in the zeta potential in that the potential slowly becomes more electronegative with increasing electrolyte concentration and then slowly decreases towards zero. The existence of this minimum has been reported previously and is not expected from double-layer theory. Currently this is an area of research (*11, 12*). The bitumen droplets also show reduced zeta potential in the presence of either Ca^{2+} or Al^{3+} ion.

Interpretation of the Energy Diagrams.   Once the values of zeta potential are determined, eq 9 can be evaluated. The results obtained are summed to those of the van der Waals interaction (eq 8) and plotted as a potential energy of interaction as a function of separation distance between the droplets. The energy is often expressed in terms of kT units, in order to better relate the energy of the interaction to the thermal energy of Brownian

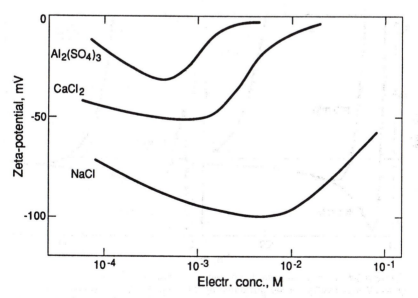

Figure 7. The zeta potentials for bitumen-in-water emulsions as a function of electrolyte concentration. (Reproduced with permission from reference 6. Copyright 1982.)

motion. DLVO theory predicts that if an energy of more than $15kT$ exists in the energy diagram, under the conditions of no flow, the emulsion would be stable. This stability is due to the small probability that one droplet would possess this much energy.

Figure 8 shows the calculated energy diagrams for the bitumen-in-water emulsions as expressed in eq 4. The figure shows that in solutions containing 20 mM CaCl$_2$ there is a net negative energy of interaction at all separation distances; that is, as bitumen droplets meet, they will coagulate and coalesce. If the droplets are immersed in a solution of 10 mM CaCl$_2$, an energy maximum of $8kT$ is shown at 3.5 nm. This energy "barrier" is not of sufficient height to result in a stable emulsion but will slow the rate of coagulation. The prediction for 3 mM CaCl$_2$ is a stable emulsion as a large positive energy is experienced at 13 nm. In solutions containing NaCl, the energy diagrams show that the emulsion will be stable in solutions containing 10 mM NaCl. As the concentration is increased, however, the large positive energy appears to shift to shorter distances, but the more interesting feature is the appearance of a energy minimum. With 300 mM NaCl, the minimum is of sufficient depth ($\sim11kT$) that droplets could become closely associated but would still remain 3–4 nm apart. In this situation the emulsion would appear to flocculate but could be easily redispersed by mechanical agitation.

To confirm the applicability of DLVO theory, coagulation tests were performed on the production samples. The results, shown photographically

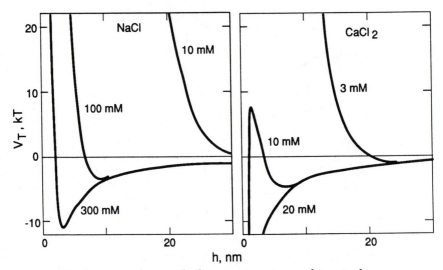

Figure 8. The energy diagram for bitumen-in-water emulsions in the presence of NaCl and CaCl$_2$. In the presence of 300 mM NaCl the emulsion should flocculate, and in the presence of the 20 mM CaCl$_2$ the emulsion will coagulate. (Reproduced with permission from reference 6. Copyright 1982.)

in Figure 9, indicate rapid coagulation in a 20 mM solution of $CaCl_2$ and flocculation and creaming in the presence of 300 mM NaCl. These results confirm the applicability of DLVO theory for bitumen-in-water emulsions.

Ionizable Surface-Group Model.   One of the mechanisms to develop interfacial charge is the ionization of functional groups on the surface of the dispersed phase. Healy and White (*13*) developed this concept into a model to predict the electric properties of interfaces. When this model was applied to the bitumen–water interface, it was assumed that the surface charge originated from the dissociation of carboxy group that belong to natural surfactants present in the oil (*14*). This work was extended in a later study (*15*) where the zeta potentials as calculated by the model were used in conjunction with DLVO theory to predict the stability of the bitumen-in-water emulsions. Agreement between the expected and observations from coagulation tests was excellent over a wide range of solution pH and electrolyte concentration.

The model can also be applied to the conventional crude-oil–water interface (*16, 17*). Those studies demonstrated that it was necessary to invoke the dissociation of up to three different types of functional groups to successfully use the model. Another interesting finding was that the electro-

Figure 9. The behavior of bitumen-in-water emulsions in the presence of 300 mM NaCl (right side) and 20 mM CaCl₂. The emulsions are behaving as predicted by DLVO theory.

phoretic mobility with pH curve for conventional crude-oil-in-water emulsions would change with time after dispersion. When applying the model, this aging process was thought to be due to certain functional groups leaving the interface.

Non-DLVO Forces.   Although DLVO theory worked very well for the electrolyte-induced coagulation of bitumen-in-water emulsions, it cannot be applied in some cases.

Polymer-Induced Flocculation.   Polymer-induced flocculation is the most commonly used technique for breaking water-in-oil emulsions in the petroleum industry. This topic will be covered in much more detail in Chapter 9, but will be briefly covered in this section.

For the polymer to be effective, it must adsorb to the interface and maintain a certain configuration. Thus the following discussion describes various experimental techniques used for the study of adsorption density and configuration of polymer at the interface. After adsorption occurs, the main mechanisms of flocculation are due to the adsorption of a single polymer molecule on separate particles, interaction through the interpenetration of adsorbed polymer, and interactions due to the loss of freedom of movement of the polymer chains.

Experimental Techniques for the Study of Polymer Adsorption. The theory of polymer adsorption and configuration is still not fully developed because the difficulties encountered in designing experiments are immense. A technique is required to measure the configuration of a polymer molecule at the interface and thus obtain concentration also.

Although the configuration of the adsorbed polymer cannot be seen directly, some experimental techniques can give an idea of the concentration and configuration. For example, bound polymer concentration can be determined through measuring the concentration of polymer in solution after the introduction of an emulsion of known surface area. Various spectroscopic techniques (infrared, electron spin resonance, nuclear magnetic resonance) may distinguish the loss of rotational and translational freedom when a polymer segment adsorbs onto an interface, and thus the amount of bound segments of a polymer can be estimated. Adsorbed layer thickness can be indirectly obtained by either determining the increase of hydrodynamic radius of the droplets or by measuring the diffusion coefficient of the emulsion droplets. This information can also be obtained by ellipsometry, assuming sufficient refractive index difference between the droplet, adsorbed layer, and bulk solution.

Interaction of Droplets with Adsorbed Polymer Layers.   In the simplest case of polymer-induced flocculation, a single polymer molecule adsorbs onto two separate droplets, and the result is flocculation. This type of situation is favored by low coverage of polymer on the surfaces and the continuous phase being a relatively good solvent. If these conditions are met, it is possible for the polymer to adsorb to the surface and remain extended into the solution in order to attach to another droplet.

As two polymer-coated droplets approach each other, the adsorbed layers will begin to interact. The extent of the interaction can be interpreted through the free energy of two different terms, the interpenetration term and the mixing term. The interpenetration term is repulsive and entropic in nature; that is, the loss of freedom of movement as the layers interpenetrate results in a repulsion term. This repulsion term can be overcome, however, by altering the mixing term, which can be repulsive or attractive in nature. The mixing term has been handled the same as for the dilution of polymer in bulk solution. The magnitude of the term is dependent upon the number of polymer molecules in the overlap region, volume fractions of the solvent and polymer in this area, and a parameter known as the Flory–Huggins interaction parameter that takes into account the enthalpy of mixing and volume of mixing effects. By altering the solution conditions, a situation can be created in which the continuous phase becomes a poorer solvent for the adsorbed polymer; flocculation results. This situation can be created by changing the temperature, adding a poorer solvent to the colloid, or adding electrolyte.

A special case to consider is the existence of what is termed depletion flocculation. This term originated from the observation that the addition of a small amount of nonadsorbing polymer will cause flocculation in a system. The reason for this effect is that, as the particles approach each other, the mobile chains of nonadsorbing polymer are squeezed out from between the particles. As the particles approach to very close distances, almost pure solvent exists between the particles, and at a given separation, the osmotic pressure that results from this pure solvent drives it out into the bulk solution and thereby causes flocculation.

Hydration and Hydrophobic Forces.   As surfaces approach each other to distances less than 10 nm, a force exists that is not accounted for in conventional DLVO theory. This force can be repulsive or attractive in nature and can be of magnitude greater than either the double-layer or van der Waals interactions. This force was discovered by Israelachivili and co-workers (*18–20*) using a unique apparatus that they developed that directly measured the force between two mica surfaces. For solid mica surfaces immersed into water, the force was repulsive and oscillatory and exponentially decayed as a function of distance from the surface (*see* Figure 10).

The origin of this force is thought to be the modification of the orientation of water molecules in the vicinity of a surface to form a structure. As surfaces approach each other, additional energy is required to decompose this structure. The oscillatory property of the force was thought to be due to

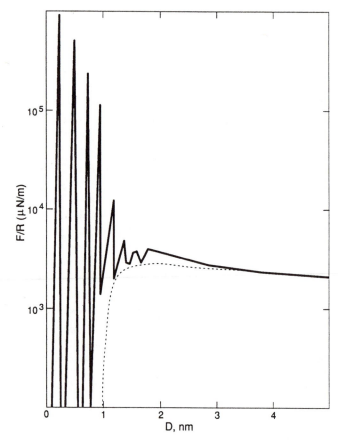

Figure 10. Schematic representation of the force as a function of distance between two mica surfaces as reported by Israelachivili and Adams (18).

the various "layers" of water molecules. A complete theory on the origin and quantitative behavior of this force is not available. It is an area of current research effort (*21, 22*).

Although the predicted bitumen-in-water emulsion stability can be accomplished without invoking this force, recent research has shown (*23*) that this force exists between liquid bilayers immersed in aqueous and nonaqueous liquids. With some types of oils, it may be important to consider this force.

Thin-Film Stability in Water-in-Oil Emulsions

Importance of Thin Films.   In order for coalescence to result in ultimate separation of the water droplets from the continuous medium, thinning and disruption of the liquid film (called "thin film" prior to rupture) between the droplets must take place. Again this process is governed by the surface forces (van der Waals, double-layer, and steric forces) that operate in the liquid lamellae between the droplets. The rupture of the thin film is usually the result of thermal or mechanical fluctuation, which results in stretching of the liquid surface with formation of surface waves that will grow in amplitude and result in droplet coalescence. Any force that "dampens" the formation of such waves will reduce or prevent coalescence. In systems containing surfactants, wave dampening occurs as a result of the so-called Marangoni effect, which arises from the presence of the surfactant film. If a film is subjected to local stretching as a result of thickness fluctuation, the consequent increase in surface area causes a local increase in interfacial tension (decrease in the surface excess of the adsorbed surface-active agent in that region) that opposes the stretching (*see* Figure 11). Because a finite time is required for surfactant molecules to diffuse to this region of the interface to restore the original surface tension (Marangoni effect), the fluctuations will tend to dampen rather than grow (i.e., rupture is prevented).

Persistent Films.   The stability of water-in-crude-oil emulsions and the factors contributing to that stability are long-standing problems of importance in the production of oil from underground reservoirs. Although a great deal of effort has been expended in the investigation of the destabilization of water-in-oil emulsions, the actual mechanisms are still not well understood. Although natural surfactants present in the crude oil can in their own right stabilize the emulsions, other types indigenous material in the oil tend to gather at the interface and play a significant role in hindering the thinning and rupture of the liquid films and act as a structural barrier to coalescence of the water droplets. Asphaltenes and porphyrinic compounds may be the stabilizing agents. Furthermore, the presence of finely divided

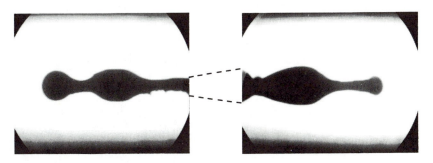

Figure 11. Photograph of bitumen drop in aqueous NaOH showing low- and high-tension sites due to periodic diffusion of natural surfactants from the oil–water interface to the aqueous phase in a spinning drop tensiometer.

solids such as sand, wax crystals, and clay particles can stabilize emulsions (24). All these "natural stabilizing agents" prevent thinning of the thin film and explain why the crude-oil films are so persistent. An example of a persistent film is shown in Figure 12, which is a photomicrograph of the bottom layer of a water-in-oil (Leduc crude) emulsion treated with demulsifier. Even after 3 days, the water droplets are still enveloped by a thin crude-oil film that will not drain any further without additional treatment.

Recent work (25–27) has shown that surfactants or medium-chain alcohols that modify the rigidity of the film, in combination with demulsifiers (which act mainly to flocculate the water droplets) can considerably speed up the separation process (see the section entitled "Effect of Demulsifier Mixture").

Use of Ultrasonic Vibration Potential To Monitor Coalescence. The complex chemical nature of crude oils makes it difficult to relate the dispersion behavior to the physicochemical properties at the crude-oil–water interface. In addition, the nonpolar and nontransparent nature of the oleic phase provides significant obstacles for studies of the interactions of the suspended water droplets in real systems. Recent development (28, 29) of electroacoustical techniques has shown considerable promise for electrokinetic measurements of colloidal systems and the direct monitoring of the rate and extent of coagulation (flocculation and coalescence) of water droplets in nontransparent water-in-oil media. The electroacoustic measurement for colloidal systems in nonpolar media is based on the ultrasound vibration potential (UVP) mode, which involves the applica-

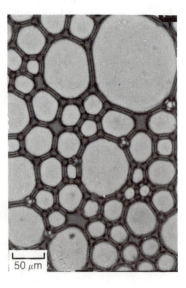

Figure 12. Photomicrograph of settled bottom layer of a water-in-oil emulsion taken 3 days after addition of demulsifier (100 ppm of Duomeen C). (Reproduced with permission from reference 27. Copyright 1990 Elsevier.)

50 μm

tion of a sonic field and the detection of an electric field. A schematic diagram of the probe and the principle of UVP are represented in Figure 13.

When voltage U_2 is applied at the transducer, a sound wave propagates into the colloid. If the densities of the dispersed and continuous phases differ, relative motion between the colloidal particles and their double layer will result. The combined relative motion will generate an electric field, which is detected as voltage U_1 between the electrodes. The measured signals are proportional to the high-frequency electrophoretic mobility $\mu(\omega)$. As derived by Babchin et al. (28), the frequency-dependent electrophoretic mobility, $\mu(\omega)$, for the case of low potentials, can be expressed by

$$\mu(\omega) = \frac{\epsilon \zeta f(\kappa R)}{\sqrt{(6\pi\eta R)^2 + \left[\frac{4}{3}\omega\rho_{\text{eff}}R^2\right]^2}} \tag{11}$$

with

$$\overline{R} = 1 + \frac{R}{\delta} \tag{12a}$$

$$\rho_{\text{eff}} = \rho_0 + \frac{9}{4R}\sqrt{\frac{2\eta\rho}{\omega}}\left[1 = \frac{2R}{9\delta}\right] \tag{12b}$$

where ϵ, η, and ρ are the dielectric permittivity, viscosity, and density of the continuous phase, respectively; ρ_0 is the density of the particle; ρ_{eff} is the effective density of a sphere in oscillatory motion; δ is the thickness of a fluid layer surrounding the particle that influences liquid flow around the particle

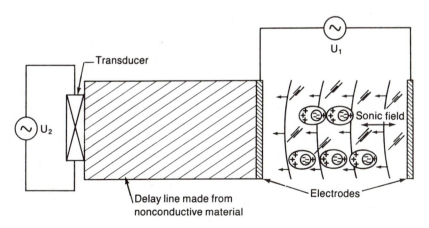

Figure 13. Schematic diagram showing the principle of UVP. (Reproduced with permission from reference 27. Copyright 1990 Elsevier.)

$[\delta = (2\eta/\rho\omega)^{\frac{1}{2}}]$; ζ is the electrokinetic potential; $f(\kappa R)$ is the Henry function; κ is the Debye–Hückel function that characterizes the extension of the double layer; R is the particle radius; and ω is the frequency.

The supplementary phase angle $\phi(\omega)$ between the applied electric field and the particle velocity response, at a fixed frequency ω is given by

$$\tan \phi(\omega) = -\frac{2}{9}\frac{\omega R^2 \rho_{eff}}{\eta \bar{R}} \tag{13}$$

The magnitude of the potential difference between the electrodes, $\Delta\Psi_0$, in the circuit U_1 is given by

$$\mathrm{UVP}(\omega) = \frac{\Phi \Delta\rho c G_f}{K^\circ} \mu(\omega) \tag{14}$$

where Φ is the volume fraction of dispersed phase, $\Delta\rho$ is the density difference between the dispersed and continuous phases, c is the sound velocity in the emulsion, K° is the complex conductivity of oil, and G_f is a geometrical factor dependent on the geometry of the electrodes.

Equations 11–14 clearly show that an increase in the effective particle radius, promoted by a coagulation process, will result in the diminution of the UVP signal and a shift in the phase angle. In addition, the low value of the complex conductivity of oil, K°, acts as a natural amplifier to provide for a significant $\Delta\Psi_0$ that makes it easy to monitor UVP even for small values of $\mu(\omega)$. Figure 14 shows that the UVP signal is sensitive to the water content in the emulsions.

Figure 15 shows the sensitivity of the UVP signal to the coagulation process. Photographs taken at 3, 12, and 24 min show that as the droplet size grows, the UVP signal decreases. Thus by measuring the UVP signal, coagulation can be monitored.

By monitoring both the UVP signal and phase angle ϕ, changes in zeta that effect only the UVP signal can be distinguished from changes in particle radius that effect both the UVP and $\phi(\omega)$ (30).

Effect of Demulsifier Mixture.   In previous studies (27) Duomeen C, which was effective in causing flocculation of the water droplets, was not very effective in breaking the interfacial film formed between the water droplets, which inhibits coalescence. (Duomeen C is a mixture of many types of surfactants; the general classification is a fatty acid ester nitrogen derivative.) However, Duomeen C in combination with docusate sodium (Aerosol OT), a hydrophilic surfactant, was much more effective in causing water separation compared to the individual chemicals. This effect is shown in Figure 16 for a 6 vol% water-in-oil (Leduc crude) emulsion in which both the UVP signal (20 min after chemical addition) and the volume

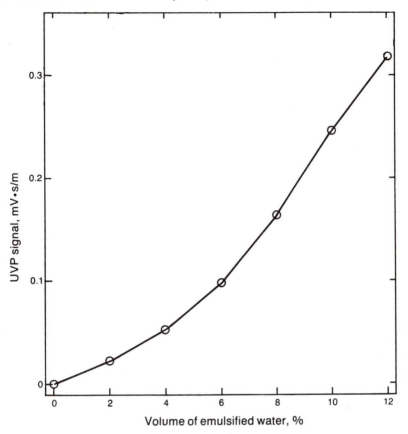

Figure 14. Sensitivity of UVP signal to water content in the emulsions.

of water recovered by centrifugation are plotted against the weight percent of Duomeen C in the mixture.

As expected, Aerosol OT, the water-soluble surfactant, by itself had practically no effect on either the UVP signal or the water separation. Duomeen C alone also had little effect on the amount of water recovered by centrifugation. The change in UVP signal therefore, likely reflected Duomeen C's ability to flocculate the droplets. The mixture of the two chemicals, however, performed in a synergistic manner, a 1:1 mixture of chemicals being most effective. Also, a direct correspondence is apparent between the minimum in UVP signal and maximum in water recovery by centrifugation.

Film Drainage and Demulsifier Adsorption.   To enhance the coagulation process, a common practice is to use chemical demulsifiers that are believed to

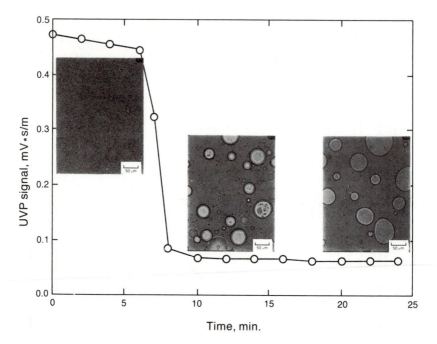

Figure 15. Sensitivity of the UVP signal to the coagulation process. Demulsifier was added at 4 min. (Reproduced with permission from reference 27. Copyright 1990 Elsevier.)

1. promote the flocculation of the droplets by weakening the repulsive forces that stabilize the emulsion

2. enhance the drainage of the interfacial film between the flocculated droplets.

The choice of chemical is usually based on trial-and-error procedures; hence, demulsifier technology is more of an art than a science. In most cases a combination of chemicals is used in the demulsifier formulation to achieve both efficient flocculation and coalescence. The type of demulsifiers and their effect on interfacial area are among the important factors that influence the coalescence process. Time-dependent interfacial tensions have been shown to be sensitive to these factors, and the relation between time-dependent interfacial tensions and the adsorption of surfactants at the oil–aqueous interface was considered by a number of researchers (27, 31–36). From studies of the time-dependent tensions at the interface between organic solvents and aqueous solutions of different surfactants, Joos and co-workers (33–36) concluded that the adsorption process of the surfactants at the liquid–liquid interface was not only diffusion controlled but that adsorption barriers and surfactant molecule reorientation were important mecha-

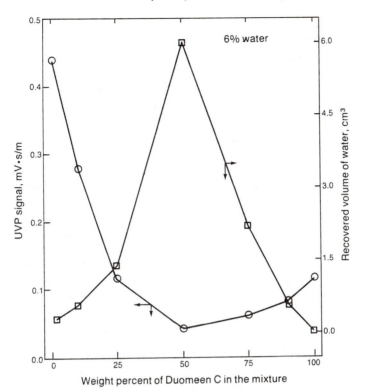

Figure 16. Comparison of the coalescence process using mixtures with a total concentration of 100 ppm.

nisms depending on the system. For surfactant adsorption from the oil phase to the oil–water interface, a reorientation process at the interface was thought to be the rate-controlling step; adsorption occurred at a much slower rate than that observed for a purely diffusion-controlled situation.

Vogler (*31*) developed a mathematical model to derive semiquantitative kinetic parameters interpreted in terms of transport and adsorption of surfactants at the interface. The model was fitted to experimental time-dependent interfacial tension, and empirical models of concentration-dependent interfacial tension were compared to theoretical expressions for time-dependent surfactant concentration. Adamczyk (*32*) theoretically related the mechanical properties of the interface to the adsorption kinetics of surfactants by introducing the compositional surface elasticity, which was defined as the proportionality coefficient between arbitrary surface deformations and the resulting surface concentrations. Although the expressions to describe the adsorption process differed from one another, it was demonstrated that the time-dependent interfacial tensions mirrored the change of surface-active substances at the interface.

For studies with real systems, Isaacs et al. (27) used the simplified approach of examining changes at the oil–water interface without specifying adsorption mechanisms or pathways. Based on measurements of time-dependent interfacial tensions, the following expression (termed the spreading rate parameter) served to characterize the relative adsorption performance of demulsifiers or demulsifier combination:

$$\text{spreading rate parameter} = \frac{\gamma_0 - \gamma_{s/e}}{\Delta t} \qquad (15)$$

where γ_0 and $\gamma_{s/e}$ are the steady-state value of oil–aqueous interfacial tension in the absence and presence of added chemical, respectively; Δt represents the time required to reach the steady-state tension, $\gamma_{s/e}$. A schematic of the technique used to measure this parameter is shown in Figure 17 together with the depiction of the adsorption of Duomeen C from the oil to the oil–water interface and Aerosol OT from the water to the water–oil interface.

To understand the reasons for the dewatering effectiveness resulting from the interactions between the two surfactants, time-dependent interfacial tensions were measured to examine the transfer of the surfactants from the bulk to the interface. Based on these measurements, Figure 18 shows a plot of the apparent spreading rate parameter, which is a measure of both

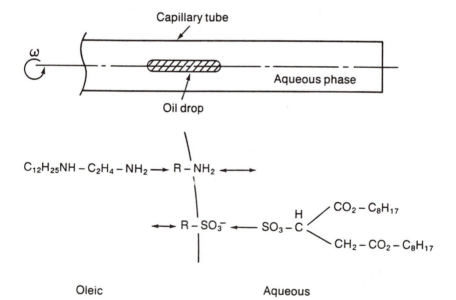

Figure 17. Schematic of the spinning drop capillary technique depicting the adsorption of oil-soluble and water-soluble surfactants from the bulk to the interface. The molecular structures of the two additives are also given.

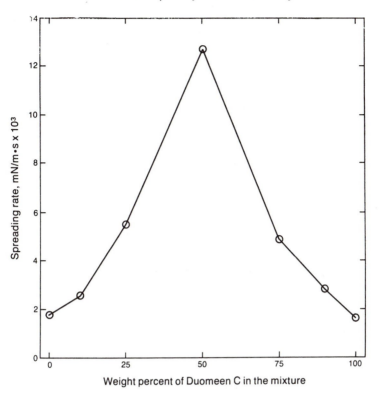

Figure 18. Apparent spreading rate as a function of the ratio of Duomeen C and Aerosol OT concentrations in the mixture. The total concentration is 100 ppm.

the ease of deformation of the interface and the speed of adsorption or mass transfer of material to the interface, as a function of the ratio of Duomeen C and Aerosol OT concentrations in the mixture. Clearly, the maximum spreading rate occurs at a 1:1 ratio of the reagents. Figure 19 shows an excellent agreement between the spreading rate parameter and both the water recovery by centrifugation and the final UVP signal. The direct correlation between results of dynamic interfacial tensions and the results of coalescence or dewatering efficiency is a new phenomenon that has the potential for quantitative analysis and tailor-making demulsifiers for a particular crude-oil emulsion system.

Conclusions

The procedures for breaking oil-in-water and water-in-oil emulsions are very different. For oil-in-water emulsions, the interfacial charge contributes to

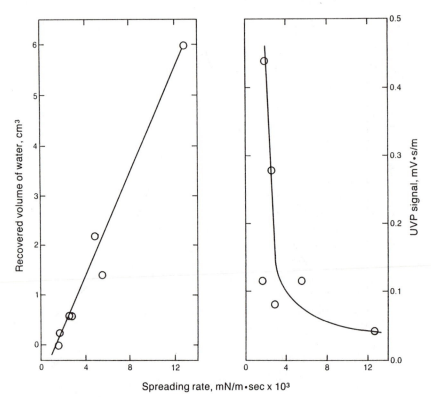

Figure 19. Comparison of electroacoustic analysis and dewatering efficiency of the mixed demulsifier as a function of the apparent spreading rate.

the stability; in water-in-oil emulsions, the strength of the interfacial film of oil that forms between the water droplets is of prime concern. In this discussion the presence of solids at the interface was not considered; if they are present, however, the additional stability in their presence would require attention.

The more common emulsion formed in the petroleum industry is the water-in-oil type. The sensitivity of electrokinetic sonic analysis to coagulation–coalescence processes in water-in-oil media is of great importance. It allows for rapid selection and optimization of different chemical demulsifiers. In addition, as a research tool, it supports the development of a fundamental understanding of chemical treatment of water-in-oil emulsions.

Acknowledgments

We thank the Alberta Research Council for financial support and Carol Hopper for help in typing this manuscript. This chapter is Alberta Research Council Contribution No. 2025.

List of Symbols

a	particle radius
A	Hamaker constant
c	sound velocity in the emulsion
dv/dz	velocity gradient
e	electronic charge
$f(\kappa R)$	Henry function
G_f	geometrical factor dependent on geometry of electrodes
k	Boltzmann constant
n_0	number of ions per unit volume
r_1, r_2	principal radii of curvature
R	particle radius
T	absolute temperature
U	electrophoretic mobility
U_1, U_2	voltage
V_{attr}	attractive interaction
V_{rep}	repulsive interaction
V_{tot}	total interaction
We_c	Weber number
z	valency of the ion

Greek

δ	thickness of fluid layer surrounding a particle
ΔA	increase in interfacial area
ΔG_{form}	free energy of formation of droplets
ΔP	Laplace pressure difference
ΔS_{conf}	increase in configurational entropy
Δt	time required to reach steady-state tension
$\Delta \gamma_{12}$	thermodynamic energy
$\Delta \rho$	density difference between dispersed and continuous phases
$\Delta \Psi_0$	potential difference between electrodes
γ_0	steady-state interfacial tension in absence of added chemical
γ_{12}	interfacial tension between two liquids
$\gamma_{s/e}$	steady-state interfacial tension in presence of added chemical

$\dot{\gamma}_e$ rate of extension
ϵ dielectric permittivity
ζ zeta potential of droplet
η viscosity of continuous phase
η_1 viscosity of continuous phase
η_2 viscosity of dispersed phase
η_1/η_2 viscosity ratio
κ Debye–Hückel function that characterizes extension of double layer
λ London wavelength
ρ density
ρ_0 particle density
ρ_{eff} effective density of a sphere in oscillatory motion
ϕ phase angle
$\phi(\omega)$ supplementary phase angle
Φ volume fraction
ω frequency

References

1. Tadros, Th. G. *L'actulaite Chim.* **1987.**
2. Lissant, K. J. *Demulsification: Industrial Applications;* Dekker: New York, 1983.
3. Chin, H. B.; Han, G. D. *J. Rheol.* **1980,** 1.
4. Heimenz, P. C. *Principles of Colloid and Surface Chemistry;* Dekker: New York, 1977.
5. Shaw, D. J. *Introduction to Colloid and Surface Chemistry,* 3rd ed.; Butterworth: London, 1980.
6. Takamura, K.; Chow, R. *Energy Process./Can.* **1982,** 9, 29.
7. Overbeek, J. Th. G. *Colloid Science;* Elsevier: Amsterdam, Netherlands, 1952.
8. Vincent, B. *J. Colloid Interface Sci.* **1973,** 42, 270.
9. Gregory, J. *J. Colloid Interface Sci.* **1981,** 83, 138.
10. O'Brien, R. W.; White, L. R. *J. Chem. Soc. Faraday Trans. 2* **1978,** 74, 1607.
11. Chow, R. S.; Takamura, K. *J. Colloid Interface Sci.* **1988,** 125, 226.
12. Hunter, R. J. *Zeta Potential in Colloid Science;* Academic Press: New York, 1981.
13. Healy, T. W.; White, L. R. *Adv. Colloid Interface Sci.* **1978,** 9, 303.
14. Takamura, K.; Chow, R. *Colloids Surf.* **1985,** 15, 35.
15. Takamura, K.; Chow, R.; Tse, D. L. In *Flocculation in Biotechnology and Separation Systems;* Attia, Y. A., Ed.; Elsevier: New York, 1987.
16. Chow, R. S.; Takamura, K. *J. Colloid Interface Sci.* **1988,** 125, 212.
17. Takamura, K.; Buckley, J.; Morrow, N. *SPE Reservoir Eng.* **1987,** 62, contribution no. 16964.
18. Israelachivili, J. N.; Adams, G. E. *J. Chem. Soc. Faraday Trans. 1* **1978,** 74, 975.
19. Israelachivili, J. N.; McGuiggan, P. M. *Science* **1988,** 241, 795.

20. Pashley, R. M.; McGuiggan, P. M.; Ninham, B. W.; Evans, D. F. *Science* **1985,** *229,* 1088.
21. van Oss, C. J.; Giese, R. F.; Costanzo, P. M. *Clays Clay Miner* **1990,** *38,* 151.
22. Leikin S.; Kornyshev, A. A. *J. Chem. Phys.* **1990,** *92,* 6890.
23. Israelachivili, J. N.; Wennerstrom, H. *Langmuir* **1990,** *6,* 873.
24. Menon, V. B.; Wasan, D. T. *Colloids Surf.* **1988,** *29,* 7.
25. Sjoblom, J.; Ming-yuan, Li; Hoiland, H.; Johansen, E. *J. Colloids Surf.* **1990,** *46,* 127.
26. Sjoblom, J.; Soderlund, H.; Lindbland, S.; Johansen, E J.; Skjarvo, I M. *Colloid Polym. Sci.* **1990,** *268,* 389.
27. Isaacs, E E.; Huang, H.; Babchin, A J.; Chow, R S. *Colloids Surf.* **1990,** *46,* 177.
28. Babchin, A J.; Chow, R S.; Sawatzky, R P. *Adv. Colloid Interface Sci.* **1989,** *30,* 111.
29. Isaacs, E E.; Huang, H.; Chow R S.; Babchin, A J. *Colloids Surf.* **1990,** *46,* 177–192.
30. Babchin, A J.; Sawatzky, R P.; Chow, R S.; Isaacs, E E.; Huang, H. Presented at the 21st Annual Fine Particle Society Meeting, San Diego, CA, 1990.
31. Vogler, E A. *J. Colloid Interface Sci.* **1989,** *133,* 228.
32. Adamczyk, Z. *J. Colloid Interface Sci.* **1989,** *133,* 23.
33. Vermeulen, M.; Joos, P. *Colloids Surf.* **1989,** *36,* 13.
34. Vermeulen, M.; Joos, P. *Colloids Surf.* **1988,** *33,* 337.
35. Hunsel, J V.; Joos, P. *Colloids Surf.* **1987,** *25,* 251.
36. Hunsel, J V.; Joos, P. *Colloids Surf.* **1987,** *24,* 139.

RECEIVED for review December 18, 1990. ACCEPTED revised manuscript June 3, 1991.

Emulsion Characterization

Randy J. Mikula

Department of Energy, Mines, and Resources, CANMET, Fuel Processing Laboratory, P.O. Bag 1280, Devon, Alberta, Canada, T0C 1E0

This chapter outlines emulsion characterization techniques ranging from those commonly found in field environments to those in use in research laboratories. Techniques used in the determination of bulk emulsion properties, or simply the relative amount of oil, water, and solids present, are discussed, as well as those characterization methods that measure the size distribution of the dispersed phase, rheological behavior, and emulsion stability. A particular emphasis is placed on optical and scanning electron microscopy as methods of emulsion characterization. Most of the common and many of the less frequently used emulsion characterization techniques are outlined, along with their particular advantages and disadvantages.

AN EMULSION IS USUALLY DEFINED as a system consisting of a liquid dispersed with or without an emulsifier in an immiscible liquid, usually in droplets of larger than colloidal sizes. In petroleum emulsions, solids play an extremely important role in both the formation and stability of emulsions. These solids can be oil-phase components such as wax crystals or precipitated asphaltenes, or mineral components that are partially oleophilic, a property that allows them to act as stabilizers between the oil and water phases.

Characterization of such emulsions therefore often involves three phases: the water phase, the oil phase, and the solids. Complete characterization of an emulsion could therefore involve detailed chemical and physical analysis of all of the emulsion components, as well as any bulk properties that might be of interest (viscosity, density, etc.). This level of detail is clearly beyond the scope of this discussion. For the purposes of this chapter, emulsion characterization will be defined as the quantification of the phases present, the determination of the nature and size distribution of

0065–2393/92/0231–0079 $13.65/0

the dispersed phase, and the measurement of the stability of the dispersed phase.

Chemical properties of the individual phases will not be discussed in this chapter because any number of conventional analytical techniques can be applied to the separated oil, water, and solids phases that might make up a typical emulsion. From a processing point of view, when treating (separating) emulsions, the main concerns usually are total oil, water, and solids in each of the feed, product, and tailings streams. As long as water in the product oil and oil in the tailings water are low, then the process is working, and detailed analysis of the composition of the emulsion components is not required. A fundamental understanding of the interactions between emulsion components that determine stability is often only required or "resorted to" when process upsets occur. Interfacial properties, film rigidity or strength, and surface tension between the various emulsion phases are extremely important in determining stability of the dispersed phase, but they will not be discussed in detail here because these measurements fall under the category of techniques used to characterize the individual emulsion components. A number of review articles and books discuss these techniques and many aspects of emulsion science and emulsion characterization (1–13).

Emulsion characterization and technology development have been driven by the medical, agricultural, food, and cosmetics industries; the petroleum and oil industries have borrowed these technologies and adapted them to their particular applications. A number of books and review articles discuss aspects of emulsion technologies specifically related to oil-field and petroleum applications (14, 15). These petroleum applications have become especially important since the advent of surfactant flooding and other tertiary oil recovery methods in which emulsions are used and/or formed.

The characterization techniques that will be discussed here are used in field situations, on-line, and in the laboratory. In order to characterize an emulsion, it is necessary to determine the amount of each phase present, the nature of the dispersed and continuous phases, and the size distribution of the dispersed phase. The stability of an emulsion is another important property that can be monitored in a variety of ways, but most often, from a processing point of view, stability is measured in terms of the rate of phase separation over time. This phenomenological approach serves well in process situations in which emulsion formation and breaking problems can be very site specific. However, emulsion stability is ultimately related to the detailed chemistry and physics of the emulsion components and their interactions, and these details cannot be completely ignored.

This chapter is structured according to the types of information provided by the various characterization techniques. Applications in the field or in the laboratory are discussed, along with advantages and disadvantages. An exception to this format is made for the microscopic techniques, which, because of their wide applicability (and because they happen to be my specialty), will be covered separately. Microscopy is seen by many as the ultimate characterization tool, at least in terms of droplet size distribution, because direct observation of a sample is simple to interpret. In spite of the perception of microscopy as the ultimate characterization method, many problems and pitfalls can be encountered in interpreting microscopic observations, and these will be discussed at length.

Two other techniques that might also warrant separate discussion, electrokinetics and viscosity determinations, are covered in detail in Chapters 2 and 4 and will only be briefly mentioned here. The figures are given with detailed captions so that they may be referred to independently of the main text.

Bulk Properties

Although surface phenomena determine the fundamental properties of emulsions in terms of size distributions and stability, the bulk properties or bulk compositions are the yardsticks by which plant operators and process personnel measure process efficiency. Accurate determination of the oil, water, and solids (if present) is therefore one of the most important aspects of emulsion characterization.

Oil-Continuous or Water-Continuous Emulsions. In most emulsion systems, the nature of the dispersed phase is quite clear. There is usually little doubt that an oil–water emulsion with 5% water is a water-in-oil emulsion, oil being the continuous phase. In many separation and treatment processes, in addition to the oil product (which might contain some emulsified water), and the water tailings (which might contain some emulsified oil), there can be an interface emulsion, a so-called rag layer, whose continuous and disperse phases are generally unknown. Figure 1 shows an optical micrograph of such an emulsion in which the oil and water are both continuous and dispersed, depending upon where in the sample one looks. Often these emulsions build to a certain level, continuously re-form and break in the separator, and never cause operational problems. Occasionally, however, they can build to such an extent that they require removal and separate treatment. Knowledge of the nature of the dispersed phase is therefore critical in determining an effective treatment.

The ratio of the oil to water alone is not sufficient to determine which is the dispersed phase because the presence of emulsifiers or solids can significantly affect the amount of dispersed phase distributed in a given amount of continuous phase. Figure 2 shows an example of a fire flood emulsion that is water-in-oil, although the emulsion contains 63% (by weight) water. Explosives are often water-in-oil emulsions with up to 92% water phase (*16, 17*).

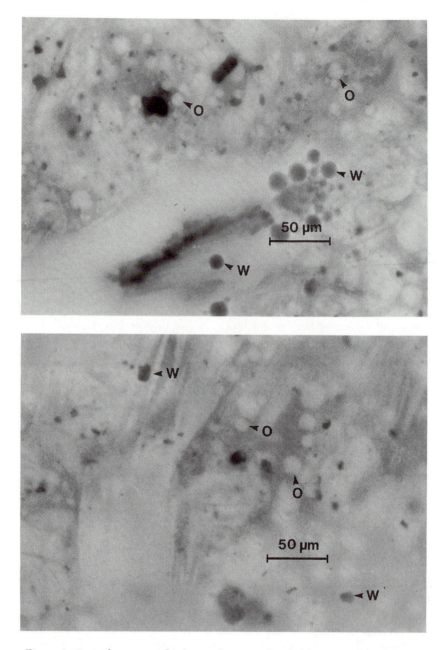

Figure 1. Optical micrograph of a rag-layer emulsion showing complex structure. In reflected mode with blue–violet light, the water component (W) is dark, and the oil component (O) fluoresces yellow (bright in this black-and-white reproduction). On a very short scale both oil-in-water and water-in-oil emulsions can be seen.

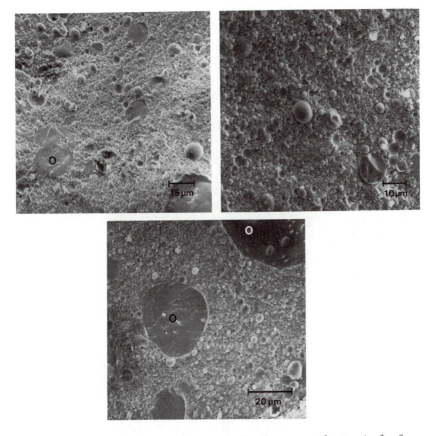

Figure 2. Scanning electron micrographs (at three magnifications) of a fire flood emulsion illustrating a case in which, although the water–oil ratio is 2.5:1, water is the dispersed phase. The composition of this emulsion is 63% water, 11% solids, and 26% oil. The compositions of the dispersed and continuous phases were determined from the X-ray signal excited in the electron microscope. The size of the dispersed water phase ranges from less than 0.1 μm up to about 10 μm. The large features labeled O are regions of oil phase that can be described as oil emulsified in a continuous phase of a water-in-oil emulsion. These complex systems are difficult to characterize with anything but microscopic methods.

Several techniques determine whether the continuous phase is oil or water. The simplest is the dilution method, in which a drop or two of the emulsion is added to water. If it is an oil-in-water emulsion it will spread and disperse. If it is water-in-oil it will remain as a drop (*18*). The dilution test can be effective, but care must be taken that sampling the emulsion does not itself determine the continuous phase. For instance, drawing a water-in-oil emulsion up through the capillary of a dropper can cause the emulsion to invert because of interactions of the water phase with the hydrophilic glass walls. This phenomenon is extremely important in the microscopic characterization of emulsions and will be discussed further in that section.

Another option is to dye the continuous phase (19, 20). Dyeing is best done under a microscope where the coloring of the continuous phase can be observed with an appropriate water- or oil-soluble dye. Several water-soluble dyes, such as methylene orange or methylene blue, can be used. A common oil-soluble dye is fuchsin. If methylene blue is mixed with the emulsion and no color change is observed, then the emulsion is most likely water-in-oil. The opaque nature of oil-field emulsions limits the applicability of these color techniques.

Electrical conductivity or capacitance of the emulsion might determine the nature of the continuous phase because water-in-oil would be much less conductive than a similar oil-in-water emulsion. This technique is useful in the laboratory to monitor emulsion inversion as a function of oil, water, or chemical addition and also is the basis of many level sensors in field situations. A significant change in the amount of solids in the oil or water phases in a process situation might give a conductivity reading that is ambiguous in terms of defining the continuous phase (21), and therefore the water–oil interface level.

Oil and Water Content with Solids.

Many methods determine the relative amounts of water and oil in emulsions. Because many emulsions of interest to the oil industry also contain solids, determination of the solids content is also important (22, 23). The Institute of Petroleum (IP), the American Petroleum Institute (API), and the American Society for Testing and Materials (ASTM) have developed standard methods for these determinations, as have most organizations or laboratories where these determinations are routinely performed (24–28). These methods are all some modification of the Dean–Stark procedure, in which the sample is placed in a porous thimble and refluxed with a suitable organic solvent.

Modified Dean–Stark Procedure.

A schematic of the apparatus for the modified Dean–Stark procedure is shown in Figure 3. The sample is held in a porous thimble suspended above the refluxing organic solvent. The water in the emulsion sample is codistilled with the solvent and is trapped in the side arm where water content can be determined directly (27, 29, 30). The organic component is dissolved in the solvent and carried to the bottom of the apparatus where it can later be quantified gravimetrically after the solvent is removed. The solids are retained in the porous thimble and are also determined gravimetrically. The size range of solids retained on the thimble is naturally related to its porosity; a common modification of the technique involves centrifuging and decanting the bitumen or oil component (in solvent) to separate the fine solids. A typical report of results would

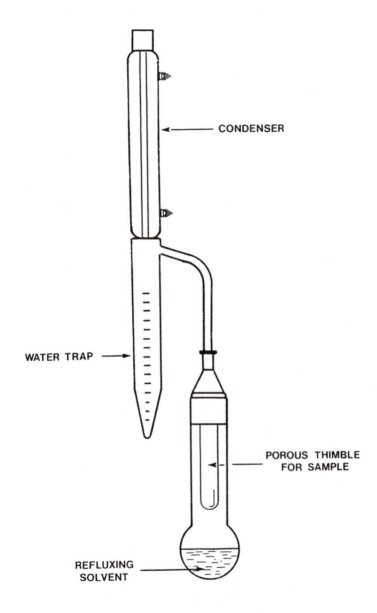

CONDENSER

WATER TRAP

POROUS THIMBLE
FOR SAMPLE

REFLUXING
SOLVENT

Figure 3. Modified Dean–Stark apparatus. The solvent (usually toluene) drips through the sample, dissolving the organic component and leaving the solids behind. The water, which codistills with the solvent, condenses and is trapped in the side arm and measured volumetrically. The solids and organic phases are determined gravimetrically after evaporating the solvent from the sample thimble and the solvent flask.

then include percent water, percent solids, percent bitumen–oil with fines, percent bitumen–oil without fines, and percent solids with fines. Often the fine solids that pass through the thimble are ignored and simply included with the bitumen or oil content. Other common modifications to the standard procedures include the type of solvent used or the refluxing time.

The major disadvantage of this technique is the time required for the refluxing and for sample workup after the extraction to complete the gravimetric determinations. Because bitumen and heavy oil may sometimes contain asphaltenes or heavy organic components that may be in solid form in the oil phase, interpretation of the fine solids component requires other information about solids composition. The advantage of the technique is that the sample can be relatively large, an important consideration in many situations where sample streams are quite heterogeneous.

Centrifugation.   Another commonly used technique for determination of oil, water, and solids is a simple centrifuge test. As with the Dean–Stark method, a standard procedure has been developed by several organizations (27, 31, 32). Basically, the test consists of diluting the emulsion with a known amount of solvent and centrifuging for a fixed time. With the specially designed centrifuge tube shown in Figure 4, the amount of water and solids in the sample can be read volumetrically. The water-and-solids is the denser phase, so the results are generally reported as BS&W (basic sediment and water) as a volumetric percent. Because it is fast and reliable, the centrifuge test is probably the most commonly used technique for field evaluation of water content in oil product streams.

The most common variation to the API, ASTM, and IP methods for field use is the addition of demulsifier or knockout drops to facilitate the separation of the phases. The demulsifier is generally added at concentrations significantly above what would be normal operational levels. The disadvantages of this technique are that it does not separate the water and solids, and it is not useful for very high-water-content streams. Filling the centrifuge tube with a representative sample can also be difficult, especially with viscous emulsions.

Oil and Water Content without Solids.   The presence of solids in an emulsion system reduces the characterization options because techniques that can quantify all three phases are not readily available. In many situations, however, only quantification of water in the oil product or oil in the tailings water is important. In other cases, the solids content is insignificant. In these situations, the range of techniques available is much more extensive and, in general, more applicable to field and on-line applications. Obviously, the methods discussed earlier also apply to these systems, along with a variety of spectroscopic and chemical analytical techniques.

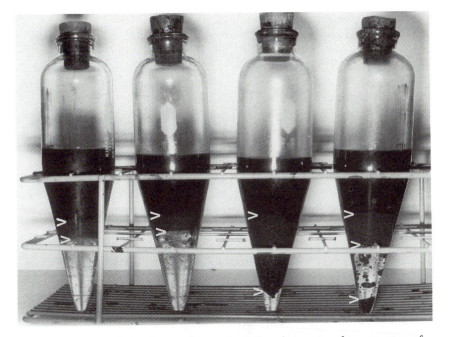

Figure 4. Centrifuge tubes are tapered to allow for precise determination of emulsion samples that have a low water content. The photograph shows that in the absence of a clear oil–water interface, it can be difficult to accurately determine water content (usually basic sediment and water). In addition, centrifugation time can significantly affect the amount of unresolved emulsion. The two centrifuge tubes on the left were centrifuged for 10 min; the two on the right were spun for only 5 min. The first three are the same sample; the tube on the right has a higher solids content. The arrows mark the unresolved emulsion layer. The tube on the far right shows the water phase with some black organic solids at the bottom of the centrifuge tube. This sedimentation can occur as a result of inadequate mixing of the oil phase (in heavy oils) with the solvent; estimation of the percentage of the oil phase relative to water and solids will be inaccurate. In this case, the black solids are oil phase closely associated with clays; therefore, they report with the bottom solids and water fraction.

Karl Fischer Titration. The Karl Fischer titration is a fast and accurate method for determining water content. Although the ASTM, API, and IP standards quote a water range of 0.02 to 2.0% (33), the technique can be successfully used at higher water contents (>10%). The technique involves titrating the emulsion sample with the Karl Fischer reagent consisting of a mixture of I_2, SO_2, and pyridine dissolved in methanol. The iodine is reduced by the sulfur dioxide in the presence of water to form HI and SO_3. These are immediately complexed by the pyridine and neutralized. Once all of the water is reacted, highly conductive free iodine appears, and the end point is identified by the increase in conductivity. Although most substances are inert to the Karl Fischer reagent, mercaptans and sulfide sulfur will interfere and must be below 500 ppm by weight. High water determinations by this method relative to those determined by Dean–Stark distillations might indicate interferences from other compounds or minerals that can react with the Karl Fischer reagent (34, 35). Process streams with high mineral or solids content can therefore be difficult to analyze accurately.

Electrical Properties.   The electrical properties of oil and water are quite different in terms of conductivity and dielectric constant (both of which can be related). These differences can be measured accurately with a capacitance probe and correlated to the amount of water in an oil stream. This type of probe is commonly used in on-line situations to monitor percent water in oil pipelines (36, 37). Generally, water and solids cannot be differentiated, so the signal is proportional to the total solids and water content. These systems have seen the greatest applications in monitoring relatively low water contents. In principle, techniques based on electrical properties can be calibrated for process streams with significant water and solids contents. However, the capacitance of the fluid changes with either an increase in solids or an increase in water, so the use of electrical properties in these situations is limited to streams where only one or the other is changing.

Other Methods.   Gamma-ray attenuation measures the density of the sample, which is related to changing oil or water content. Gamma-ray density meters are quite common in process monitoring, but they are useful for emulsion characterization only in cases where the solids content is known to be zero or completely constant (38). Otherwise the density information obtained cannot be reliably related to oil or water content.

Microwave-based meters have also been used to monitor water content in emulsions (39). Microwave techniques can be used in two ways: Either the attenuation of the microwave radiation due to absorption by the water phase is measured, or capacitance or resonance changes in a microwave cavity are noted. The capacitance-change method is much more sensitive, although both, like the gamma-ray absorption method, are limited in that solids content must be constant or zero in order to accurately interpret the information obtained. Both of these techniques are applicable to field situations and on-line monitoring.

In special cases where the continuous phase is reasonably transparent, absorbance of light or simple turbidimetry can be related to oil or water content in an emulsion.

Rheology.   Viscosity and other fluid-flow parameters of emulsions are important, not just for establishing pumping and handling protocols, but because they relate to other emulsion properties, such as size distribution of the dispersed phase, the presence of solids or emulsifiers, and the nature of the continuous phase. Many commercial instruments perform rheological determinations, and many thorough reviews cover various aspects of viscosity and other rheological parameters (*40–43*). The operating principles of the various instruments and the characterization techniques available will be covered in Chapter 4.

Emulsion instability and phase separation during viscosity measurements limit the applicability of many of the measurement techniques. This phenomenon is illustrated in Figure 5. The reproducible peak in shear stress with increasing shear rate is related to bitumen separating from this emulsion at the rotor–plate interface of a conventional viscometer.

Dispersed-Phase Properties

The chemical and physical nature of the dispersed phase is generally the primary consideration in order to define or to characterize an emulsion.

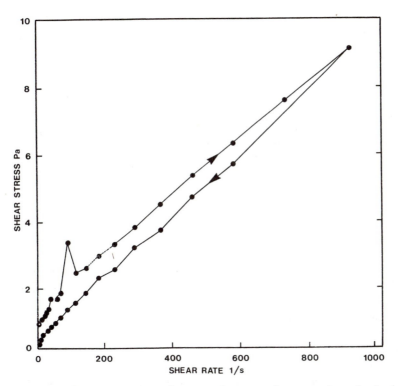

Figure 5A: Shear stress versus shear rate for an emulsion sample with a high solids content. The peak at low shear is reproducible and is due to oil separating from the emulsion onto the rotors. In rheometers that cannot measure at such low shear rates, this peak can be incorrectly attributed to stress overshoot or yield strength.

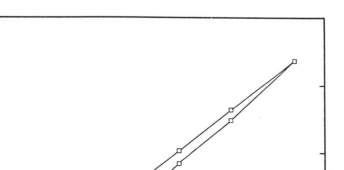
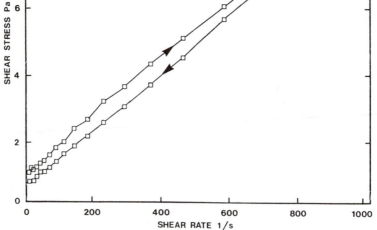

Figure 5B: Shear stress versus shear rate for the same sample with this easily separable bitumen removed. Now the peak at low shear rate does not appear. The area enclosed by the upper and lower curves represents thixotropic (or shear-thinning) behavior in this emulsion.

After all, the stability and size distribution of this phase determine most bulk emulsion properties. Fixed proportions of oil, water, and solids can be combined in various ways to produce emulsions having different size distributions of the dispersed phase, given only small differences in emulsifier or ion additions to the water or oil phases. These physical differences can lead to significantly different viscosity and stability in emulsions with nominally identical bulk composition.

The selection of optimum treatment protocols may depend significantly on determination of the size distribution of the dispersed phase. For instance, centrifugation might not be effective in a system with high viscosity and a very small size distribution of dispersed phase. Stokes' law can be used to predict the residence time needed if size distribution and viscosity are known. The smaller the average size of the dispersed phase, the larger the residence time required. In fact, the residence time increases as the inverse of the square of the diameter of the dispersed phase.

This section will focus mainly on characterization of the physical nature of the dispersed phase or its size distribution. Electrokinetic characterization techniques, which determine the electric double-layer properties of the dispersed phase, will be only briefly mentioned. Again, electrokinetic properties, their significance, and their measurement have been covered in review articles (*44, 45*).

Aside from microscopy, the techniques for determining the size distribution of the dispersed phase in emulsion systems can be broadly divided into three categories: techniques that depend upon the differences in electrical properties between the dispersed and continuous phases, those that effect a physical separation of the dispersed droplet sizes, and those that depend upon scattering phenomena due to the presence of the dispersed phase. Overviews of these types of techniques are found elsewhere (*1–4, 13, 46–49*).

Size Distribution Using Electrical Properties. As mentioned earlier, for water content determinations, techniques that depend upon differences in electrical properties do not distinguish between water and solids. This property limits their applicability to systems in which solids are negligible or the process stream is so well defined that the differences in signal can be attributed to differences in size distribution and not to total water and solids (*50, 51*). Clearly these measurements require extensive calibration and are not generally applicable to oil-field emulsions.

A technique that is widely used in spite of these drawbacks, however, is performed with the automated (Coulter) counter (*51–53*). In this instrument, the emulsion droplets (or particles) are diluted in an electrolyte and passed through a fine capillary that connects two larger chambers containing immersed electrodes. A potential difference is applied between the electrodes. The resistance change that occurs when an oil droplet passes through the orifice between the plates is proportional to the amount of electrolyte displaced and therefore to the size of the particle. Figure 6 shows a schematic of a typical experimental setup. A wide range of orifice sizes is available to cover size ranges from 0.4 μm to about 500 μm.

This technique, or variations of it that might measure voltage, current, or capacitance changes, is also known as a sensing-zone technique. These methods always require calibration and are limited to oil-in-water emulsions because the technique depends upon the displacement of electrolyte in the sensing zone. Dilution of the emulsion is often required because the appearance of two particles in the sensing zone at one time would be measured as a single larger particle. The size range analyzed in a single capillary is also limited because particles on the order of 40% of the capillary diameter lead to blockage, and particles smaller than about 2% of the capillary diameter do not produce a signal above the noise (background) and are effectively invisible.

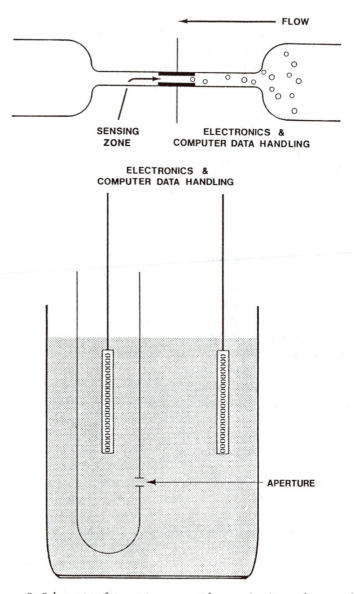

Figure 6. Schematic of a sensing-zone technique (top). As the particle or emulsion droplet (suspended in an electrolyte) passes through the sensing zone, the capacitance or resistance changes in proportion to the size of the particle. These signals can be sorted and interpreted as a size distribution by using an equivalent spherical diameter. Multiple droplets can be mistakenly interpreted as a single larger particle, but several alternative designs minimize this problem. The signal is proportional to the amount of electrolyte displaced; consequently, solids and emulsion droplets cannot be distinguished. These types of techniques are applicable only to oil-in-water emulsions. Bottom: The most common instrument for this technique, the Coulter counter.

Size Distribution Using Scattering Properties. The size range probed by the various scattering techniques is a function of the wavelength; neutron-scattering, X-ray-scattering, and light-scattering techniques are related in terms of the physical interaction between the radiation and the particles and cover sizes from 0.4 nm to hundreds of micrometers.

Light Scattering. The most common commercially available sizing instruments depend upon light scattering to obtain size information. The availability of inexpensive, well-defined light from laser sources has resulted in a wide variety of scattering techniques using light (*16, 54–59*).

Light scattering can be broadly divided into time-averaged scattering, with which either spatial distribution or intensity is measured, and time-fluctuation scattering, which includes photon correlation spectrometry, in which scattering is correlated to the microscopic motion of individual scattering centers. These techniques have been discussed in detail in several reviews (*59–63*). Only a brief overview of the most common time-averaged methods will be given here. These methods include Fraunhofer diffraction and light scattering at larger angles, (Mie scattering), which are the basis of many commercially available sizing instruments.

Quasi-elastic light scattering or photon correlation spectrometry, Fraunhofer diffraction, and other techniques that depend upon light have the same drawback; namely, the opacity of most oil production samples makes them unsuitable for use. Typical problems with the theory and subsequent data reduction of the scattering information to a size distribution include an assumption of the nature of the size distribution (typically lognormal, although software is available with other options) and an inability to distinguish aggregates from large single particles. Solids and the dispersed phase of the emulsion cannot be distinguished from each other, a disadvantage shared with the sensing-zone techniques. In addition, the sample must be dilute enough to minimize multiple scattering.

Figures 7 and 8 illustrate the experimental setup and examples of the signal observed in Fraunhofer diffraction for monodisperse particles and for a polydisperse sample, respectively. The detection system in most commercial instruments is either an array of intensity sensors or a single detector with a moving mask that measures intensity differences of the overlapping concentric rings that are not discernible in Figure 8. Although no calibration is necessary for monodisperse spherical systems, the data are output as an equivalent spherical diameter and the range of applicability is generally for sizes exceeding 10 μm. For many emulsion systems, the refractive index is close to that of water; therefore, the practical lower limit is 10 μm (*13*). For solid particulate systems, the refractive index is generally large; consequently, the applicability of this technique can be extended to smaller sizes.

Most instrument manufacturers use scattering at larger angles, as well as diffraction, to probe the smaller sizes (smaller than about 10 μm). Scattering at larger angles involves a distinct dependence upon refractive index, and

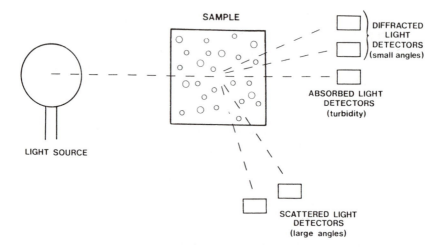

Figure 7. Schematic of a light-scattering apparatus. Three techniques are illustrated. First, attenuation of the incident light or turbidimetry can indicate the amount of dispersed phase but offers no information about the size distribution. Second, diffraction of the incident beam (Fraunhofer diffraction) offers size information for relatively large sizes when the particles are on the order of, or larger than, the wavelength of the incident light. Third, scattering through larger angles (Mie scattering) occurs with particles smaller than the wavelength of the incident light. This large-angle scattering can be affected by the refractive index of the scattering centers. Computer data handling reduces these signals to a size distribution. Variations of these basic techniques involve detection of scattered light as a function of angle, correlation of the scattered photons (photon correlation spectroscopy), and detection of scattering as a function of wavelength or polarization of the incident light.

various manufacturers use either the position or wavelength dependence of the scattered light at larger angles. Assumptions about an "average" refractive index or the nature of the size distribution (bimodal or log-normal, for instance) must be made to determine a size distribution from the light-scattering information at larger angles.

X-ray and Neutron Scattering. Small-angle X-ray and neutron scattering also can be used to probe emulsion size distributions, but at a much smaller resolution, down to the molecular level (about 4-A resolution with neutron scattering) (*64–67*). This level of detail in determining molecular aggregates is clearly not applicable to emulsions commonly encountered in oil extraction and processing situations, although the principles are the same as for light-scattering phenomena.

Size Distribution by Physical Separation. Size distributions obtained by physical separations generally involve systems of solid particles

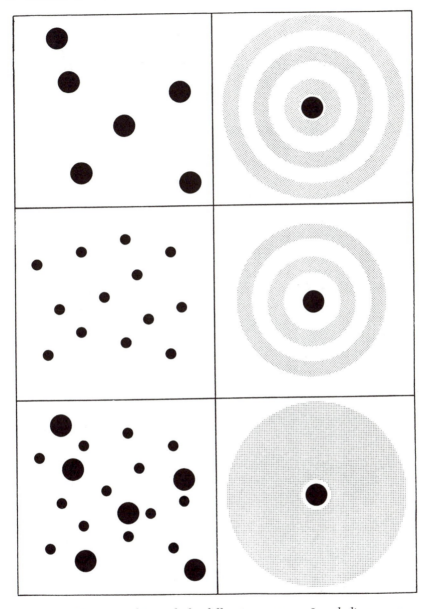

Figure 8. Three typical Fraunhofer diffraction patterns. In polydisperse systems, the interpretation of the relationship between these patterns and the size distribution can be difficult and requires sensitive photomultipliers. The transmitted beam is blocked out, and the detectors are arranged outward from the center. In some cases a single detector has a movable mask to measure diffracted light intensity as a function of position. Subtle differences in size distribution (i.e., log-normal vs bimodal, etc.) cannot be distinguished, and generally some assumption must be input to the data reduction programs.

that are not affected by the handling required. However, some new developments in hydrodynamic chromatography, size exclusion chromatography, and field-flow fractionation may have applications to microemulsion systems and perhaps to the larger emulsions more commonly encountered in the oil field. Detailed descriptions of these techniques can be found in the literature (68).

Chromatographic Techniques. Hydrodynamic and size exclusion chromatography are similar in that they depend upon conventional chromatographic principles of flow (of particles or droplets in this case) in a carrier fluid. In size exclusion chromatography, the larger particles exit the system first because they are not slowed by interactions with pores in the packing material. In hydrodynamic chromatography, the larger particles exit first because they are too big to stay in the slow carrier-fluid velocity zones near the packing or at the walls of the chromatographic column.

Figure 9 illustrates these techniques. These techniques are most applicable to the separation of microemulsions, micellar systems, or large molecules. Typical size ranges up to 1 μm, although hydrodynamic chromatography has been applied to larger systems when used without packing (up to 60 μm) (69–72). Unstable systems cannot be characterized with these techniques because of interactions with the column or packing material. The major advantage of the techniques is the physical separation of the size fractions for further characterization.

Sedimentation Techniques. Other techniques that effect a physical separation include gravitational or centrifugal sedimentation, in which particles or emulsion droplets are separated on the basis of size and density. The separation that occurs can be quantified by monitoring X-ray or light absorbance as a function of position. Stokes' law then can be used to determine the particle size distribution from the absorbance data as a function of the sedimentation time (73, 74).

Field-Flow Fractionation. Field-flow fractionation (Figure 9) is like the other physical separation techniques except that a field is applied at right angles to the flow; particles or droplets are thereby separated depending upon their interaction with the field. The field can be electric, magnetic, gravitational, thermal, or whatever force might interact with the particles. Fractionation using either gravity or centrifugal force is the principle behind some commercially available instruments (75–81).

Emulsion Stability

Determining emulsion stability is one of the most important tests that can be performed on an emulsion. The ease with which the oil and water phases

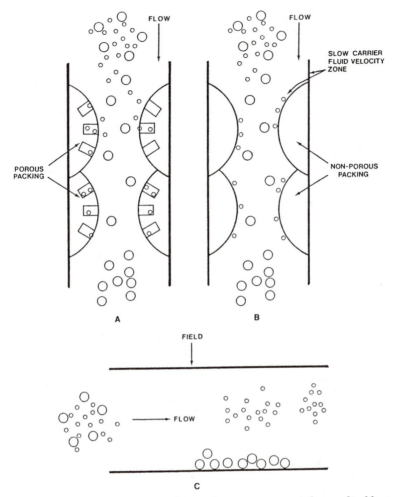

Figure 9. Three chromatographic techniques are mainly applicable to microemulsions and very stable systems because they depend upon interactions at surfaces and in pore spaces. A: In size exclusion chromatography, the sample is injected and is segregated in the column by virtue of the fact that the smaller particles interact and get held up in the small pore spaces, while the larger particles elute through more quickly. B: Hydrodynamic chromatography works because smaller particles and droplets can approach the column or capillary wall to near the boundary where carrier flow is essentially zero. The larger droplets are more strongly affected by the flow and elute more quickly than the smaller ones. C: Field-flow fractionation depends upon separation of flowing emulsion droplets in an applied field. A sample is injected into a carrier with an applied field perpendicular to the flow. The field is often just gravity but might be electrical or magnetic depending upon the nature of the emulsion. Several commercial instruments use sedimentation or gravity as the applied field. Subsequent detection of the droplets is often via a light-scattering type of technique.

separate establishes the treatment protocol for the emulsion and ultimately determines the cost of treatment. Determination of the most effective demulsifier for a particular process is generally done off-line, and the effectiveness of a demulsifier depends upon the degree of destabilization. The destabilization is most often monitored by simply observing the phase separation as a function of time. When a clear interface is present between the oil and water layers, this method can be very effective, but when the separation is not so distinct, operator bias can affect the results. To overcome this bias, centrifugation can be used to clarify the interface (82, 83), light scattering to automate the determination of separation (84, 85), or microscopic techniques to monitor droplet coalescence (86–88).

Bottle Tests.   The most common method of determining relative emulsion stability is the simple bottle test. There are probably as many different bottle test procedures as there are people who routinely use them. In general they involve dilution of the emulsion with a solvent (to reduce viscosity), shaking to homogenize the emulsion or to mix in the demulsifier to be evaluated, and a waiting and watching period during which the extent of phase separation is monitored along with the clarity of the interface and the turbidity of the water phase. Depending upon the viscosity of the original emulsion, the test may be done at elevated temperatures or with varying amounts of diluent. Separation might also be enhanced by centrifugation, although usually not when diluent is also added.

Figure 10 shows a bottle test as a function of time in which the upper oil phase steadily increases in volume. This type of test provides a significant amount of information relating to both the stability of the emulsion phase and the clarity of the separated water.

Centrifugation.   Centrifugation is a modification of the bottle test in which the sedimentation force is artificially increased to effect separation. Diluent may or may not be added depending upon the stability of the emulsion. Specially designed stroboscopic centrifuges monitor phase separation as a function of time; they provide information exactly analogous to the bottle test previously described (at higher gravity, or g, forces) while continuously observing the sample interface. In both the simple bottle test and the centrifuge test, settling and separation of the oil and water phases are dependent upon the size of the dispersed phase and the viscosity of the continuous phase (Stokes' law). Therefore, the upper part of the oil phase often contains significantly less water than the layer near the oil–water interface. Depending upon the process conditions, the water content as a function of depth and time in the oil layer might be a very important parameter. The water content as a function of depth in the oil layer can then be determined by spin or centrifuge tests (at high speed for water content determination) or by Dean–Stark tests.

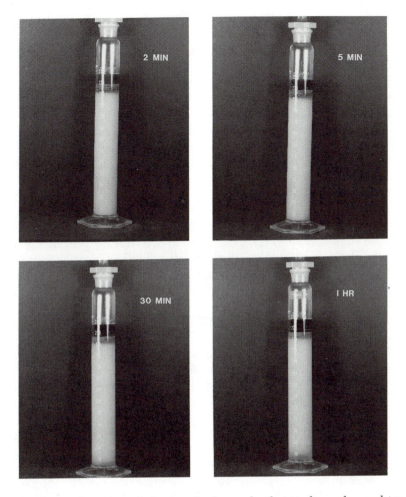

Figure 10. This series of photographs from a bottle test shows the emulsion separating over time as evidenced by the steady increase in the upper oil phase. The lower water phase contains most of the solids but does not change in volume significantly. The interface emulsion in between the oil and water steadily decreases (destabilizes) in volume and resolves into the oil and water phases. As shown in Figure 1, these interface emulsions can have a complex morphology or structure.

Electrokinetics. Bottle tests and centrifugation may be somewhat crude, but they do offer a relative measure of emulsion stability that combines, to some extent, all of the factors that affect stability. Electrokinetic measurements are somewhat more elegant because they allow direct measurement of the degree of electrostatic stability in an emulsion system. The zeta potential, or relative magnitude of the electric charge on the surface, is related to emulsion stability, as is the thickness of the double layer (the diffuse layer of compensating charge). For electrostatically stabilized systems, the double-layer thickness can be as important in determining relative stability as the surface potential as measured by the zeta potential.

A wide variety of electrokinetic measurements or instruments can be used to quantify the electrostatic stability of the dispersed phase. These measurements will only be summarized here.

A dispersed emulsion droplet (electrostatically stabilized) can be thought of as a charged center with a pliable cover of compensating charge due to orientation of the water molecules (or counterions) around the droplet. When exposed to an electric field, the droplet will move according to the surface charge.

The motion of the droplet in an applied electric field distorts the relationship between the charged center and the outer layer to create a dipole. With this simplistic view, it is possible to understand the principle behind electrophoretic mobility, whereby the relative motion of particles or emulsion droplets is measured with an applied electric field. These measurements often depend upon microscopic observation of the droplet motion in the applied electric field and a calculation of droplet velocities to determine their electrophoretic mobility. Figure 11 is a schematic of a typical experimental setup.

By observing many droplets, the average electrophoretic mobility or charge on the emulsion can be determined. For highly charged systems, it may be possible to destabilize the dispersion by adjusting pH. Compression of the double layer by changing the concentration of counterions can also destabilize emulsions or dispersions that are electrostatically stabilized. Alternatively, the addition of surface-active agents can bring the system to its zero point of charge, or electrostatically destabilized state. The methods for making these measurements range from direct observation, which can be very tedious, to some fairly automated systems that count particle by particle to give a distribution of electrophoretic mobilities (44, 45). These automated instruments are invaluable in cases where mixtures of dispersed droplets and solids might have differing electrophoretic mobilities.

Figure 12 shows the electrophoretic mobility of a population of oil droplets that have a significant average negative charge. The photograph in the same figure shows the behavior of these electrostatically stabilized particles. The electrophoretic mobility of the oil droplets can be modified in a variety of ways by adding other ions or polymers to affect the surface charge or to neutralize it.

Figure 13 illustrates the effect of a cationic polymer that neutralizes the charge somewhat and brings the droplets close to the zero point of change. The accompanying photomicrograph shows the treated particles coagulating.

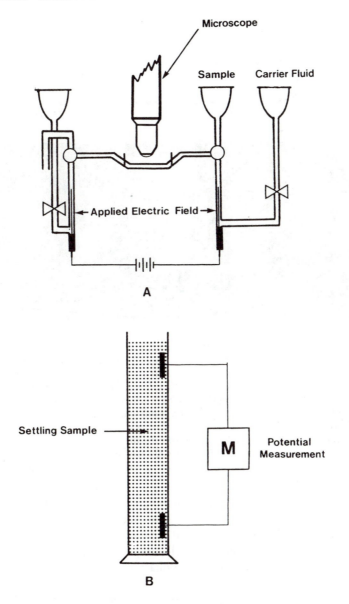

Figure 11. Schematic representation of the electrophoretic mobility (A) measurement showing the major components. In an applied electric field, emulsion droplets move according to their surface charge. These charges can electrostatically stabilize an emulsion system by preventing the droplets from coming into contact and coalescing. The motion of the droplets is visually observed, and the electrophoretic mobilities of a number of particles are measured to determine zeta potential. The sedimentation potential (B) is also illustrated.

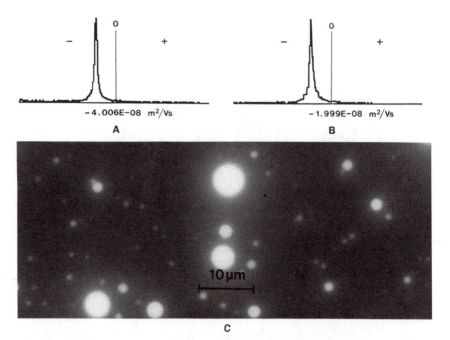

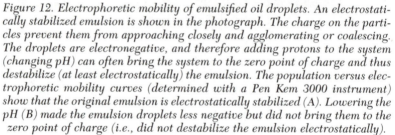

Figure 12. Electrophoretic mobility of emulsified oil droplets. An electrostatically stabilized emulsion is shown in the photograph. The charge on the particles prevent them from approaching closely and agglomerating or coalescing. The droplets are electronegative, and therefore adding protons to the system (changing pH) can often bring the system to the zero point of charge and thus destabilize (at least electrostatically) the emulsion. The population versus electrophoretic mobility curves (determined with a Pen Kem 3000 instrument) show that the original emulsion is electrostatically stabilized (A). Lowering the pH (B) made the emulsion droplets less negative but did not bring them to the zero point of charge (i.e., did not destabilize the emulsion electrostatically).

In spite of the droplets being destabilized electrostatically, no evidence of droplet coalescence is seen. By the same token, an electrostatically stabilized emulsion might still coalesce and separate, sediment, or cream if other destabilizing forces overbalance the electrostatic component. Creaming refers to concentration of the dispersed phase without completely separating the oil and water phases.

Film stability and interfacial forces are important in determining emulsion stability and the likelihood of creaming or complete separation of the phases. Characterization of these interfacial effects is an important factor in determining the fundamental properties that might ultimately determine coalescence kinetics. Some relevant papers and reviews have been published elsewhere (54, 89–96).

The presence of the electric double layer and its distortion when droplets move is the principle behind several related methods such as the sedi-

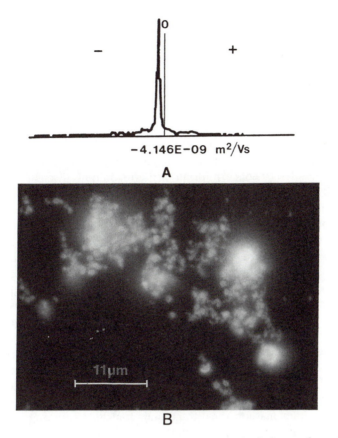

Figure 13. Electrophoretic mobility (Pen Kem 3000) of the emulsion from Figure 12 after cationic polymer addition (A). The cationic polymer has neutralized the oil droplet surface charge and electrostatically destabilized the emulsion. The photomicrograph (B) shows this destabilized emulsion that has begun to flocculate or agglomerate but that is not coalescing. This electrostatic destabilization is not the only factor affecting emulsion stability. Factors such as interfacial tension and film strength can prevent coalescence of the emulsion droplets, even though they can now closely approach each other and agglomerate.

mentation potential techniques (also represented in Figure 11). One fairly new technique that deserves special mention is the electrosonic amplifier, which can measure electrical currents in an emulsion that is sonically agitated (and thereby create dipoles in the electric double layer) or measure the sound wave generated when the emulsion is electrically stimulated. The advantage of this technique is that neither dilute solutions nor transparent systems are required (to allow for direct observation of the dispersed phase). Unfortunately, the relationship between the measured signal and electrophoretic mobility of the emulsion is not as straightforward as in conventional microelectrophoresis that employs direct observation (97).

Microscopy

Microscopy is often the last word in the determination of the size distribution of dispersed systems (98–101). Throughout the literature, distributions obtained by various particle and emulsion sizing techniques are compared to the values determined by microscopy (13, 102–107). Establishing a representative sample is a concern for all of the techniques discussed and is not necessarily a particular problem for microscopic observation, although this criticism is often given for the microscopic methods. Indeed, many of the sample handling concerns discussed in this section apply equally to samples prepared for other techniques.

In solid particulate systems, direct observation is justifiably the last word. In emulsions where creaming, sedimentation, and coalescence can change the nature of the sample, microscopic observation has unique sample handling problems. If these special sampling problems are addressed, then microscopy can indeed provide the benchmark for the physical characterization of the dispersed phase in emulsion systems.

Figure 14 shows a multiple emulsion easily characterized with optical microscopy in the fluorescent mode. Other techniques are not capable of distinguishing this emulsion from a simple oil-in-water emulsion with a much larger size distribution.

The complex mathematical treatments for light-scattering experiments and the experimental complexities of some of the other characterization techniques mean that, in general, greater care is taken in the interpretation of the results and operators are aware of potential data reduction problems. In microscopy, because "seeing is believing", the tendency is to ignore sampling problems and to reach conclusions that are sometimes based on sampling artifacts or peculiarities of the microscopic observation technique.

As long as the possible problems are known, microscopy can be regarded as the single most important emulsion characterization tool. In the appropriate circumstances it can give information about the relative amounts of oil, water, and solids in an emulsion system; their interactions or associations; the size distribution of the dispersed phase; and the rate of coalescence of the dispersed droplets. Various microscopic techniques can be used to define not only the physical nature of the sample, but also the chemical composition, both mineral and organic.

Optical Microscopy.   Optical microscopy involves the use of transmitted light, reflected light, polarized light, fluorescence, and more recently, techniques such as confocal microscopy. Each of these variations has particular strengths and applicability.

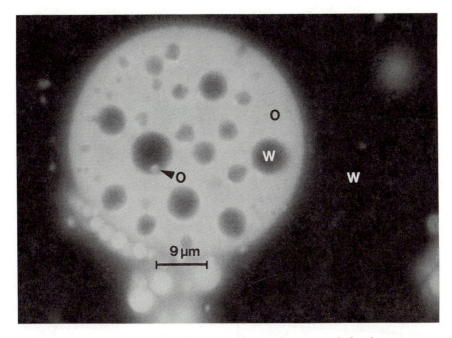

Figure 14. Optical micrograph using reflected fluorescent light showing a multiple emulsion that is extremely difficult to characterize by conventional techniques. The continuous water phase (W, dark) shows a large dispersed oil droplet (O, bright) that contains a water droplet that also contains emulsified oil. The arrow points out an oil-in-water in oil-in-water emulsion droplet. Characterization of these multiple emulsions can be accurately carried out only with microscopic techniques.

Transmitted-light microscopy requires a sample sufficiently thin to allow light to pass through it. This requirement is often accomplished by simply smearing the emulsion sample on a slide. Care must be taken to ensure that the slide is properly prepared to accept the continuous phase. A hydrophilic glass surface, for instance, can invert an oil-continuous emulsion to a water-continuous one. Correct determination of something as basic as the nature of the continuous phase can therefore be difficult with emulsions that are unstable. Careful observation of emulsion behavior using both hydrophilic and oleophilic sample holders is sometimes required to determine the effect of emulsion interactions with the sample holders. When emulsion instability makes sample collection and observation difficult, fast freezing the emulsion and subsequent observation of the frozen sample can avoid emulsion changes due to sample preparation and handling.

The transmitted-light technique is limited by the opaque nature of most oil samples and, in cases where the sample cannot be made thin enough, an alternative technique using reflected light is available.

Figure 15 is a schematic of the experimental setup for reflected-light microscopy. By use of reflected light, the sample can simply be put in a small

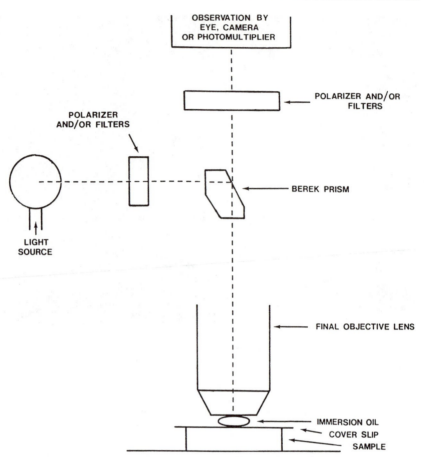

Figure 15. Schematic of the optical microscope in reflected-light mode. Air or oil immersion objectives may be used. Oil immersion objectives require a glass cover slip over the sample and an oil drop of appropriate refractive index to bridge the gap between the objective lens and the sample cover glass; this setup has the advantage of much higher resolution. The light source can be plain or polarized white light (tungsten lamp) for observation of solids, or appropriately filtered blue–violet light (high-pressure mercury lamp) to excite fluorescence of the oil phase. To investigate fluorescence behavior, the reflected blue–violet or ultraviolet light is filtered out, and only the fluorescent light (longer wavelength) is returned to the detector. Other techniques such as dark-field illumination allow particles to be counted and not sized. The droplets are seen only as points of light on a dark background.

container or well slide. Often a cover slip is put over the sample, and an oil immersion lens is used. This setup allows one to focus beyond the cover slide and observe the sample past the level of the cover-slip–sample interface where air bubbles are commonly entrapped. Figure 16 shows air bubbles

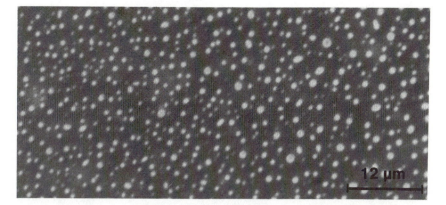

Figure 16. Photomicrograph (white light, reflected mode) illustrating trapped air bubbles at the cover-slip–sample interface. These can easily be mistaken for the dispersed phase, especially in transmitted-light mode.

trapped at the sample–glass interface that might be mistaken for dispersed-phase droplets, especially if transmitted light is used.

Clays and other solids are often transparent to white light, and in these cases, polarized light can be used to observe the clays. To enhance observation of the oil, the fluorescence behavior of the organic phase is used. This approach involves incident light of violet or ultraviolet wavelength and observation of the fluorescent light in the visible region. The incident reflected beam is filtered out, and the returning light is due to the fluorescent behavior of the oil phase. Figure 17 shows the fluorescence of bitumen associated with clays in contrast to the behavior of bitumen that is relatively clay free. The differing fluorescent behavior might be indicative of a particular oil component that preferentially associates with the mineral phase.

This fluorescence technique is mainly applicable to oil-in-water emulsions in which the oil phase appears as bright spots in a dark background (because the water does not fluoresce). The effect is illustrated in Figures 18 and 19. Figure 18 shows an oil-in-water emulsion under white light. Although the clays and other solids are visible, there is little or no evidence of the dispersed oil phase. Figure 19 illustrates that the oil phase appears bright and is much easier to resolve with fluorescent light. The oil droplets can be seen to be quite distinct from the continuous water phase (although the clays are now invisible). This type of image is particularly suitable for image analysis and automated droplet counting and size characterization.

A potential problem with optical microscopy, especially with high-intensity mercury vapor lamps (for blue–violet incident light) is localized sample heating. With some marginally stable emulsions, the heating effect could be enough to break the emulsion. This effect is illustrated in the series of photographs in Figure 20. The first image is of a multiple emulsion (water-

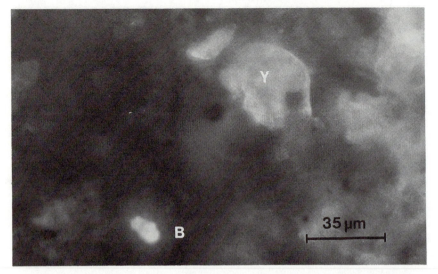

Figure 17. Fluorescence behavior of the organic phase can be an important indicator of the composition of the oil phase or of oil–solid interactions. These oil components (bitumen associated with clays versus bitumen that is clay-free) clearly differ in fluorescence behavior, a result indicating different organic-phase compositions or different oil–solid interactions. In this black-and-white reproduction, the oil component that fluoresces blue (B) appears brighter than the component that is fluorescing yellow (Y). The free organic component exhibits the blue fluorescence; the organic material associated with clays fluoresces yellow.

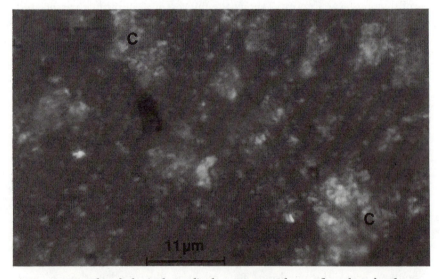

Figure 18. White-light (polarized) photomicrograph in reflected mode of an oil-in-water emulsion with a significant solids content. With polarized light, the clays (C) appear bright, but the oil droplets cannot be seen at all.

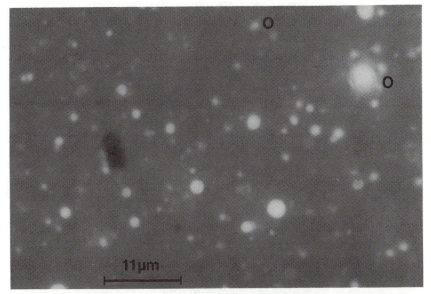

Figure 19. Reflected-light photomicrograph of the same field of view as Figure 18 in the fluorescence mode showing bright oil droplets in a dark water-continuous phase. In this photograph the clays cannot be seen. This type of image with high contrast between the phases is ideal for automated analysis. However, droplets not exactly in focus (O) may be incorrectly sized.

in-oil-in-water) under white light that is broken after a short observation period under blue–violet light (fluorescence mode).

The availability of low-cost computing and image analysis compatibility has helped to reduce the time involved in quantification of microscopic analysis to determine size distribution. The comparison of images to rulers photographed under the same conditions and the use of split-image micro-scopes (*105–107*) have largely been replaced by automated image-analysis techniques. Most suppliers of image-analysis equipment offer programs or routines to separate or "deagglomerate" the spheres when oil droplets are touching or agglomerated. These programs and the ability to automatically size emulsion droplets greatly reduces the tedium of size analysis by micro-scopic methods.

To be confident of an average size analysis within 10%, approximately 150 particles should be sized. To increase the confidence level to 5%, approximately 740 particles should be sized (*104*). Of course these numbers are only a rough guide, and the actual confidence levels will depend upon the nature of the size distribution. Figure 21 shows a problem that can occur when too much reliance is placed on automated image analysis, namely, inaccurate sizing of droplets that are slightly out of focus. The fields of view to be analyzed, either manually or with automated methods, have to be chosen carefully.

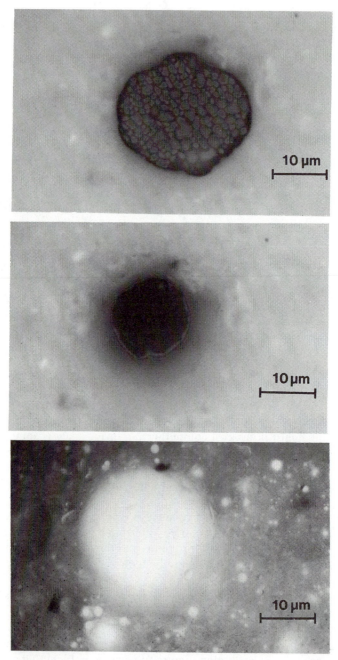

Figure 20. Heating of an emulsion sample under blue–violet light. The top photograph shows a water-in-oil-in-water emulsion under white light in the reflected mode. After a short observation period under blue–violet light (middle), the multiple emulsion is broken and only the oil-in-water component remains (bottom).

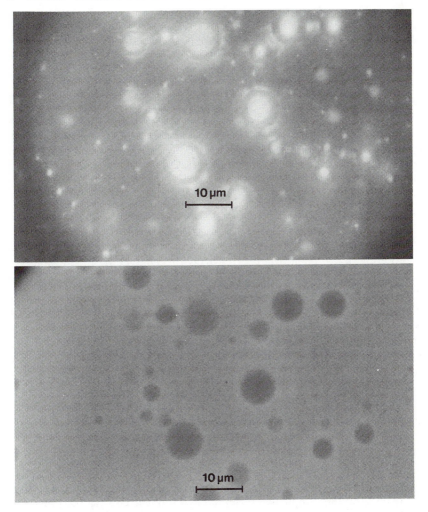

Figure 21. White-light (top) and blue-light fluorescence mode (bottom) photomicrographs of a water-in-oil emulsion. With white light the water droplets have internal reflections that lead to a halo effect and an incorrect size estimate. With incident blue–violet light to excite oil-phase fluorescence, the emulsified water droplets appear as dark circles in a bright oil background and are significantly easier to size. However, droplets that are above or below the plane of focus will still be incorrectly sized.

Because of the interactive nature of the microscopic technique, in other words, the human factor, there can be differences in size analyses by different operators, and operator bias to either small or large particles. Missing large particles affects the mass distribution, and neglecting small particles affects the number distribution. Of course these concerns are not limited to microscopic analysis, if size distributions are biased.

Particles smaller than about 0.5 μm begin to approach the resolution limit of the optical microscope, and often particles can be recognized but not sized properly because of limitations in both the resolution and the depth of field or focus in the optical system.

The confocal microscope solves some of these problems and adds a new dimension to optical microscopic analysis. The confocal microscope digitizes the intensity information in a field of view and, by adjusting the focus, makes it possible to reconstruct an image that is in focus over a significant depth in a sample. Through this reconstruction, the optical image that can be produced provides more information about associations between the water, oil, and solid phases. Figure 22 shows a series of confocal microscope photographs of a typical interface emulsion. The diameter of the oil droplet increases as the plane of focus passes in sections through the droplet.

Electron Microscopy. Both scanning and transmission electron microscopy have been used extensively to characterize emulsion systems (*107–110*). Transmission electron microscopy is somewhat less common and almost invariably involves the observation of replicas or metal reproductions of the emulsion sample. Scanning electron microscopy (SEM) is much more analogous to conventional optical microscopy. The scanning electron microscope offers significant advantages over the optical microscope in terms of depth of field and resolution. However, the vacuum environment and energy deposited by the electron beam means that sample handling and preparation are much more difficult.

The two main techniques commonly discussed in the literature are known as direct observation (or frozen hydrated observation) and the observation of replicas. Both techniques involve the fast freezing of the sample in a cryogen such as liquid nitrogen, propane, or freon. The frozen sample is then fractured to reveal the interior features. This fractured surface can be coated with a metal film or observed directly. Often, the metal film is removed from the sample and observed as a replica. This type of procedure allows the creation of a permanent archive of the samples prepared, and the observation is the same as with any other electron microscope sample with no concern about contamination of the microscope or beam damage to the sample.

Transmission electron microscopy, with few exceptions, involves the creation of replicas because it depends upon the electron beam passing through the sample, with regions of low or high density appearing as bright and dark areas. When replicas are used in transmission electron microscopy, the metal or carbon replica is shadowed with a second metal to accentuate the topography of the emulsion on the replica. These methods are outlined in Figure 23.

Scanning electron microscopy is relatively simple compared to transmission electron microscopy, and the images obtained are significantly easier to

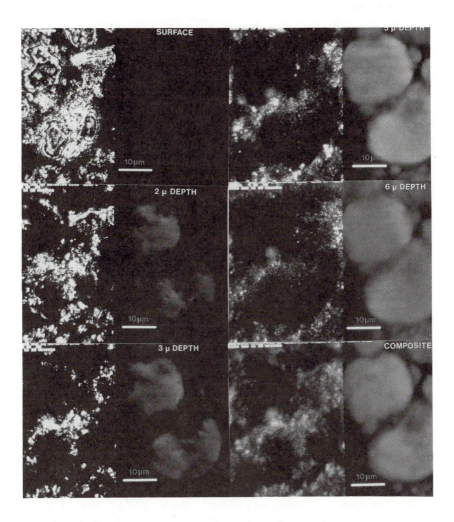

Figure 22. This series of confocal micrographs are composites of both polarized white light (on the left) to show the clays and of fluorescent light (on the right) to show the oil component. The confocal microscopic technique allows digitization of the data from images focused at discrete depths in the sample. This feature gives the effect of observing slices of the sample at successive intervals of depth. The increase in apparent size of the large oil droplet is due to slices being taken progressively closer to the center of the droplet. This technique makes it easier to characterize the relationship between the solids and the dispersed phases. Computer reconstruction of the slices can give a three-dimensional effect (greater depth of field) similar to that obtained with scanning electron microscopy. (Photographs taken by V. A. Munoz and W. W. Lam of CANMET, at the Ontario Laser and Light Wave Research Centre.)

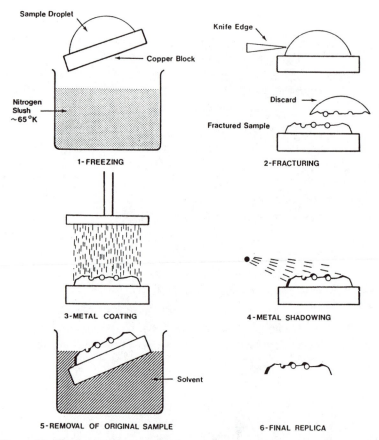

Figure 23. The freeze-fracture sample preparation technique and the six main steps in preparing a sample for electron microscopy. The first step is freezing the sample rapidly enough to prevent large ice crystal formation and resulting sample distortion. After the sample is frozen, it is fractured to reveal the interior features. This fractured sample can then be put into an electron microscope with appropriate cryogenic capability (frozen hydrated or direct observation). More commonly a replica of the sample is created by coating the fractured surface with metal, shadowing with a second metal, dissolving away the original sample, and then observing the replica in a conventional electron microscope. For transmission electron microscopy, in which the image formation depends upon differences in sample density, the replica must be shadowed, or coated directionally with metal to provide a density contrast in the peaks and valleys of the emulsion replica. During the freezing and metal coating steps, the sample must be kept in a vacuum environment to keep it frozen and to prevent frost deposition. Light frost can sometimes be mistaken for the dispersed phase. With direct observation, the sample can be carefully warmed (to about 130–150 K), and any frost layer can be sublimed away. This step obviously cannot be done with a replica. Another advantage of direct observation is the X-ray information that can help identify the composition of the various emulsion components.

interpret. In addition, the direct observation option gives the potential for much more information about the chemical composition of the emulsion. Figures 24 and 25 show typical water-in-oil emulsions studied by frozen hydrated observation in a scanning electron microscope.

With direct observation, the sample must be kept cold in the electron microscope, and care is required to prevent sample damage in the beam and to prevent microscope contamination. In addition, these frozen samples are often difficult to image because of charging effects that distort the image. The benefit of this extra care in sample handling, however, is that electron beam interactions with the sample produce characteristic X-ray signals that allow identification of components of the emulsion being observed. This technique has been refined to the point where, in special cases, chemical compositional differences at the emulsion interface can be identified, as well as the composition of the dispersed and continuous phases (*109, 110*).

Figure 26 shows an oil-in-water emulsion with corresponding X-ray spectra of the continuous and dispersed phases and of the interface itself. Figure 27 illustrates the resolution improvements of SEM observation relative to optical microscopy. Figure 28 shows corresponding X-ray spectra that identify the droplets as oil and suggest that they may be stabilized by fine clay particles.

High-speed computers and the ability to digitize electron signals at video rates mean that, in spite of poor initial image quality in dealing with direct observation of frozen hydrated samples, several relatively noisy im-

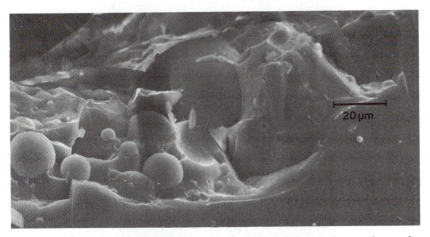

Figure 24. Electron micrograph showing the relatively featureless surface and the fractured interior of a water-in-oil emulsion. This image was prepared with a metal-coated frozen sample, a modification of direct observation in which the sample is coated sufficiently to prevent sample charging but not enough to produce a replica. This technique still requires an electron microscope with cryogenic capability.

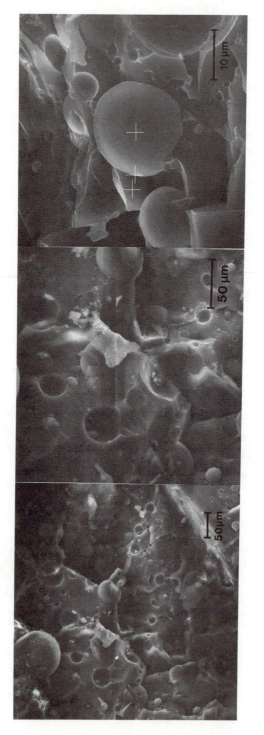

Figure 25. Electron micrographs showing three typical views of a water-in-oil emulsion by direct observation. The resolution and depth of field are significantly better than can be achieved via optical microscopy.

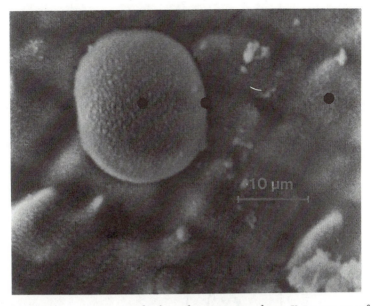

Figure 26A. Electron micrograph of an oil-in-water emulsion. X-ray spectra of the dispersed and continuous phases and the interface are shown in Figures 26B and 26C. The chemical composition of the interface itself can also be characterized. The resolution of the image is much greater than the resolution of the X-ray information. The spots marking the X-ray acquisition points approximately represent the area where the X-rays are produced.

ages can be averaged to reduce the noise level and produce an image that is close in quality to that obtained from replicas.

The optical and electron microscopic techniques are quite complimentary in terms of the information that they can provide. Optical microscopy, in fluorescence mode or with polarized light, can provide information about the organic phases in the emulsion. Electron microscopy, through the X-rays excited in the sample, can provide information about the inorganic or mineral phases present.

The practical lower limit of emulsion sizing with optical microscopy is on the order of 0.5 μm. This limit is much lower with electron microscopy, on the order of 0.1 μm or less with direct observation of frozen samples in a scanning electron microscope, and 0.01 μm or less with replicas and transmission electron microscopy. Sizes smaller than these lower limits can be recognized with each of these techniques, but quantification of the size distribution becomes difficult. Furthermore, at levels of about 0.01 μm, it is extremely difficult to avoid artifacts and subsequent misinterpretations. As mentioned earlier, sample preparation is an extremely important consideration in both optical and electron microscopic techniques. With optical

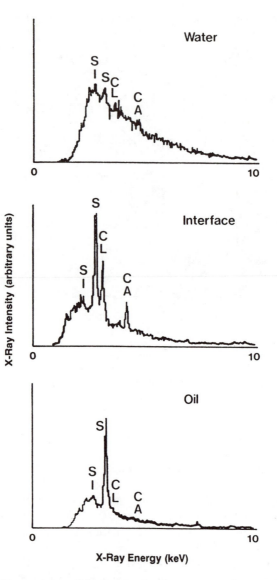

Figure 26B. X-ray spectra of the dispersed and continuous phases and interface of the oil-in-water emulsion in Figure 26A doped with calcium. The top X-ray spectrum of the continuous phase shows no X-ray peaks, only background bremsstrahlung radiation, because this particular detector is not sensitive to the oxygen in the water phase. The bottom spectrum shows only a sulfur peak typical of many bitumens and heavy oils. The middle spectrum is of the interface and clearly shows chlorine and calcium (in this part) or iron (in Figure 26C), which are not present in either the dispersed or continuous phases. The chlorine is present in the emulsifier that was used to prepare this emulsion.

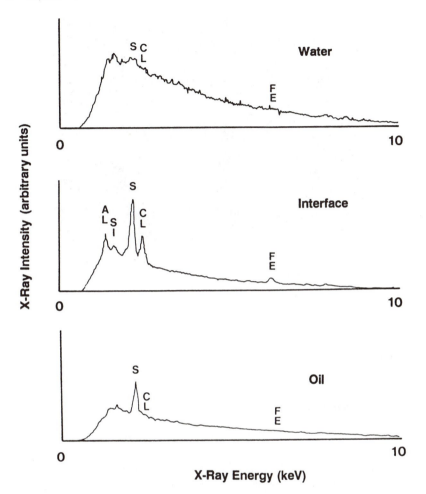

Figure 26C. X-ray spectra of the dispersed and continuous phases and interface of the oil-in-water emulsion in Figure 26A doped with iron. See also the caption to Figure 26B.

microscopy, interactions of the sample with the sample holder can affect which phase is observed as continuous and, with electron microscopy, artifacts due to the freezing process can affect interpretation of the results (*108*).

Microscopic techniques offer the potential for complete emulsion characterization because they are capable of quantifying volumetrically the relative amounts of oil, water, and solids present, determining the size distribution of the dispersed phase, and determining some chemical compositional information about both the organic and inorganic components.

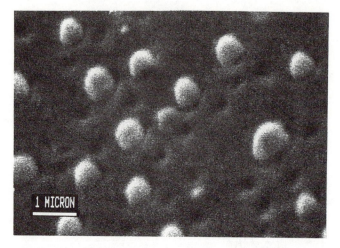

Figure 27. Electron micrograph of discrete oil droplets on the interior surface of an air bubble. These oil droplets are less than 1 μm in diameter and illustrate the high resolution possible using frozen hydrated observation.

However, the researcher must be aware of the limitations of this technique in terms of sample handling and preparation, and of the very real danger of overinterpreting images once they are acquired.

New Developments

New developments in emulsion characterization can simply mean recent applications of well-established technologies to emulsion systems or the application of unconventional methods that, although not in widespread use, may be well established in particular operations. Several of the techniques discussed previously could have been assigned to this section; conversely, some of those discussed here might not be regarded as new by those who may be using these techniques extensively. Admittedly, the distinction is partly a reflection of my own bias.

Nuclear Magnetic Resonance Spectroscopy. Nuclear magnetic resonance (NMR) spectroscopy offers several intriguing possibilities to identify water "structure", or the ordered arrangement of water molecules at an emulsion or solid surface. This type of information might help in understanding the differences between emulsified water and continuous-phase water, especially in those emulsions that contain portions of both. Information on the range of structure in the water phase helps in understanding the effective size of some dispersed and solid phases in emulsion systems and therefore some of the factors affecting their stability.

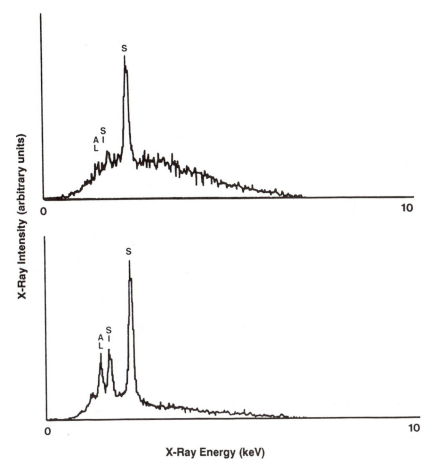

Figure 28. X-ray spectra of the small oil droplets in Figure 27. The upper spectrum was acquired with an incident beam energy of 10 keV. At this energy the electrons do not significantly penetrate the dispersed oil phase, and the X-ray signal shows a high sulfur component typical of a heavy oil. At 15 keV the electrons penetrate the droplet, and the X-ray signal comes from behind the oil droplet. This spectrum shows significant Al and Si; hence, fine clays may play a role in stabilizing this emulsion.

Nuclear magnetic resonance spectroscopy can probe bulk properties of either the water (via proton NMR spectroscopy) or the organic phases (via carbon-13 NMR spectroscopy). On-line sensors have been developed to determine oil and water content in certain emulsion systems, although the NMR technique requires a magnetic field and radio frequency generators to produce the signal, which means that it is not rugged enough for many on-line applications (*111*). However, as a quick laboratory test for oil or water content, it has a wider range of applicability. The signal can be affected by magnetic or paramagnetic species, and to a lesser extent, by the solid phase, if for example, the organic species are absorbed on a solid surface.

Near-Infrared Spectroscopy. Near-infrared (NIR) spectroscopy is a technique that has been around for some time but, like NMR spectroscopy, has only recently been improved and developed for on-line applications. Near-infrared analysis (NIRA) is a nondestructive technique that is versatile in the sense that it allows many constituents to be analyzed simultaneously (*112, 113*). The NIR spectrum of a sample depends upon the anharmonic bond vibrations of the constituent molecules. This condition means that the temperature, moisture content, bonding changes, and concentrations of various components in the sample can be determined simultaneously. In addition, scattering by particles such as sand and clay in the sample also allows (in principle) the determination of particle size distributions by NIRA. Such analyses can be used to determine the size of droplets in oil–water emulsions.

To determine the oil, water, and solids contents simultaneously, sophisticated statistical techniques must usually be applied, such as partial least-squares analysis (PLS) and multivariate analysis (MVA). This approach requires a great deal of preparation and analysis of standards for calibration. Near-infrared peaks can generally be quantified by using Beer's law; consequently, NIRA is an excellent analytical tool. In addition, NIRA has a fast spectral acquisition time and can be adapted to fiber optics; this adaptability allows the instrument to be placed in a control room somewhat isolated from the plant environment.

In Figure 29, the spectrum of an oil-sand sample shows the fundamental C–H peaks at 3.5 μm. From the two peaks in this region, one could determine the aromatic–aliphatic ratio of the hydrocarbons present in the sample. The fundamental water vibration is at approximately 3 μm (this peak would be substantially larger in a conventional emulsion sample), and the fundamental vibrations due to clays are at approximately 2.8 μm. The shape of the clay peaks indicates that kaolinite and a small amount of swelling clays such as bentonite are present in this sample.

Differential Scanning Calorimetry. Differential scanning calorimetry (DSC) is a technique with the potential to determine the relative amounts of free and emulsified water. The freezing, or more correctly, the supercooling behavior of emulsified water is very different from that of free water, so the amount of free versus emulsified water in a sample can be characterized. This parameter is important in the characterization of produced fluids and interface emulsions in which water might exist simultaneously as both continuous and emulsified phases.

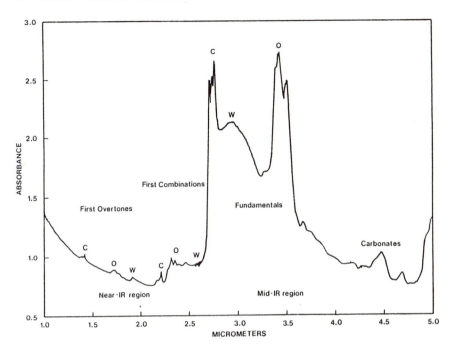

Figure 29. NIR spectrum of an oil-sand sample illustrating peaks due to clay minerals (C), oil (O), and water (W) phases. These spectra can be obtained via fiber optics; therefore, this technique has the potential for on-line quantification of oil, water, and mineral emulsion components. Either fundamentals, first combinations, or first overtones can be used to quantify particular emulsion components. The method requires calibration with standards that can be difficult, given complex field emulsions.

Figure 30 shows a plot of the amount of heat absorbed by the sample as a function of time for a sample that has both emulsified and free water. The emulsified water, or dispersed-phase water, is in small volumes, a condition that lessens the likelihood of a nucleation site for the beginning of the freezing process and results in supercooling of the dispersed phase (*114*). The latent heat of crystallization or solidification is then taken up at a lower temperature than for the free water, which freezes at a higher temperature. The degree of supercooling is greater for emulsions of smaller size distributions, and, in special cases, the technique can also give an indication of the size distribution of the emulsion. These latent heats and freezing temperatures can also be affected by solutes that depress the freezing point and by fine solids that provide nucleation sites. In spite of these potential interferences, the technique is one of the few that will allow a determination of free versus emulsified water.

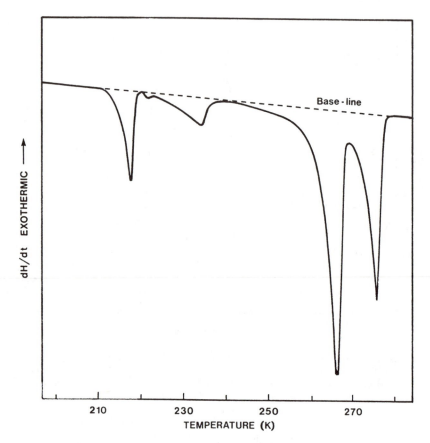

Figure 30. Freezing behavior of an emulsion characterized by differential scanning calorimetry. The free water will freeze at approximately 273 K. Emulsified water will supercool and freeze at lower temperatures, depending upon size distribution. The smallest droplets freeze last because of the smaller volume, and so fewer nucleation sites are available for ice crystal formation and water freezing. The different freezing behavior of free versus emulsified water gives this technique the potential to quantify the relative proportions of these two types of water. (Reproduced with permission from reference 114. Copyright 1984.)

Summary

To select and use the emulsion characterization technique best suited to the application at hand, it is necessary to develop an understanding of the unique capabilities and limitations of each method. Oil-field emulsion characterization requirements are generally fairly straightforward. Operations, production, and research personnel are generally interested in determining the extent of change from normal operation in terms of some baseline data. In most cases, this determination involves simple analysis for total water, oil, and solids. When operations begin to drift, a wide range of corrective actions can be taken to remedy the situation; the most important factor is therefore speed of analysis so that drift in product or tailings quality can be quickly recognized. This requirement for speed means reliance on on-line sensors or, more commonly, on a regular sampling protocol and centrifuge tests.

Some of the more sophisticated techniques offer detailed information or levels of accuracy that are not required in day-to-day operations. However, when operational upsets cannot be handled by normal methods, details of the emulsion properties have to be understood. For example, subtle changes in the size distribution of the dispersed phase (while total oil, water, and solids remain constant) can be important in determining process performance. An oil-in-water or water-in-oil emulsion can invert during processing as one or the other phase is removed, and the point in the process when this inversion occurs can have implications for the efficiency of the operation. The addition of diluent to reduce oil-phase viscosity, for instance, is much more efficient if oil is the continuous phase.

The efficiency of any water-removal steps depends upon the size distribution of the dispersed water and the stability of the emulsion. Emulsion formation may be exacerbated by inappropriate pumping speed or other process variables. An evaluation of the chemical and physical factors that determine emulsion size distribution or emulsion bulk properties is essential to optimize emulsion breaking efficiency.

The detailed emulsion characterization methods discussed herein can be used to help resolve operational upsets only if a base line of data exists for normal operation. In fact, without a thorough characterization of the normal emulsion properties such as size distribution and mineral and organic composition, the techniques for detailed characterization may actually hinder the understanding and ultimate solution of a particular processing problem by introducing extraneous information. When a base line of data exists, detailed information on the size distribution and the relationship between the dispersed, continuous, and solid phases is invaluable.

Acknowledgments

Thanks to K. C. McAuley for preparation of the figures and photographs and to V. A. Munoz and W. W. Lam, who contributed many of the photographs related to the optical and electron microscopy. Thanks are also due to C. K. Preston and J. C. Donini for the contribution on NIRA, C. A. Angle for the electrokinetic data, R. Zrobok for the rheology scans, and H. A. Hamza for useful comments and discussion on characterization technologies in general.

References

1. *Encyclopedia of Emulsion Technology;* Becher, P., Ed.; Dekker: New York, 1983; Vol. 1.
2. *Encyclopedia of Emulsion Technology;* Becher, P., Ed.; Dekker: New York, 1985; Vol. 2
3. *Encyclopedia of Emulsion Technology;* Becher, P., Ed.; Dekker: New York, 1988; Vol. 3.
4. Sherman, P. *Emulsion Science;* Academic Press: London, 1968.
5. Robb, I. D. *Microemulsions;* Plenum: New York, 1982.
6. Prince, L. M. *Microemulsions: Theory and Practice;* Academic Press: New York, 1977.
7. Becher, P. *Principles of Emulsion Technology;* Reinhold: New York, 1955. 2nd Ed.; ACS Monograph Series 162; Reinhold: New York, 1965.
8. Becher P.; Yudenfreund M. N. *Emulsions, Latices, and Dispersions;* Dekker: New York, 1978.
9. Smith, A. L. *Theory and Practice of Emulsion Technology;* Academic Press: London, 1976.
10. Lissant, K. J. *Emulsions and Emulsion Technology;* Surfactant Science Series 6; Dekker: New York, 1974; Vol. 1.
11. Lissant, K. J. *Emulsions and Emulsion Technology;* Surfactant Science Series 6; Dekker: New York, 1974; Vol. 2.
12. Lissant, K. J. *Emulsions and Emulsion Technology;* Surfactant Science Series 6; Dekker: New York, 1984; Vol. 3.
13. Barth, H. G. *Modern Methods of Particle Size Analysis;* Wiley Interscience: New York, 1984.
14. Shah, D. O. *Surface Phenomena in Enhanced Oil Recovery;* Plenum: New York, 1981.
15. Shah D. O.; Schechter, R. S. *Improved Oil Recovery by Surfactant and Polymer Flooding;* Academic Press: New York, 1977.
16. Goulden, J. D. S. *Trans. Faraday Soc.* **1958,** *54,* 941.
17. Walstra, P. J. *J. Colloid Interface Sci.* **1968,** *27,* 493.
18. Briggs, T. L. *J. Phys. Chem.* **1914,** *18,* 34.
19. Newmann, F. *J. Phys. Chem.* **1914,** *18,* 34.
20. Tronnier H.; Bussius, H. *Seifen-Oele-Fette-Wachse* **1960,** *86,* 747.
21. Bhatnagar, S. S. *J. Chem. Soc.* **1920,** *117,* 542.
22. Pickering, S. U. *J. Chem. Soc.* **1934,** 1112.
23. Scarlett, A. J.; Morgan, W. L.; Hildebrand, J. H. *J. Phys. Chem.* **1927,** *31,* 1566.
24. *Standard Methods for Analysis and Testing of Petroleum and Related Products 1988;* Institute of Petroleum, London, Wiley: Chichester, United Kingdom; 1988; Vol. 2.
25. *API Manual of Petroleum Measurement Standards;* American Petroleum Institute: Washington, DC; 1987.
26. *Annual Book of ASTM Standards;* American Society for Testing and Materials: Philadelphia, PA, 1987.
27. *Syncrude Analytical Methods for Oil Sand and Bitumen Processing;* Syncrude Canada, Ltd.: Edmonton, Alberta, Canada, 1979.
28. Hearst, P. J. *Determination of Oil in Water by Organic Carbon Analysis;* U.S. Navy Civil Engineering Laboratory: Port Hueneme, CA, 1979.
29. *Water in Petroleum Products and Bituminous Materials by Distillation;* IP74/82; ASTM D95–83; API 10.2, American Petroleum Institute: Washington, DC.

30. *Determination of Water in Crude Oil by Distillation;* IP358/82; ASTM D4006–81; API 10.2, American Petroleum Institute: Washington, DC.

31. *Emulsion Stability of Water-in-Oil Emulsions at Ambient Temperature;* IP290/84, Institute of Petroleum: London.

32. *Water and Sediment in Crude Oil by the Centrifuge Method;* IP 359/82; ASTM D4007-81; API 10.3, American Petroleum Institute: Washington, DC.

33. *Water in Crude Oil by Karl Fischer Volumetric Titration;* IP356/87; ASTM 4377–87, American Society for Testing and Materials: Philadelphia, PA, 1987.

34. Mitchel, J., Jr.; Smith, D. M. *Aquametry Part II (The Karl Fischer Reagent): A Treatise of Methods of the Determination of Water;* 2nd Ed.; Wiley: New York, 1980.

35. Smith, D. M.; Bryant, W. M. D.; Mitchell, J. *J. Am. Chem. Soc.* **1939,** *61,* 2407–2412.

36. Kapff, S. F. *Oil Gas J.* **1979,** *77(7),* 133–135.

37. Bailey, S. J. *Control Eng.* **1975,** 34–37.

38. Ellis, D. V. *IEEE Trans. Nucl. Sci.* **1988,** *35(1),* 806–811.

39. Thompson, F. *Meas. Control* **1989,** *22(7),* 210–215.

40. Whorlow, R. W. *Rheological Techniques;* Ellis Horwood, Ltd.: Chichester, United Kingdom, 1980.

41. Walters, K. *Rheometry;* Chapman and Hall: London, 1975.

42. Barnes, H. A.; Hutton, J. F.; Walters, K. *An Introduction to Rheology;* Elsevier: Amsterdam, Netherlands, 1989.

43. Ferry, J. D. *Viscoelastic Properties of Polymers;* Wiley: New York, 1980.

44. Hunter, R. J. *Zeta Potential in Colloid Science: Principles and Applications;* Academic Press: London, 1981.

45. Riddick, T. M. *Control of Colloid Stability Through Zeta Potential;* Livingston: Wynnewood, PA, 1968; Vol. 1.

46. Allen, T. *Particle Size Measurement,* 3rd Ed.; Chapman and Hall: London, 1981.

47. Groves, M. J. *Pharm. Technol.* **1980,** *4(5),* 80–94.

48. Orr, C., Jr. In *Particle Size Analysis;* Groves, M. J., Ed.; Heyden: London, 1978; pp 77–100.

49. Singh, H. N.; Swarup, S.; Singh R. P.; Saleem, S. M. *Ber. Bunsenges. Phys. Chem.* **1983,** *87(12),* 1115–1120.

50. Tausk, R. J. M.; Stassen, W. J. M.; Wilson, P. N. *Colloids Surf.* **1981,** *2(1),* 89–99.

51. Rabinovich, F. M. *The Application of Conductimetric Particle Counters to Medicine;* Medical: Moscow, U.S.S.R., 1972.

52. Rabinovich, F. M. *Conductimetric Methods of Dispersion Analysis;* Khimya Leningrad Otd.: Leningrad, U.S.S.R., 1970.

53. Ross, S. *Chemistry and Physics of Interfaces;* American Chemical Society: Washington, DC, 1965.

54. van der Waarden, M. *J. Colloid Sci.* **1954,** *9,* 215–230.

55. Mie, G. *Ann. Phys.* **1908,** *25,* 377–385.

56. Robillard F.; Patitsas, A. J. *Can. J. Phys.* **1974,** *52,* 1571–1576.

57. Robillard, F.; Patitsas, A. J.; Kaye, B. *Powder Technol.* **1974,** *9,* 307–315.

58. Lacharojana S.; Caroline, D. *NATO Adv. Study Inst. Ser., Ser. B* **1977,** *B23,* 499–514.

59. Nicholson J. D.; Clarke, J. H. R. *Surfactants Solution* **1984,** *3,* 1663–1674.

60. Sjoblom E.; Friberg, S. *J. Colloid Interface Sci.* **1978,** *67(1),* 16–30.

61. Daniels, C. A.; McDonald, S. A.; Davidson, J. A. In *Emulsions, Latices, and Dispersions;* Becher, P.; Yudenfreund, M. N., Eds.; Dekker: New York, 1978; pp 175–197.

62. Van der Hulst, H. C. *Light Scattering by Small Particles;* Chapman and Hall: London, 1957.
63. Kaler, E. W.; Davis, H. T.; Scriven, L. E. *J. Chem. Phys.* **1983,** *79(11),* 5685–5692.
64. Kaler, E. W.; Bennett, K. E.; Davis, H. T.; Scriven, L. E. *J. Chem. Phys.* **1983,** *79(11),* 5673–5684.
65. Herbst, L.; Hoffmann, H.; Kalus, J.; Thurn, H.; Ibel, K. *Neutron Scattering in the Nineties;* International Atomic Energy Agency Bulletin: Vienna, Austria, 1985.
66. Gunier, A.; Fournet, G. *Small Angle Scattering of X-rays;* Wiley: New York, 1955.
67. Caldwell, K. D.; Li, J. *J. Colloid Interface Sci.* **1989,** *132(1),* 256–268.
68. McHugh, A. J. *Crit. Rev. Anal. Chem.* **1984,** *15(1),* 63.
69. Small, H. *J. Colloid Interface Sci.* **1974,** *48,* 147–161.
70. Small, H.; Saunders, F. L.; Solc, J. *Adv. Colloid Interface Sci.* **1976,** *6,* 237–266.
71. Silebi C. A.; McHugh, A. J. *J. Appl. Polym. Sci.* **1979,** *23,* 1699–1721.
72. Groves M. J.; Freshwater, D. C. *J. Pharm. Sci.* **1968,** *57,* 436–453.
73. Groves, M. J.; Yalabik, H. S. *Powder Technol.* **1975,** *12,* 233–238.
74. Caldwell, K. D.; Karaiskakis, G.; Myers, M. N.; Giddings, J. C. *J. Pharm. Sci.* **1981,** *70,* 1350.
75. Caldwell, K. D.; Karaiskakis, G.; Giddings, J. C. *Colloids Surf.* **1981,** *3,* 233.
76. Kirkland, J. J.; Rementer, S. W.; Yau, W. W. *Anal. Chem.* **1981,** *53,* 1730.
77. Giddings, J. C.; Karaiskakis, G.; Caldwell, K. D.; Myers, M. N. *J. Colloid Interface Sci.* **1983,** *92,* 66.
78. Kirkland, J. J.; Dilks, C. H., Jr.; Yau, W. W. *J. Chromatrogr.* **1983,** *255,* 255.
79. Feng-Shyang, Y.; Caldwell, K. D.; Myers, M. N.; Giddings, J. C. *J. Colloid Interface Sci.* **1983,** *93,* 115.
80. Yau, W. W.; Kirkland, J. J. *Anal. Chem.* **1984,** *56,* 1461.
81. Bailey, E. D.; Nichols, J. B.; Kraemer, E. O. *J. Phys. Chem.* **1936,** *40,* 1149.
82. Nichols, J. B.; Bailey, E. D. In *Physical Methods of Organic Chemistry,* 2nd ed.; Weissberger. A., Ed.; Interscience: New York, 1949.
83. Lips, A.; Smart, C.; Willis, E. *Trans. Faraday Soc.* **1971,** *67,* 2979.
84. Lips, A.; Willis, E. *J. Chem. Soc. Faraday Trans. 1* **1973,** *69,* 1226.
85. Lissant, K. J.; Mayhan, K. J. *J. Colloid Interface Sci.* **1973,** *42,* 201.
86. Eley, D. R.; Hey, M. J.; Symonds, J. D.; Willison, J. H. R. *J. Colloid Interface Sci.* **1976,** *54,* 462.
87. Davis, S. S.; Purewal, T. S.; Burbage, A. S. *J. Pharm. Pharmacol.* **1976,** *28,* 60.
88. Jeffreys, G. V.; Davies, G. A. *Recent Advances in Liquid/Liquid Extraction;* Pergamon: Elmsford, NY, 1971, p 495.
89. Woods, D. R.; Burril, K. A. *J. Electroanal. Chem.* **1972,** *37,* 191.
90. Liem, A. J. S.; Woods, D. R. *AIChE Symp. Ser.* **1974,** *70(144),* 8.
91. Flummerfelt, R. W.; Catalano, A. B.; Tong, C. H. In *Surface Phenomena in Enhanced Oil Recovery;* Proc. Stockholm Symp. Aug. 1979, Shah, D. O. Ed.; Plenum: New York, 1981; p 571.
92. Buscall, R.; Ottewill, R. H. *Colloid Sci.* **1975,** *2,* 191.
93. Aveyard, R.; Vincent, B. *Prog. Surf. Sci.* **1977,** *2,* 59.
94. de Vries, A. J. *Proc. Int. Congr. Surf. Act. 3rd,* **1960,** *2,* 566.
95. Sonntag, H.; Strenge, K. *Coagulation and Stability of Disperse Systems;* Halstead-Wiley: New York, 1969.
96. Pesheck, P. S.; Scriven, L. E.; Davis, H. T. *Scanning Electron Microscopy;* Scanning Electron Microscopy, Inc.: Chicago, IL, 1981; Vol. I, pp 515–524.
97. Lianos, P.; Lang, J.; Sturm, J.; Zana, R. *J. Phys. Chem.* **1984,** *88(4),* 819–822.

98. Kano, K.; Yamaguchi, T.; Ogawa, T. *J. Phys. Chem.* **1984,** *88(4),* 793–796.
99. Eley, D. D.; Hey, M. J.; Symonds, J. D. *J. Colloid Interface Sci.* **1976,** *54,* 462.
100. Slattery, J. *AIChE J.* **1974,** *20,* 1145.
101. Watson, H. H.; Mulford, D. J. *Br. J. Appl. Phys.* **1954,** *3,* S105.
102. Montgomery, D. W. *Rubber Age* **1964,** *45,* 759.
103. Ross, W. D. *Filtr. Sep.* **1973,** *10,* 587.
104. Dixon, W. J.; Massey, F. J., Jr. *Introduction to Statistical Analysis,* 3rd Ed.; McGraw-Hill: New York, 1969, p 550.
105. Loveland, R. P. *Photomicrography;* Wiley: New York, 1970.
106. Klein, M. V. *Optics;* Wiley: New York, 1970.
107. Schott, H.; Royce, A. E. *J. Pharm. Sci.* **1983,** *72,* 313.
108. Robards, A. W.; Sleytr, U. B. *Low Temperature Methods in Biological Electron Microscopy;* Elsevier: Amsterdam, Netherlands, 1985.
109. Mikula, R. J. *Colloids Surf.* **1987,** *23,* 267–271.
110. Mikula, R. J. *J. Colloid Interface Sci.* **1988,** *121(1),* 273–277.
111. Gambhir, P. N.; Agarwala, A. K. *J. Am. Oil Chem. Soc.* **1985,** *62(1),* 103–108.
112. Walling, P. L. *J. Soc. Cosmet. Chem.* **1988,** *39(3),* 191–199.
113. Rostaing, B.; Delaquis, P.; Guy, D.; Roche, Y. *S.T.P. Pharma.* **1988,** *4(6),* 509–515.
114. Senatra, D.; Guarini, G. G. T.; Gabrielli, G.; Zoppi, M. *J. Phys.* **1984,** *45(7),* 1159–1174.

RECEIVED for review December 18, 1990. ACCEPTED revised manuscript April 24, 1991.

4

Rheology of Emulsions

R. Pal[1], Y. Yan, and J. Masliyah[*]

Department of Chemical Engineering, University of Alberta, Edmonton, Alberta, Canada T6G 2G6

This chapter presents a brief review of the rheological classification of fluids and instruments used for viscosity measurements. A discussion of the rheology of suspensions and how it relates to that of emulsions is given. Predictive correlations for emulsion viscosity are discussed in detail. The effect of added solids to an emulsion is fully treated, and its relation to a bimodel system is discussed.

Rheological Classification of Fluids

A fluid, that is, a liquid or a gas, is a substance that undergoes continuous deformation under the action of an applied shear force or stress. In other words, when a fluid is subjected to shear, it flows. On the other hand, a solid deforms under the action of an applied shear force and retains its original shape upon the cessation of the applied shear force (1).

The manner by which a fluid obeys a given shear-stress–shear-rate relationship determines its class within the rheological classification of a fluid.

Newtonian Fluids. A fluid is said to be Newtonian when it obeys Newton's law of viscosity, given by

$$\tau = -\eta\dot{\gamma} \tag{1}$$

where τ is the shear stress (Pa or N/m², force per unit area), $\dot{\gamma}$ is the shear rate exerted on the fluid (s⁻¹), and η is constant and is referred to as the fluid dynamic or the shear viscosity (kg/m·s or Pa·s).

---

[1]Current address: Department of Chemical Engineering, University of Waterloo, Waterloo, Ontario, Canada N2L 3G1
[*]Corresponding author

0065–2393/92/0231–0131 $11.00/0

A rheological instrument such as a viscometer can be used to evaluate τ and $\dot{\gamma}$ and hence obtain a value for the shear viscosity, η. Examples of Newtonian fluids are pure gases, mixtures of gases, pure liquids of low molecular weight, dilute solutions, and dilute emulsions. In some instances, a fluid may be Newtonian at a certain shear-rate range but deviate from Newton's law of viscosity under either very low or very high shear rates (2).

To be more precise, the general tensor equation of Newton's law of viscosity should be obeyed by a Newtonian fluid (2); however, for one-dimensional flow, the applicability of eq 1 is sufficient. For a Newtonian fluid, a linear plot of τ versus $\dot{\gamma}$ gives a straight line whose slope gives the fluid viscosity. Also, a log–log plot of τ versus $\dot{\gamma}$ is linear with a slope of unity. Both types of plots are useful in characterizing a Newtonian fluid. For a Newtonian fluid, the viscosity is independent of both τ and $\dot{\gamma}$, and it may be a function of temperature, pressure, and composition. Moreover, the viscosity of a Newtonian fluid is not a function of the duration of shear nor of the time lapse between consecutive applications of shear stress (3).

Fluids that do not obey Newton's law of viscosity can be broadly grouped into time-independent and time-dependent non-Newtonian fluids. Subclassifications for each group are convenient (3).

Time-Independent Non-Newtonian Fluids.   Time-independent non-Newtonian fluids are characterized by having the fluid viscosity as a function of the shear rate (or shear stress). However, the fluid viscosity is independent of the shear history of the fluid. Such fluids are also referred to as "non-Newtonian viscous fluids". Figure 1 shows a typical shear diagram for the various time-independent non-Newtonian fluids.

Pseudoplastic Fluids.   A pseudoplastic or a shear-thinning fluid is one of the most commonly encountered non-Newtonian fluids. The variation of the shear stress, τ, versus the shear rate, $\dot{\gamma}$, for a pseudoplastic fluid is shown in Figure 2. A plot of τ versus $\dot{\gamma}$ is characterized by linearity at very low and very high shear rates. The slope at very low shear rate gives the

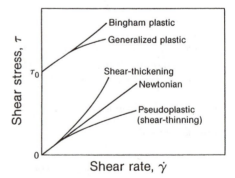

Figure 1. Shear diagram for the various type of fluids.

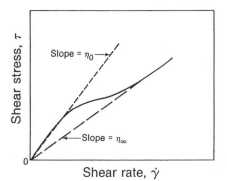

Figure 2. Shear diagram for a pseudo-plastic fluid.

viscosity at zero shear rate, η_0. The slope of the linear portion of the curve at high shear rates gives the viscosity at infinite shear, η_∞. In the intermediate range of shear rate, the viscosity is a variable, and it decreases with shear rate. In this region of the τ versus $\dot{\gamma}$ curve, the Ostwald–de Waele equation (4), commonly called the power law, is usually used to correlate the shear stress and shear rate. This power law is given as

$$\tau = -K\dot{\gamma}[\dot{\gamma}]^{n-1} \tag{2}$$

where K and n are constants. When $n = 1$, a Newtonian fluid is obtained. For a pseudoplastic fluid, n is less than unity. In the region where the power law is valid, a log–log plot of τ versus $\dot{\gamma}$ gives a straight line whose slope is n.

The power law does not describe the regions of the viscosity curve near $\dot{\gamma} = 0$ and $\dot{\gamma} \to \infty$. To this end, the Ellis model at low shear rates and the Sisko model at high shear rates can be used (2). The models are given by

$$\text{Ellis: } \dot{\gamma} = -\tau \left[\frac{1}{\eta_0} + K_1 \mid \tau \mid^{\alpha-1} \right] \tag{3}$$

$$\text{Sisko: } \tau = -\dot{\gamma} \left(\eta_\infty + K_2 \mid \dot{\gamma} \mid^{\delta-1} \right) \tag{4}$$

where K_1, K_2, α, and δ are adjustable parameters. Both the Ellis and Sisko models contain Newton's law and the power law as limiting forms. Each of these models contains three adjustable parameters in contrast to two for the power law. Sisko's model finds many uses for viscosity correlations for greases at high shear rates (5).

Reiner-Philippoff's model (6) is normally used to fit the entire pseudoplastic curve:

$$\tau = -\dot{\gamma} \left[\eta_\infty + \frac{\eta_0 - \eta_\infty}{1 + \tau^2/A} \right] \tag{5}$$

where η_0 and η_∞ are the fluid viscosities in the limits of $\dot{\gamma} \to 0$ and $\dot{\gamma} \to \infty$, respectively; and A is an adjustable parameter.

Most macromolecular fluids and concentrated emulsions are pseudoplastic fluids. They can also exhibit viscoelastic characteristics.

Dilatant Fluids. Dilatant fluids or shear-thickening fluids are less commonly encountered than pseudoplastic (shear-thinning) fluids. Rheological dilatancy refers to an increase in the apparent viscosity with increasing shear rate (3). In many cases, viscometric data for a shear-thickening fluid can be fit by using the power law model with $n > 1$. Examples of fluids that are shear-thickening are concentrated solids suspensions.

Fluids with a Yield Stress. Both pseudoplastic and dilatant fluids are characterized by the fact that no finite shear stress is required to make the fluids flow. A fluid with a yield stress is characterized by the property that a finite shear stress, τ_0, is required to make the fluid flow. A fluid obeying

$$\tau = \tau_0 - \eta_B \dot{\gamma} \tag{6}$$

is called a Bingham plastic. Here the parameter η_B is a constant. η_B is not a real viscosity but a viscosity defined after the $\dot{\gamma}$-axis is shifted to τ_0 (*see* Figure 1). For most practical fluids with a yield stress, the plastic viscosity, η_B, is a function of shear rate. Such fluids are referred to as generalized plastic. Drilling mud is a good example of a generalized plastic fluid.

Time-Dependent Non-Newtonian Fluids.

Time-dependent non-Newtonian fluids are characterized by the property that their viscosities are a function of both shear rate and shear history.

Thixotropic Fluids. Thixotropic fluids are characterized by a decrease in their viscosity with time at a constant shear rate and fixed temperature. When shear rate is steadily increased from 0 to a maximum value and then immediately decreased toward 0, a hysteresis loop is formed, as shown in Figure 3. The shape of the hysteresis loop is also a function of the rate by which the shear rate, $\dot{\gamma}$, is changed. Oil-well drilling muds, greases, and food materials are examples of thixotropic fluids.

Rheopectic Fluids. Rheopectic fluids are characterized by an increase in their viscosity with time at a constant shear rate and fixed temperature. As for a thixotropic fluid, a hysteresis loop is also formed with a rheopectic fluid if it is sheared from a low to a high shear rate and back to a low shear rate. However, a different rate is usually followed upon lowering the shear rate, as is shown in Figure 3. Bentonite clay suspensions and sols are typical examples of rheopectic fluids (3).

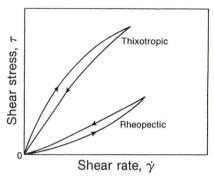

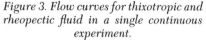

Figure 3. Flow curves for thixotropic and rheopectic fluid in a single continuous experiment.

Viscoelastic Fluids. A full description of viscoelastic fluids was given by Skelland (3):

> These materials exhibit both viscous and elastic properties. In a purely Hookean elastic solid, the stress corresponding to a given strain is independent of time, whereas for viscoelastic substances the stress will gradually dissipate. In contrast to purely viscous liquids, on the other hand, viscoelastic fluids flow when subjected to stress, but part of their deformation is gradually recovered upon removal of the stress.

Flour dough, Napalm, jellies, and concentrated emulsions are typical examples of viscoelastic fluids.

In the previous sections, the non-Newtonian viscosity (η) was used to characterize the rheology of the fluid. For a viscoelastic fluid, additional coefficients are required to determine the state of stress in any flow. For steady simple shear flow, the additional coefficients are given by

$$\tau_{11} - \tau_{22} = -\psi_1(\dot{\gamma})\dot{\gamma}_{21}^2 \tag{7a}$$

$$\tau_{22} - \tau_{33} = -\psi_2(\dot{\gamma})\dot{\gamma}_{21}^2 \tag{7b}$$

The functions ψ_1 and ψ_2 are known as the primary and secondary normal stress coefficients (7). Subscripts 1, 2, and 3 for τ and $\dot{\gamma}$ refer to the flow direction, shear axis, and neutral axis, respectively. In general, the primary and secondary normal stress coefficients are strong functions of the shear rate $\dot{\gamma}$, and ψ_2 is usually about 10% of ψ_1. Moreover, ψ_1 is a positive quantity, whereas ψ_2 is a negative quantity. Nonzero values of ψ_1 and ψ_2 give rise to the die-swell phenomenon, fluid climbing up a rotating shaft (Weissenberg effect), and secondary flow between moving surfaces.

For steady-state shearing flows, the relationship between the shear-stress tensor and the shear-rate tensor is given by Criminale–Ericksen–

Filbey equation (7). For cases of small deformation and deformation gradients, the general linear viscoelastic model can be used for unsteady motion of a viscoelastic fluid. Such a model has a memory function and a relaxation modulus. Bird and co-workers (6, 7) gave details of the available models.

In the previous sections, fluids were classified in distinct categories. However, some emulsions cannot be fitted into any one category. Some dispersions or emulsions exhibit various non-Newtonian behaviors depending on the level of the shear rate. For example, dispersions of latex particles exhibit a Newtonian behavior at very low shear rate; a random three-dimensional structure is formed, and the Brownian motion is dominant. This condition is shown as a region A in Figure 4. At still a higher shear rate, the three-dimensional structure changes to a two-dimensional one and the viscosity decreases with shear rate. In this regime, the fluid exhibits a shear-thinning behavior. This is region B. At still higher shear rate, the three-dimensional structure lends itself to a completely two-dimensional structure and the suspension behaves as a Newtonian fluid (regime C). When the shear rate is increased further, the two-dimensional ordered structure becomes unstable and the suspension viscosity increases (8, 9). In this regime (D), the fluid has a shear-thickening behavior. As will be discussed in the section "Viscosity of Emulsion with Added Solids", this behavior was observed for a bitumen-in-water emulsion with added solids.

References 10 and 11 give more details of the rheological properties of fluids.

Viscosity Measurements: Instruments

The directly measured variables are not the fluid properties such as viscosity, but the forces, torques, and rate of rotation pertaining to a given apparatus (7). In some cases, the measured shear stress, τ, and shear rate $\dot{\gamma}$ data can be used to construct a model or use an established model to fit the data.

Some of the typical apparatus used to measure the rheological behavior of fluids will be discussed in the following section. A detailed compilation of commercial rheological equipment is given in reference 12.

Capillary Tube. Basically, for these devices, the frictional pressure drop associated with the laminar flow of the fluid of interest is measured in a capillary tube. Normally, various tube lengths and diameters are used to eliminate end effects (13, 14). Appendix 1 (taken directly from ref. 7) gives the interrelationship between the measured flow rate, pressure drop, shear stress, and shear rate (equations A-1 to A-3).

The capillary tube viscometer is not suitable for settling suspensions. This instrument suffers from the fact that the shear rate is not constant across the tube radius and the fluid cannot be sheared as long as desired. It is

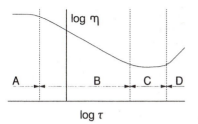

log τ

Figure 4. Shear diagram for a typical suspension.

a one-pass system. However, a fairly high shear rate can be achieved in a capillary tube viscometer.

Rotational Viscometers. Rotational viscometers are the most widely used instruments for the measurement of the rheological properties of a fluid (e.g., a pure liquid, emulsion, or suspension). The test fluid is placed in a gap formed by either two coaxial rotating cylinders, two flat discs, or a flat disc and a cone. The major advantages of the rotational viscometers are

1. Ease of use.
2. Commercial availability.
3. The shear rate can be changed in a preset manner.
4. A wide range of viscosity can be handled.
5. The test fluid can be sheared to any length of time.
6. A fairly uniform shear stress is imposed on the fluid, mainly because of a narrow gap or the use of an appropriate design.
7. They can be used for relatively nonsettling suspensions.
8. They can be adopted to measure normal stresses for viscoelastic fluids.
9. They can be adopted for use at high temperature and pressure.

One of the major disadvantages is that a very viscous test fluid cannot be subjected to a prolonged period of high shear rate without extensive temperature rise within the test fluid. This effect is due to the fact that the temperature control system cannot cool the test fluid fast enough to counteract the internal heat generation due to fluid flow (*15*).

Concentric Cylinder Rotary Viscometer. These instruments are designed to have the test fluid in the annulus between two concentric cylinders that are subjected to shear due to the rotation of either the inner or the outer cylinder. The torque required to rotate one of the cylinders is a measured quantity. From the geometry of the cylinders and the rotational speed, the shear rate can be evaluated. Appendix 1 gives a summary of the stress–shear relationships in equations D-1 to D-3. Figure 5 shows a simple arrangement for a concentric cylinder viscometer.

The concentric cylinder viscometers are supplied with different inner and outer cylinders such that various gap widths can be formed. For rheological measurements of emulsions and suspensions, care must be taken to ensure a gap width of at least 20 times the suspended particle size in order to avoid "wall effects". Moreover, experiments should be conducted with different gap widths to ensure the absence of any wall slip that is usually encountered in emulsion viscosity measurements (16). However, uniformity of shear rate can be achieved only when the ratio of the gap width to the inner cylinder radius is small.

Cone and Plate Viscometer.   A cone and plate viscometer is shown in Figure 6. The test fluid is placed in the gap between the cone and the plate. The cone angle can be from 0.3° to 10°. The cone is made to rotate. In instruments such as a rheogoniometer, the flat bottom plate is provided with pressure-sensing devices for the measurement of normal stresses for viscoelastic fluids (quantities ψ_1 and ψ_2 of Appendix 1). For small cone angles (e.g., 1°), the shear rate within the gap is very uniform. However, for the larger angles (e.g., 5°), the shear rate within the gap is not very uniform, and it is not suitable for non-Newtonian fluids. Because of the small gap, this

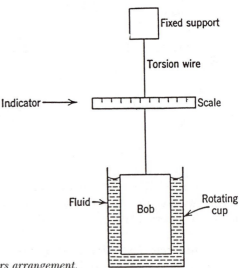

Figure 5. Concentric cylinders arrangement.

type of a viscometer is not very suitable for viscosity measurement of suspensions. Such viscometers can give fairly low and high shear rates.

The pertinent rheological expressions for the cone and plate are given in Appendix 1, equations B-1 to B-5.

Parallel Plate Viscometer.   This instrument resembles the cone and plate viscometer, except that it has a flat horizontal rotating plate in place of the cone. The shear rate within the narrow gap of the two plates is not as uniform as for the cone and plate viscometer. The limiting shear rates for the parallel plate viscometer are similar to those of the cone and plate instrument. This type of a viscometer is suitable for rheological measurements of suspensions and emulsions.

The equations defining the various rheological quantities are given in Appendix 1 by equations C-1 to C-5.

Rheology of Emulsions

The rheological behavior of an emulsion can be Newtonian or non-Newtonian depending upon its composition. At low to moderate values of dis-

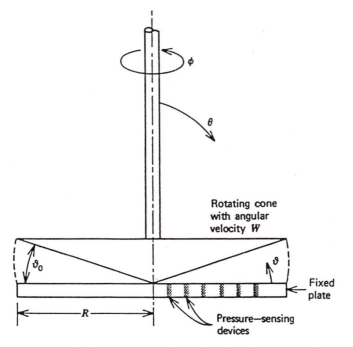

Figure 6. Cone and plate viscometer. (Reproduced with permission from reference 7. Copyright 1987 Wiley.)

persed-phase concentration, emulsions generally exhibit Newtonian behavior. In the high-concentration range, emulsions behave as shear-thinning fluids (17–21).

Figure 7 shows the rheograms (shear stress versus shear rate plots) of a typical emulsion system at different values of dispersed-phase concentration. The volume–surface mean diameter of the oil droplets for the system shown is 10 μm. For a given concentration, the variation of log τ against log $\dot{\gamma}$ is linear, a result indicating that the emulsions follow a power law (eq 2).

The rheograms shown in Figure 7 indicate that the emulsions considered are Newtonian up to a dispersed-phase concentration of 40% by volume as the slope of the rheograms is unity. At dispersed-phase concentrations above 40%, the slope of the rheograms is less than unity, a result indicating a pseudoplastic or shear-thinning behavior; that is, the apparent viscosity (the ratio of the shear stress to the shear rate) decreases with an increase in the shear rate. In the concentration range in which emulsions behave as non-Newtonian fluids, the deviation of the slope of the rheograms from unity increases with the increase in the dispersed-phase concentration. Also, quite likely at very high concentrations of the dispersed phase, emulsions may develop a yield stress (yield stress is the stress that must be exceeded before any flow of material can take place).

Han and King (22) had shown that concentrated emulsions can exhibit viscoelastic behavior even though the dispersed and the continuous phases are both Newtonian. For a given shear stress, the primary normal stress

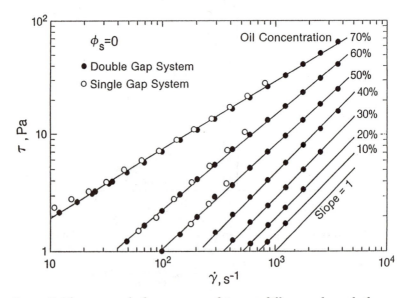

Figure 7. Rheograms of oil-in-water emulsions at different values of oil concentration. (Reproduced with permission from reference 21. Copyright 1990.)

coefficient, ψ_1, was maximum at a dispersed-phase volume fraction of about 0.3. The system used was Indopol L100–glycerine (22).

Viscosity of Emulsions. The viscosity of an emulsion, defined as the ratio of shear stress to shear rate, depends upon several factors:

1. the viscosity of the continuous phase
2. the volume fraction of the dispersed phase
3. the viscosity of the dispersed phase
4. the average particle size and particle size distribution
5. shear rate
6. the nature and the concentration of the emulsifying agent
7. temperature

The viscosity of an emulsion is directly proportional to the continuous-phase viscosity (η_c), and therefore, all the viscosity equations proposed in the literature are written in terms of the relative viscosity (η_r). If an emulsifying agent is present in the continuous phase, as is the case with emulsions, η_c is then the viscosity of the emulsifier solution rather than the viscosity of the pure fluid phase (i.e., oil or water alone). When an emulsion is prepared, some of the emulsifying agent becomes adsorbed at the oil–water interface; this adsorption tends to lower the original concentration of emulsifier in the continuous phase and cause an associated decrease in η_c. However, the amount of emulsifier adsorbed is usually very low compared with the total amount present, and therefore any decrease in concentration of the emulsifier can easily be neglected (23).

The volume fraction of the dispersed phase is the most important factor that affects the viscosity of emulsions. When particles are introduced into a given flow field, the flow field becomes distorted, and consequently the rate of energy dissipation increases, in turn leading to an increase in the viscosity of the system. Einstein (24, 25) showed that the increase in the viscosity of the system due to addition of particles is a function of the volume fraction of the dispersed particles. As the volume fraction of the particles increases, the viscosity of the system increases. Several viscosity equations have been proposed in the literature relating viscosity to volume fraction of the dispersed phase. We discuss these equations in a later section.

Unlike a solid-in-liquid suspension, the viscosity of an emulsion may depend upon the viscosity of the dispersed phase. This dependence is especially true when internal circulation occurs within the dispersed droplets. The presence of internal circulation reduces the distortion of the flow field around the droplets (26), and consequently the overall viscosity of an emulsion is lower than that of a suspension at the same volume fraction. With the increase in dispersed-phase viscosity, the internal circulation is reduced, and consequently the viscosity of an emulsion increases. The phenomenon of internal circulation is important only when the emulsifier is not present at the droplet surface. The presence of an emulsifier greatly inhibits internal circulation (27), and the emulsion droplets behave more like rigid particles. Thus, in the presence of emulsifiers, the effect of dispersed-phase viscosity on the overall emulsion viscosity is negligible.

For monodisperse or unimodal dispersion systems (emulsions or suspensions), some literature (28–30) indicates that the relative viscosity is independent of the particle size. These results are applicable as long as the hydrodynamic forces are dominant. In other words, forces due to the presence of an electrical double layer or a steric barrier (due to the adsorption of macromolecules onto the surface of the particles) are negligible. In general the hydrodynamic forces are dominant (hard-sphere interaction) when the solid particles are relatively large (diameter >10 μm). For particles with diameters less than 1 μm, the colloidal surface forces and Brownian motion can be dominant, and the viscosity of a unimodal dispersion is no longer a unique function of the solids volume fraction (30).

In systems where Brownian motion is significant, the relative viscosity decreases with the increase in the particle size. Figure 8 shows Krieger's data (31) for a 50% monodispersion of polystyrene spheres in benzyl alcohol in the absence of both steric and electroviscous forces. At a given shear stress, the relative viscosity decreases with the increase in the particle size. This result implies that the importance of the Brownian motion decreases with increase in particle size. Krieger (31) showed that the effect of the

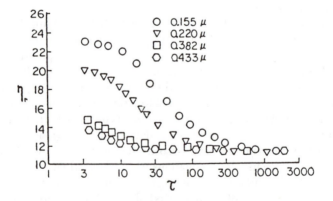

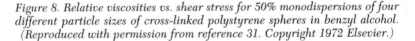

Figure 8. Relative viscosities vs. shear stress for 50% monodispersions of four different particle sizes of cross-linked polystyrene spheres in benzyl alcohol. (Reproduced with permission from reference 31. Copyright 1972 Elsevier.)

Brownian motion can be taken into account by plotting η_r against the reduced shear stress τ_r, given as

$$\tau_r = \frac{\tau a^3}{kT} \tag{8}$$

where k is the Boltzmann constant and T is the absolute temperature. The data for the different sized particles then fall on a single curve, as shown in Figure 9.

Figure 10 shows the variation of the relative viscosity with the counterion molarity at different reduced shear stress values for monodisperse polystyrene latex having a diameter of 0.192 μm at dispersed-phase volume fraction $\phi = 0.509$. Clearly, η_r is a function of the electrolyte concentration in addition to the reduced shear stress.

Figure 11 shows the relative-viscosity–concentration behavior for a variety of hard-sphere suspensions of uniform-size glass beads. Even though the particle size was varied substantially (0.1 to 440 μm), the relative viscosity is independent of the particle size. However, when the particle diameter was small (~1 μm), the relative viscosity was calculated at high shear rates, so that the effect of Brownian motion was negligible. Figure 8 shows that η_r becomes independent of the particle size at high shear stress (or shear rate).

The effect of particle size distribution on the viscosities of suspensions and emulsions has been investigated (28, 32–35). Most of these studies indicate that the effect of particle size distribution is of enormous magnitude

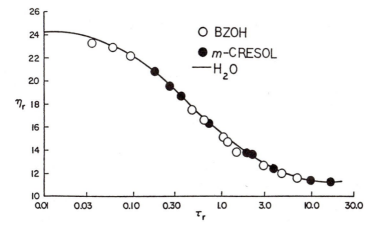

Figure 9. Relative viscosities vs. reduced shear stress for 50% monodispersions of polystyrene spheres of various sizes in different media. Points are taken from Figure 8. (Reproduced with permission from reference 31. Copyright 1972 Elsevier.)

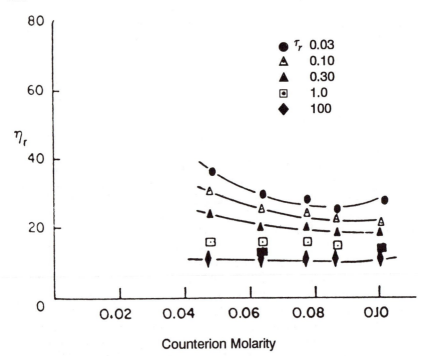

Figure 10. Variation of relative viscosity of monodisperse polystyrene latex with electrolyte concentration at different reduced shear stress. (Reproduced with permission from reference 31. Copyright 1972 Elsevier.)

at high values of dispersed-phase concentration. At low concentrations, however, the effect is small.

Figure 12 shows Chong et al. (28) data for monodisperse and bidisperse (bimodal) suspension systems. In a bidisperse suspension, the volume fraction of small spheres (diameter d) in the mixture is kept constant at 25% of the total solids. The figure shows that the viscosity of a bidisperse suspension is a strong function of the particle size ratio, d/D, where D is the diameter of the large particles. The viscosity decreases substantially by decreasing d/D at a given total solids concentration. The data for the unimodal system fall well above the bimodal suspensions. Also, the effect of particle size distribution decreases at lower values of total solids concentration.

Figure 13 illustrates another very interesting point. Here the relative viscosity of a bimodal suspension is plotted as a function of volume percent of small spheres in total solids. At any given total solids concentration, the relative viscosity decreases initially with the increase in volume percent of small spheres, and then it increases with further increase in small spheres. The minimum observed in the relative viscosity plots of a bimodal suspension is quite typical. There are no fundamental reasons why a similar behavior would not be true for emulsions.

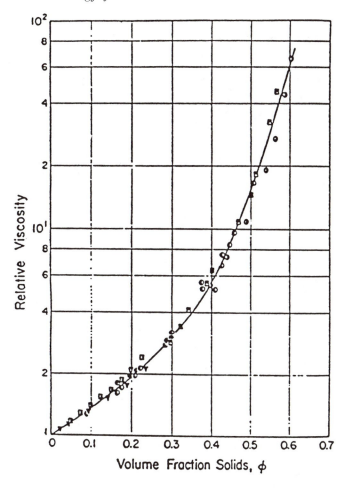

Figure 11. Relative viscosity vs. concentration behavior for suspensions of spheres having narrow size distributions. Particle diameters range from 0.1 to 440 μm. (Reproduced with permission from reference 30. Copyright 1965 Academic Press.)

A similar study was conducted by Poslinski et al. (36) on the effect of a bimodal size distribution of solids. They confirmed the findings of Chong et al. (28) in that the relative shear viscosity can exhibit a minimum for a plot of relative viscosity versus volume percent of small particles in total solids. Moreover, the primary normal stress also exhibited a minimum. Poslinski et al. showed that the relative viscosity can be predicted from the knowledge of the maximum packing volume fraction of the bimodal solids systems.

Shear rate influences the viscosity of emulsions quite significantly when their behavior is non-Newtonian. As discussed earlier, in the low ϕ range, emulsions exhibit a Newtonian behavior, and consequently shear rate does

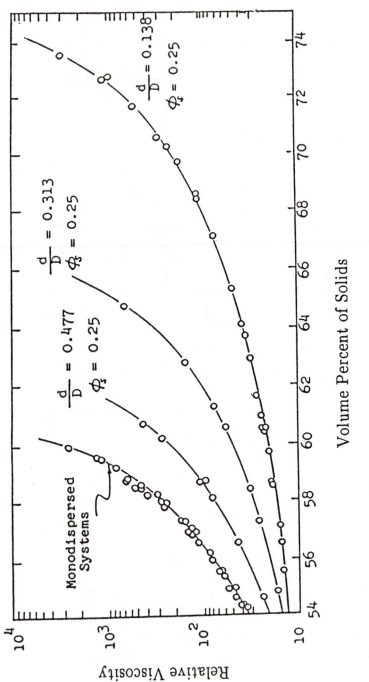

Figure 12. Relative viscosity as a function of solids concentration and particle size distribution. (Reproduced with permission from reference 28. Copyright 1971.)

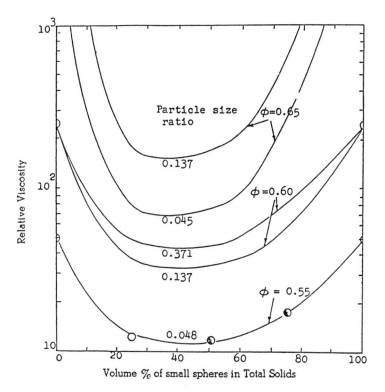

Figure 13. Dependence of relative viscosity upon particle size ratio for bimodal suspensions of spheres. (Reproduced with permission from reference 28. Copyright 1971.)

not affect the viscosity of emulsions. In the high ϕ range where emulsions exhibit non-Newtonian behavior, the apparent viscosity decreases significantly with the increase in the shear rate.

The chemical nature and the concentration of an emulsifying agent also play a role in determining the viscosity of emulsions (37). The average particle size, particle size distribution, and the viscosity of the continuous phase (to which an emulsifier is normally added) all depend upon the properties and concentration of emulsifying agent. Also, ionic emulsifiers introduce electroviscous effects, leading to an increase in the emulsion viscosity.

The viscosity of emulsions is a strong function of temperature; it decreases with the increase in temperature (18). The decrease in emulsion viscosity that occurs with raising the temperature is mainly due to a decrease in the continuous-phase viscosity. The increase in temperature may also affect the average particle size and particle size distribution. When the apparent viscosity of an emulsion (at a given shear rate) is plotted as a function of $1/T$ (*see* Figure 14), a linear relationship is normally observed. The data follow the Arrhenius type relationship:

$$\eta = A \exp (B/T) \tag{9}$$

where A and B are constants dependent upon the system and shear rate and T is the absolute temperature.

Emulsion Viscosity Equations.   Only a few viscosity–concentration equations have been developed purely for emulsion systems. Most equations have been adapted by analogy from studies with suspensions of solid particles (*20*).

For suspensions of nondeformable spherical globules at very low concentrations, Einstein (*24, 25*) derived the following relationship between the viscosity of the suspension and the volume fraction:

$$\eta_r = \frac{\eta}{\eta_c} = 1 + 2.5\phi \tag{10}$$

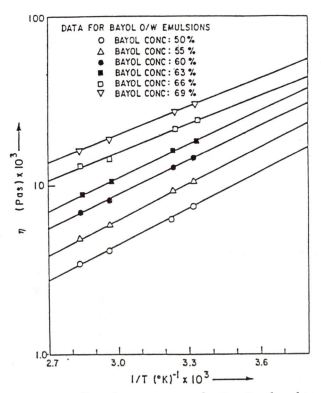

Figure 14. Effect of temperature on the viscosity of emulsions.

The main assumptions made in arriving at eq 10 were (1) no interaction between the globules and (2) no slippage at the particle–fluid interface. Most emulsions of practical interest exceed the concentration for which eq 10 is valid. With increasing concentration, the hydrodynamic interaction between the globules increases, and eventually mechanical interference occurs between the particles as packed-bed concentrations are approached. To take into account the increased hydrodynamic interaction, many investigators (30, 38–43) have expanded eq 10 into the polynomial form:

$$\eta_r = 1 + K_1\phi + K_2\phi^2 + K_3\phi^3 + \ldots \tag{11}$$

where K_1, K_2, K_3 . . . are constants.

Several closed-form equations have also been proposed. Using Einstein's equation (eq 10), Roscoe (42) and Brinkman (43) independently developed the following equation for the higher concentration suspension:

$$\eta_r = 1 / (1 - \phi)^{2.5} \tag{12}$$

This equation reduces to Einstein's equation in the limit of $\phi \to 0$.

On the basis of the analogy with the influence of variable pressure on a material that obeys Hooke's law, Richardson (44) calculated the compressibility of an emulsion whose dispersed-phase volume concentration is increased from ϕ to $\phi + \Delta\phi$. From this calculation he derived the following expression for the relative viscosity of concentrated emulsions:

$$\ln \eta_r = K\phi \tag{13}$$

where K is a constant that depends on the system.

Broughton and Squires (45) later modified Richardson's equation to

$$\ln \eta_r = K_1\phi + K_2 \tag{14}$$

where K_1 and K_2 are constants that depend on the system. This form showed better agreement with their experimental data.

Hatschek (46) developed the following equation for concentrated emulsions:

$$\eta_r = 1 / (1 - \phi^{1/3}) \tag{15}$$

which was later modified by Sibree (47, 48) to

$$\eta_r = 1 / [1 - (h\phi)^{1/3}] \tag{16}$$

where h is a hydration factor that depends on the emulsion system.

Eilers (49, 50) proposed the following empirical equation to fit his data on polydisperse bitumen emulsions:

$$\eta_r = \left[1 + \frac{1.25 \, \phi}{1 - K\phi} \right]^2 \tag{17}$$

where the constant K varied from 1.28 to 1.30. For uniform spheres Eilers proposed $K = 1.35$.

Using functional analysis, Mooney (51) derived the following equation for concentrated suspensions:

$$\eta_r = \exp \left[\frac{2.5\phi}{1 - K\phi} \right] \tag{18}$$

where K is a constant. This equation is a solution of the following functional equation:

$$\eta_r(\phi_1 + \phi_2) = \eta_r \left[\frac{\phi_1}{1 - K\phi_2} \right] \eta_r \left[\frac{\phi_2}{1 - K\phi_1} \right] \tag{19}$$

which Mooney arrived at on the basis of theoretical arguments. Equation 18 is widely quoted in the emulsion literature.

Barnea and Mizrahi (52) modified Mooney's equation to

$$\eta_r = \exp \left[\frac{2.66\phi}{1 - \phi} \right] \tag{20}$$

Equation 20 was found to fit the viscosity data of a variety of suspensions.

Using functional analysis similar to Mooney, Kreiger and Dougherty (53) derived the following equation:

$$\eta_r = \frac{1}{(1 - \phi / \phi_m)^{[\eta]\phi_m}} \tag{21}$$

where $[\eta]$ is the intrinsic viscosity and ϕ_m is the maximum packing concentration.

In the region of high concentration levels, that is, near maximum packing concentration, Frankel and Acrivos (54) derived the following theoretical equation on the basis of lubrication theory:

$$\eta_r = \frac{(9/8)(\phi / \phi_m)^{1/3}}{1 - (\phi / \phi_m)^{1/3}} \tag{22}$$

This equation and all the other equations discussed so far are valid only for Newtonian emulsions. As mentioned earlier, emulsions exhibit non-Newtonian behavior at high concentrations. This behavior means that the viscosity equations must incorporate the shear rate effect.

The general tendency in the available literature is to apply the Newtonian viscosity equations to non-Newtonian emulsions at high shear rate. According to Sherman (55), Mooney's equation can be applied to non-Newtonian emulsions at high shear rates.

Pal and Rhodes (18, 20) recently used an empirical approach to correlate the viscosity data of emulsions. On the basis of the extensive amount of experimental data collected, the following correlation was proposed:

$$\eta_r = \left[1 + \frac{(\phi / \phi^\circ)}{1.187 - (\phi / \phi^\circ)} \right]^{2.49} \tag{23}$$

where ϕ° is the dispersed-phase concentration at which the relative viscosity becomes 100. Both the Newtonian and non-Newtonian emulsions were correlated by eq 23. Figure 15 shows comparison of the correlation with the experimental data. As can be seen, the correlation describes the experimental data quite adequately. The theoretical basis for this correlation also was given (20).

The normalization of the volume fraction ϕ with either ϕ_m as in eq 22 for suspensions or by ϕ° as in eq 23 for emulsions tends to suggest that the relative viscosity, η_r, can be better correlated by the ratio of (ϕ/ϕ_m) or (ϕ/ϕ°) than by simply the use of the volume fraction ϕ alone. Such a normalization takes into account the effect of forces other than the hydrodynamic forces. Such plots are given by Poslinski et al. (36) and Pal and Rhodes (20).

Summary.

1. Emulsions can exhibit Newtonian, shear-thinning, and viscoelastic behaviors.

2. In the absence of colloidal forces for a monodispersion, the relative viscosity is independent of the particle size.

3. When Brownian motion and/or colloidal forces are present, the relative viscosity of a dispersion becomes a function of the particle size.

4. The viscosity of a bimodal dispersion is a strong function of the particle size ratio of the two particle fractions.

5. Mooney- or Pal–Rhodes-type correlations can be used to correlate emulsion relative viscosities with the dispersed-phase volume fraction.

Viscosity of Emulsion with Added Solids

Little work has been done on the rheology of emulsions with added solids, despite the fact that handling of mixtures of emulsions and solids is encoun-

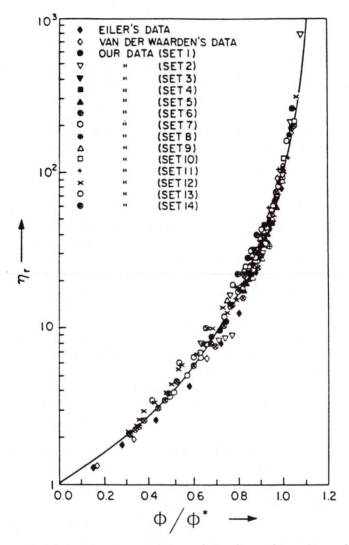

Figure 15. Relative viscosity vs. ϕ/ϕ° correlation for emulsions. (Reproduced with permission from reference 20. Copyright 1989 Wiley.)

tered in many engineering processes. For example, heavy crude oil is often produced in the form of oil-in-water (O/W) or water-in-oil (W/O) emulsions. The produced emulsions often contain some solids despite the use of mechanical filters, though the concentration of solids is generally low [less than 5% by weight (56)]. As will be discussed, solids behave quite differently from the oil droplets in the emulsion with respect to the rheological properties. The concentration of the solids and that of the oil in an emulsion are generally not "additive" for the evaluation of the viscosity. This condition is particularly true when the solids are of different size and shape than the spherical oil droplets.

In this section, we discuss the effects of solids addition on the rheology of oil-in-water emulsions, in particular, the effects of solids size (size distribution) and shape (spherical versus irregular). Because the type of the oil used to form an emulsion is important in determining the viscosity of the oil-in-water emulsion, the rheology of the emulsion–solids mixtures is also influenced by the type of oil. Thus, two distinct emulsion systems with added solids will be discussed: (1) synthetic (Bayol-35) oil-in-water emulsions (*21, 57*) and (2) bitumen-in-water emulsions (*58*). The synthetic oil has a viscosity of 2.4 mPa·s, whereas the bitumen has a viscosity of 306,000 mPa·s at 25 °C. The Sauter mean diameter of the oil droplets is 10 μm for synthetic oil, and 6 μm for bitumen-in-water emulsions. The synthetic O/W emulsions are fairly shear-thinning, whereas the bitumen O/W emulsions are fairly Newtonian.

We also discuss the conditions under which emulsions can be considered as a continuous phase toward the added solids, where the prediction of the emulsion–solids viscosity is possible.

Effect of Solids Addition. Figure 16 shows the variation of the viscosity with shear stress when solids are added to a synthetic O/W emulsion. The solids are sand particles with a Sauter mean diameter of 9 μm. The

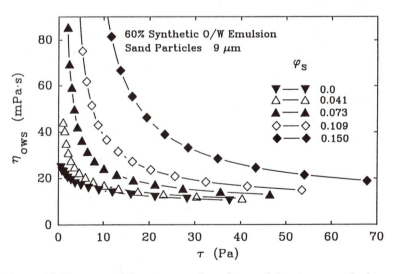

Figure 16. Variation of the viscosity of emulsion–solids mixtures with shear stress at different solids volume fraction. (Reproduced with permission from reference 57. Copyright 1991 Pergamon Press.)

oil concentration (solids-free basis) is 60% by volume. The addition of solids to the emulsion increases the viscosity of the emulsion, and the viscosity of the emulsion–solids mixtures decreases with increasing τ (shear thinning). The addition of solids to emulsions may also introduce yield stress. The addition of smaller solids was found (57, 58) to result in more pronounced yield stress than that of larger solids.

The effect of solids addition to a bitumen-in-water emulsion is shown in Figure 17; the bitumen concentration (solids-free basis) is about 59% by volume, and the added solids are the 9- and the 33-μm silica sand (Figures 17a and 17b, respectively). The addition of solids to the emulsion increases the viscosity for both the 9- and the 33-μm silica sand. Figure 17a shows that after a rapid decrease, the viscosity tends to increase with increasing shear stress; that is, the emulsion–solids mixture exhibits dilatant behavior. The dilatant behavior becomes more noticeable as the solids volume fraction increases. Comparison of Figures 17a and 17b indicates that the larger size solids result in a more pronounced shear-thickening behavior. The sharp increase in the viscosity at low shear stress with decreasing τ is indicative of a yield stress (Figure 17a). However, the yield stress is not obvious with the 33-μm solids (Figure 17b). Shear-thickening behavior is also observed with spherical glass beads. The degree of shear thickening increases with solids volume fraction. Also, the onset of shear thickening occurs at a lower shear stress for the larger particles. Similar conclusions have been made by Barnes (59) in a review on shear thickening in suspensions.

Various explanations in the literature concerning shear thickening all tend to rely on some kind of change in the relative spatial disposition of the particles. According to Goodwin (60) and Hoffman (61), a change in the flow pattern takes place when dilatancy occurs. At a very low shear rate, the particles tend to arrange themselves to be as far as possible from one another. When the shear rate increases, the hydrodynamic forces increase, and the particles are forced into layers. In other words, the particles follow the fluid streamlines with a minimum contact. This process is characterized by a decrease in the viscosity as the shear rate is increased. This shear-thinning behavior was observed in the study reported here for both the solids-free emulsions and the emulsion–solids mixtures.

When the shear rate increases up to a critical value, the layers begin to disrupt, and the two-dimensional layering is disturbed. With further increase in shear rate, a three-dimensional random arrangement is obtained. Thus, the viscosity begins to increase with the shear rate, (i.e., shear-thickening behavior) (8, 9).

Shear thickening was not observed for synthetic oil emulsions with added solids. This observation is in agreement with Barnes (59) that almost all solids suspensions showed shear thickening, whereas shear thickening was never observed in emulsions. Considering the dramatic difference in the viscosity of the synthetic oil and the bitumen, the interparticle interaction

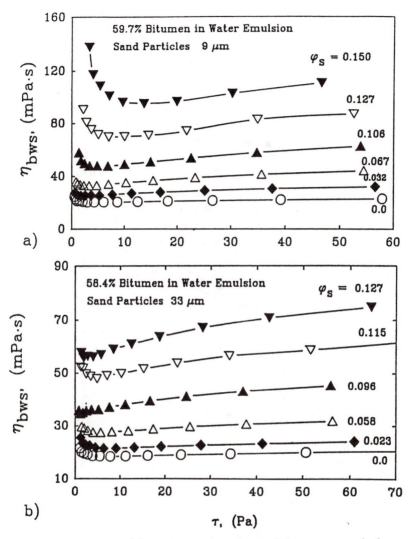

Figure 17. Variation of the viscosity of emulsion–solids mixtures with shear stress. (Reproduced with permission from reference 58. Copyright 1991.)

between the solids and the synthetic oil droplets is much weaker. However, the bitumen droplets behave more like solids compared with the synthetic oil droplets; this behavior thus causes dilatancy.

Effects of Solids Shape. The viscosities of emulsion–solids mixtures are compared when irregular-shaped silica sand and spherical glass beads are added separately to an oil emulsion. The results are shown in Figure 18 for different sizes of glass beads and silica sand for synthetic oil.

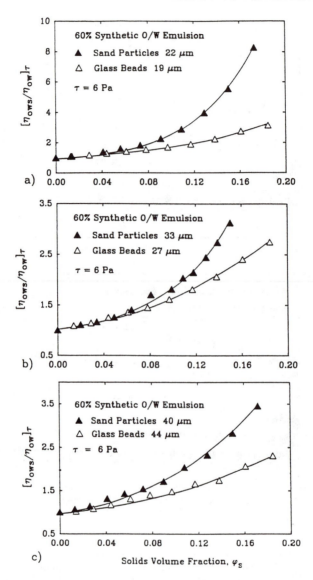

Figure 18. Effect of solids shape on the viscosity of emulsion–solids mixtures. (Reproduced with permission from reference 57. Copyright 1991 Pergamon Press.)

Because the emulsion–solids mixtures are non-Newtonian fluids, a shear stress or a shear rate must be specified when any comparison is made. In Figure 18a, comparison is made between the 22-μm silica sand and the 19-μm glass beads at a shear stress of 6 Pa. The differences in the solids mean diameter and in their size distributions are small. The degree of influence on the viscosity is therefore mainly due to the solids shape. Similar results are shown in Figures 18b and 18c for larger solids sizes. Evidently, the addition of the irregular-shaped silica sand results in a much higher viscosity than that of the spherical glass beads.

For the bitumen-in-water emulsions, different types of flow behavior such as shear thinning and shear thickening were observed over the shear-stress range studied; therefore, comparison of the viscosity is made at three typical shear stresses, 1, 14, and 50 Pa. At a shear stress of about 1 Pa, the mixtures acted as a shear-thinning fluid, whereas at shear stresses of 14 and 50 Pa, the mixtures acted as a dilatant fluid. Results (58) showed that the irregular-shaped silica sand always gave a higher viscosity than the smooth spherical glass beads.

The effect of particle shape has received little attention in the published literature, though rodlike particles with different aspect ratios have been studied (62). The difficulty with the irregular-shaped solids lies in the characterization of the irregularity of the surface and the sharpness of the edges of the surface. However, generally, the more the solids shape deviates from that of a sphere, the higher is the viscosity of the suspension (63). The rotation of nonspherical particles in shear flow may increase the frequency of contact and trap layers of liquid on to their surfaces. Both of these effects result in an increase in the effective solids volume fraction. Furthermore, according to Clarke (63), irregular-shaped solids may also interlock and scrape harshly together rather than making simple impact; thus irregular shape causes high energy loss.

Effects of Solids Size. The effect of solids size on the viscosity of the emulsion–solids mixtures is shown in Figure 19 for synthetic O/W emulsions. The oil concentration (solids-free basis) is 60% by volume, and the solids used are silica sand. The comparison is made at shear stresses of 6 and 14 Pa. The viscosity is expressed as the relative viscosity $(\eta_{ows}/\eta_{ow})_\tau$, that is, the viscosity of the emulsion–solids mixture divided by the viscosity of the solids-free emulsion. At low solids volume fraction (<0.1), solids size has little effect.

However, at higher solids volume fractions, the effect of solids size becomes significant, especially at low shear rates. The smallest sand particles (9 μm) give the highest viscosity. At a solids volume fraction of about 0.17, the addition of the 40-μm sand increases the viscosity of the emulsion by a factor of 3 (Figure 19a), whereas the 9-μm sand increases the viscosity by a factor of 25. Furthermore, the effect of solids size on the viscosity becomes less pronounced at a higher shear stress, as shown in Figure 19b. This result is consistent with the fact that the effect of the yield stress becomes weaker with increasing τ. The same trends were observed when different sizes of glass beads were added to the synthetic O/W emulsions (57).

The effect of solids size on the viscosity of bitumen-in-water emulsions containing solids is shown in Figure 20. The shear stresses at which the comparison is made are 1, 14, and 50 Pa. At a very low shear stress (Figure

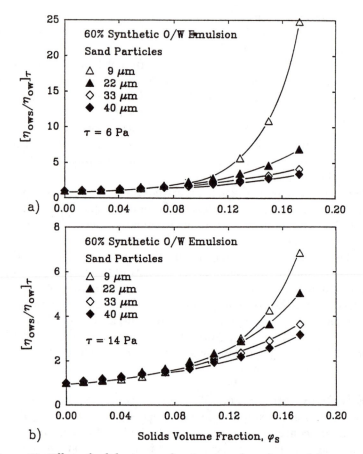

Figure 19. Effect of solids size on the viscosity of emulsion–solids mixtures. (Reproduced with permission from reference 57. Copyright 1991 Pergamon Press.)

20a) the addition of the 9-μm sand gives a higher viscosity than that of the 33-μm sand, particularly at high solids volume fraction. However, when comparison is made at shear stresses of 14 and 50 Pa, the two fractions yield approximately the same viscosity, and the size effect becomes negligible (Figures 20b and 20c). The same conclusion was reached for the spherical glass beads (58).

Points of Special Interest. The addition of solids to an emulsion forms a bimodal system. If the coarse particles are sufficiently larger than the fine particles in a bimodal system, then the viscosity of the bimodal system can be predicted from the unimodal viscosity data (35). In the emulsion–solids mixtures, if the solids are much larger than the oil droplets, then

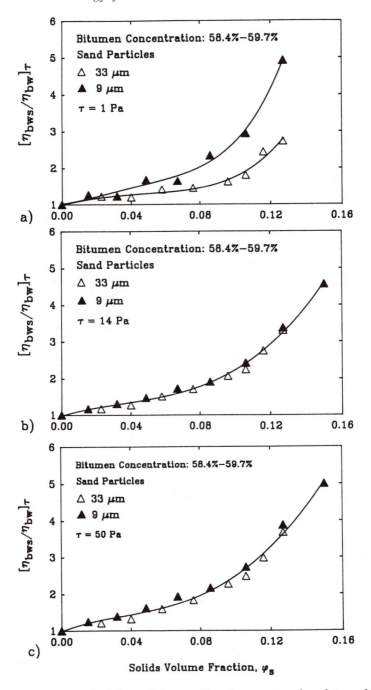

Figure 20. Effect of solids size (silica sand) on the viscosity of emulsion–solids mixtures. (Reproduced with permission from reference 58. Copyright 1991.)

the emulsions behave as a homogeneous or a continuous phase toward the solids. Thus, the viscosity of emulsion–solids mixtures can be predicted from the viscosity data of the pure emulsion and the pure solids suspensions.

The conditions under which emulsions can be considered as a continuous phase toward the solids are determined by analyzing the dependence of η_{ows}/η_{ow} on the oil concentration. Figure 21 shows the variations of the relative viscosity $(\eta_{ows}/\eta_{ow})_\tau$ with solids volume fraction for different values of d_p/d_o, where d_p is the mean diameter of solid particles and d_o is the mean diameter of oil droplets. The addition of the solids was made to the emulsions having oil concentrations of 40–60%. Figure 21a, where d_p/d_o is about 1, shows that at the same solids volume fraction the relative viscosity is higher for the emulsions with higher oil concentrations. This finding indicates that an interaction between the solids and the oil droplets gives rise to a higher viscosity.

As the size ratio of the sand particle to the oil droplets d_p/d_o increases to about 2, there is less dependence on the oil concentration, as shown in Figure 21b. When the size ratio increases to about 3, as shown in Figure 21c, the relative viscosity becomes independent of the oil concentration; this result indicates that the emulsions act as a continuous phase toward the solids. Under this condition, the solids and the droplets behave independently, and no interparticle interaction occurs between the solids and the droplets. Yan et al. (64) showed that when the emulsions behave as a continuous phase toward the solids, the viscosity of the mixtures can be predicted quite accurately from the viscosity data of the pure emulsions and the pure solids suspensions. The viscosity of an emulsion–solids mixture having an oil concentration of β_0 (solids-free basis) and a solids volume fraction of ϕ_s (based on the total volume) can be calculated from the following equation:

$$\eta_{ows}(\beta_0, \phi_s) = \eta_{ow}(\beta_0) \frac{\eta_{sw}(\phi_s)}{\eta_w} \tag{24}$$

where $\eta_{ow}(\beta_0)$ is the viscosity of the pure emulsion having an oil concentration of β_0, $\eta_{sw}(\phi_s)$ is the viscosity of the pure solids suspension having a solids concentration of ϕ_s, and η_w is the viscosity of water.

The critical size ratio above which emulsions can be regarded as a continuous phase toward solids is not a unique value for the following reasons. First, the size distributions of the solids and the oil droplets are most likely to be different from one system to another. Second, the viscosity of the oil used to form the emulsions also affects this critical size ratio. Evidence of this dependence on oil viscosity is the fact that in the bitumen (viscosity 306 Pa·s)-in-water emulsions, the solids do not "see" the emulsions as a continuous phase even at a size ratio of 6. If solids are regarded as having an infinite viscosity, Farris (35) found that the critical size ratio is about 10 for a bimodal solids suspension.

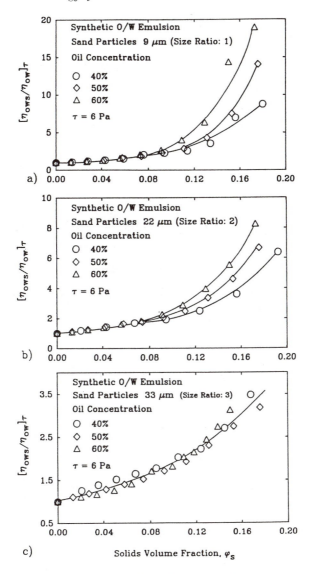

Figure 21. Variation of $[\eta_{ows}/\eta_{ow}]_\tau$ with solids volume fraction. (Reproduced with permission from reference 57. Copyright 1991 Pergamon Press.)

As was just shown, the emulsions act as a continuous phase toward the solids when the solids are much larger than the emulsion droplets. However, when the added solids are of a size similar to the emulsion droplets, the oil droplets cannot be treated as solids. Figure 22 shows the variation of $(\eta_{ows}/\eta_w)_\tau$ with the total particulate volume fraction, that is, the sum of the solids volume fraction and that of the oil in the emulsions. The added solids were

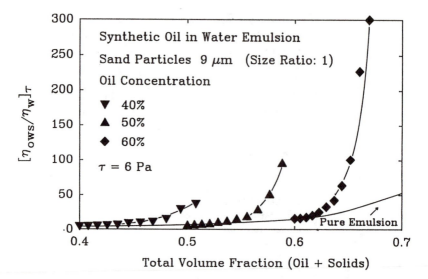

Figure 22. Variation of $[\eta_{ows}/\eta_w]_\tau$ with total volume fraction of the dispersed phases for 9-μm silica sand. (Reproduced with permission from reference 57. Copyright 1991 Pergamon Press.)

silica sand (9 μm). Both the average diameter ($d_p/d_o = 1$) and the size distribution of this fraction of sand is close to the oil droplets. Figure 22 shows that as the solids concentration increases, the curves for the emulsion–solids mixtures deviate significantly from the solids-free emulsion curve, a result indicating that the solids do not behave in the same manner as the oil droplets, even though the sizes of the solids and the oil droplets are similar. In other words, the oil and the solids concentrations are not additive. A mixture of emulsion–solids always has a higher viscosity than the solids-free emulsion at the same total concentration of the dispersed phases. Similar conclusions were also made with respect to bitumen-in-water emulsions.

Summary.

1. The non-Newtonian nature of emulsion–solids mixtures depends on the nature of the pure emulsions. Addition of solids to a highly shear-thinning emulsion also results in a shear-thinning mixture. However, addition of solids to a fairly Newtonian bitumen emulsion results in a more complex mixture that can exhibit different non-Newtonian behaviors at different shear stress or shear rate.

2. The addition of irregularly shaped solids always gives a higher viscosity than smooth spherical solids.

3. The effects of solids size (size distribution) on the rheology of emulsion–solids mixtures is also significant. Smaller solids tend to give a higher yield stress, whereas larger solids tend to induce more pronounced shear-thickening behavior. When the emulsion–solids mixtures are of shear-thinning nature, smaller solids yield a higher viscosity than larger solids. However, when shear thickening occurs, the effect of solids size on the viscosity of the mixtures seems negligible.

4. When the solids are sufficiently larger than the oil droplets in an emulsion, the solids "see" the emulsion as a continuous phase. The viscosity of the mixture can be predicted from the viscosity data of the pure emulsion and the pure solids suspension. The critical size ratio for this to occur depends on the viscosity of the oil used to form the emulsion, and it lies between 3 and 10.

5. The solids cannot be treated as oil droplets even if the size and size distribution of the solids are similar to the oil droplets. The addition of solids to an emulsion generally gives a higher viscosity than the pure emulsion at the same total concentration.

List of Symbols

a	particle diameter
A, B	Arrhenius constant
d	diameter of the small particles in a bimodal suspension
D	diameter of the large particles in a bimodal suspension
d_o	Sauter mean diameter of oil droplets (μm)
d_p	Sauter mean diameter of solid particles (μm)
h	hydration factor
k	Boltzmann constant
K	power law constant
n	power law constant
O/W	oil-in-water emulsion
T	absolute temperature (K)
W/O	water-in-oil emulsion

Greek

β_0	oil concentration in emulsion–solids mixture, solids-free basis
$\dot{\gamma}$	shear rate (s^{-1})
η	viscosity (Pa·s)

$[\eta]$	intrinsic viscosity
η_0	viscosity at zero shear rate
η_∞	viscosity at infinite shear rate
η_B	plastic viscosity
η_{bw}	viscosity of bitumen-in-water emulsions (Pa·s)
η_{bws}	viscosity of bitumen-in-water emulsions with added solids (Pa·s)
η_c	continuous-phase viscosity
η_{ow}	viscosity of O/W emulsions (Pa·s)
η_{ows}	viscosity of emulsion–solids mixtures (Pa·s)
η_r	relative viscosity
η_{sw}	viscosity of solids in water suspensions (Pa·s)
η_w	viscosity of water (Pa·s)
τ	shear stress (Pa)
τ_r	reduced shear stress
ϕ	volume fraction of dispersed phase
ϕ°	dispersed-phase concentration at which relative viscosity becomes 100
ϕ_m	maximum packing concentration
ϕ_o	volume fraction of oil in O/W emulsions
ϕ_s	volume fraction of solids in emulsion–solids mixtures or volume fraction of small spheres in a mixture
ψ_1, ψ_2	primary and secondary normal stress coefficients

Subscripts

0	value evaluated at $\dot{\gamma} \to 0$
∞	value evaluated at $\dot{\gamma} \to \infty$
bw, bws	bitumen–water, bitumen–water–solids
B	Bingham plastic
c	continuous phase
m	value evaluated at maximum packing
ow, ows	oil–water, oil–water–solids
p	particles
r	relative viscosity, viscosity of an emulsion or dispersion normalized with the continuous-phase viscosity, η_c
sw	solids–water
τ	at constant shear stress

Appendix 1: Relations for Determining Viscometric Functions in Standard Experimental Arrangements

A. Capillary Viscometer

Q = Volume rate of flow

$\Delta\mathscr{P}$ = Pressure drop through tube

R = Tube radius

L = Tube length

$\dot\gamma_R$ = Shear rate at tube wall

τ_R = Shear stress at tube wall

$$\eta(\dot\gamma_R) = \frac{\tau_R}{(Q/\pi R^3)}\left[3 + \frac{d\ln(Q/\pi R^3)}{d\ln\tau_R}\right]^{-1} \tag{A-1}$$

$$\dot\gamma_R = \frac{1}{\tau_R^2}\frac{d}{d\tau_R}(\tau_R^3 Q/\pi R^3) \tag{A-2}$$

$$\tau_R = \Delta\mathscr{P}R/2L \tag{A-3}$$

B. Cone-and-Plate Instrument

R = Radius of circular plate

ϑ_0 = Angle between cone and plate (usually less than 4°)

W_0 = Angular velocity of cone

$\mathscr{T}$ = Torque on plate

$\mathscr{F}$ = Force required to keep tip of cone in contact with circular plate

$\pi_{\theta\theta}(r)$ = Pressure measured by flush-mounted pressure transducers located on plate

p_a = Atmospheric pressure

$$\eta(\dot\gamma) = \frac{3\mathscr{T}}{2\pi R^3\dot\gamma} \tag{B-1}$$

$$\Psi_1(\dot\gamma) = \frac{2\mathscr{F}}{\pi R^2\dot\gamma^2} \tag{B-2}$$

$$\Psi_1(\dot\gamma) + 2\Psi_2(\dot\gamma) = -\frac{1}{\dot\gamma^2}\frac{\partial\pi_{\theta\theta}}{\partial\ln r} \tag{B-3}$$

$$\Psi_2(\dot\gamma) = \frac{p_a - \pi_{\theta\theta}(R)}{\dot\gamma^2} \tag{B-4}$$

$$\dot\gamma = W_0/\vartheta_0 \tag{B-5}$$

Continued

Appendix—Continued

C. Parallel-Disk Instrument

R = Radius of disks
H = Separation of disks
W_0 = Angular velocity of upper disk

$$\eta(\dot\gamma_R) = \frac{(\mathscr{T}/2\pi R^3)}{\dot\gamma_R}\left[3 + \frac{d\ln(\mathscr{T}/2\pi R^3)}{d\ln\dot\gamma_R}\right] \quad (C\text{-}1)$$

$\mathscr{T}$ = Torque required to rotate upper disk
$\mathscr{F}$ = Force required to keep separation of two disks constant

$$\Psi_1(\dot\gamma_R) - \Psi_2(\dot\gamma_R) = \frac{(\mathscr{F}/\pi R^2)}{\dot\gamma_R^2}\left[2 + \frac{d\ln(\mathscr{F}/\pi R^2)}{d\ln\dot\gamma_R}\right] \quad (C\text{-}2)$$

$\dot\gamma_R$ = Shear rate at edge of system

$$\Psi_1(\dot\gamma_R) + \Psi_2(\dot\gamma_R) = \frac{1}{\dot\gamma_R^2}\frac{d\pi_{zz}(0)}{d\ln\dot\gamma_R} \quad (C\text{-}3)$$

$\pi_{zz}(0), \pi_{zz}(R)$ = Normal pressure measured on disk at center and at rim

$$\Psi_2(\dot\gamma_R) = \frac{p_a - \pi_{zz}(R)}{\dot\gamma_R^2} \quad (C\text{-}4)$$

p_a = Atmospheric pressure

$$\dot\gamma_R = \frac{W_0 R}{H} \quad (C\text{-}5)$$

D. Couette Viscometer

R_1, R_2 = Radii of inner and outer cylinders
H = Height of cylinders
W_1, W_2 = Angular velocities of inner and outer cylinders

$$\eta(\dot\gamma) = \frac{\mathscr{T}(R_2 - R_1)}{2\pi R_1^3 H|W_2 - W_1|} \quad (D\text{-}1)$$

$$\Psi_1(\dot\gamma) = \frac{-[\pi_{rr}(R_2) - \pi_{rr}(R_1)]R_1}{(R_2 - R_1)\dot\gamma^2}$$

Couette Viscometer—Continued

$$+ \frac{\rho R_1^2}{3\dot{\gamma}^2} (W_1^2 + W_2^2 + W_1 W_2) \qquad \text{(D-2)}$$

$\mathcal{F}$ = Torque on inner cylinder

$$\dot{\gamma} = \frac{|W_2 - W_1| R_1}{R_2 - R_1} \qquad \text{(D-3)}$$

$\pi_{rr}(R_1), \pi_{rr}(R_2)$ = Normal pressures measured on inner and outer cylinders

E. Axial Annular Flow

$$(\pi_{rr})_1 - (\pi_{rr})_2 = -\int_{R_1}^{R_2} \frac{\Psi_2 \dot{\gamma}^2}{r} \, dr \qquad \text{(E-1)}$$

R_1, R_2 = Radii of inner and outer cylinders

$(\pi_{rr})_i$ = Reading of a flush-mounted pressure transducer at R_i

F. *Torsional Flow between Two Disks, the Upper One of Which is Rotating and Attached to a Vertical Tube*

$$h = \frac{W_0^2}{\rho g H} \int_{R_1}^{R_2} (\Psi_1 + \Psi_2) r \, dr - \frac{W_0^2}{6g} (R_2^2 - R_1^2) \qquad \text{(F-1)}$$

h = Height of rise of fluid in tube
R_1 = Radius of tube
R_2 = Radius of disks
W_0 = Angular velocity of tube-disk assembly
H = Gap between disks
g = Gravitational acceleration

Continued

Appendix—Continued

G. *Truncated Cone-and-Plate Instrument*

R = Radius of circular plate

R_0 = Radius of truncated section of cone

ϑ_0 = Angle between cone and plate (usually less than 4°)

H = Gap between truncated section of cone and circular plate

$\pi'_{\theta\theta}(r)$ = Pressure measured by transducers mounted at the bottom of holes along bottom disk

$\pi_{\theta\theta}(r)$ = Normal stress on bottom circular disk

W = Angular velocity of truncated cone

$\dot\gamma_0$ = Shear rate for $R_0 \leq r \leq R$

p^* = Hole pressure

p_a = Atmospheric pressure

$$\pi_{\theta\theta}(0) = \pi'_{\theta\theta}(0) \tag{G-1}$$

$$\Psi_1(\dot\gamma_0) + 2\Psi_2(\dot\gamma_0) = -\frac{1}{\dot\gamma_0^2}\frac{\partial \pi'_{\theta\theta}}{\partial \ln r} \quad (R_0 \leq r \leq R) \tag{G-2}$$

$$\Psi_1(\dot\gamma_0) + \Psi_2(\dot\gamma_0) = \frac{1}{\dot\gamma_0}\frac{d}{d\dot\gamma_0}\left\{(\pi_{\theta\theta}(0) - p_a)\right\} \tag{G-3}$$

$$p^*(\dot\gamma_0) = p_a - \pi'_{\theta\theta}(R) - \Psi_2(\dot\gamma_0)\dot\gamma_0^2 - [\Psi_1(\dot\gamma_0) + 2\Psi_2(\dot\gamma_0)]\dot\gamma_0^2 \ln\frac{R}{R_0} \tag{G-4}$$

$$\dot\gamma_0 = \frac{WR_0}{H} \tag{G-5}$$

Acknowledgments

The authors thank Alberta Oil Sands Technology and Research Authority and the Natural Sciences and Engineering Research Council of Canada for financial support.

References

1. Gerhart, P. M.; Gross, R. J. *Fundamentals of Fluid Mechanics;* Addison-Wesley: Reading, PA, 1985.
2. Brodkey, R. S.; Hershey, H. C. *Transport Phenomenon;* McGraw-Hill: New York, 1988.
3. Skelland, A. H. P. *Non-Newtonian Flow and Heat Transfer;* Wiley: New York, 1967.
4. Ostwald, W.; Auerback, R. *Kolloid-Z.* **1926,** *38,* 1926.
5. Sisko, A. *Ind. Eng. Chem.* **1958,** *50,* 1789.
6. Bird, R.; Steward, W.; Lightfoot, E. *Transport Phenomenon;* Wiley: New York, 1960.
7. Bird, R.; Armstrong, R.; Hassager, D. *Dynamics of Polymeric Liquids,* 2nd ed.; Wiley: New York, 1987.
8. Tadros, Th. F. lecture notes–special seminar; Edmonton, Alberta, Canada, 1989.
9. Hoffman, R. L. *J. Colloid Interface Sci.* **1974,** *46(3),* 491–506.
10. Walters, K. *Rheometry;* Chapman and Hall: London, 1975.
11. Gittus, J. *Creep, Viscoelasticity, and Creep Fracture in Solids;* Wiley: New York, 1975.
12. van Wazer, J. R.; Lyons, J. W.; Kim, J. W.; Cowell, R. E. *Viscosity and Flow Measurements;* Interscience: New York, 1963.
13. Whorlow, R. W. *Rheological Techniques;* Ellis Horwood, Ltd.: Chichester, United Kingdom, 1980.
14. Fredrickson, A. G. *Principles and Applications of Rheology;* Prentice-Hall: Englewood Cliffs, NJ, 1964.
15. Sukarek, P. C.; Laurence, R. L. *AIChE J.* **1974,** *24,* 474.
16. Yang, M. C.; Scriven, L. E.; Macosko, C. W. *J. Rheol.* **1986,** *30(5),* 1015.
17. Sherman, P. In *Encyclopedia of Emulsion Technology;* Becher, P., Ed.; Dekker: New York, 1983; Vol. 1, p 416.
18. Pal, R.; Rhodes, E. *J. Colloid Interface Sci.* **1985,** *107(2),* 301.
19. Pal, R.; Bhattacharya, S. N.; Rhodes, E. *Can. J. Chem. Eng.* **1986,** *64,* 3.
20. Pal, R.; Rhodes, E. *J. Rheol.* **1989,** *33(7),* 1021.
21. Pal, R.; Masliyah, J. *Can. J. Chem. Eng.* **1990,** *68,* 24.
22. Han, C. D.; King, R. G. *J. Rheol.* **1980,** *24,* 213.
23. Sherman, P. *Rheol. Acta.* **1962,** *2(1),* 74.
24. Einstein, A. *Ann. Phys.* **1906,** *19,* 289.
25. Einstein, A. *Ann. Phys.* **1911,** *34,* 591.
26. Sherman, P. In *Encyclopedia of Emulsion Technology;* Becher, P., Ed.; Dekker: New York, 1983; Vol. 1, p 415.
27. Sherman, P.; *Industrial Rheology;* Academic Press: London, 1970, p 138.
28. Chong, J. S.; Christiansen, E. B.; Baer, A. D. *J. Appl. Polym. Sci.* **1971,** *15,* 2007.
29. Metzner, A. B. *J. Rheol.* **1985,** *29(6),* 739.

30. Thomas, D. G. *J. Colloid Sci.* **1965,** *20,* 267.
31. Krieger, I. M. *Adv. Colloid Interface. Sci.* **1972,** *3,* 111.
32. Richardson, E. G. *J. Colloid Sci.* **1950,** *5,* 404.
33. Richardson, E. G. *J. Colloid Sci.* **1953,** *8,* 367.
34. Eveson, G. F. In *Rheology of Disperse Systems;* Mill, C. C., Ed.; Pergamon: London, 1959.
35. Farris, R. J. *Trans. Soc. Rheol.* **1968,** *12,* 281.
36. Poslinski, A. J.; Ryan, M. E.; Gupta, R. K.; Seshadri, S. G.; Frechette, F. J. *J. Rheol.* **1988,** *32,* 751.
37. Sherman, P. *Emulsion Science;* Academic Press: New York, 1968, p 316.
38. Guth, E.; Simha, R. *Kolloid-Z.* **1936,** *74,* 266.
39. Saito, N. *J. Phys. Soc. Jpn.* **1950,** *5,* 4.
40. Vand, V. *J. Phys. Colloid Chem.* **1948,** *52,* 277.
41. Manley, R. J.; Mason, S. G. *Can. J. Chem.* **1955,** *33,* 763.
42. Roscoe, R. *Br. J. Appl. Phys.* **1952,** *3,* 267.
43. Brinkman, H. C. *J. Chem. Phys.* **1952,** *20,* 571.
44. Richardson, E. G. *Kolloid-Z.* **1933,** *65,* 32.
45. Broughton, J.; Squires, L. *J. Phys. Chem.* **1938,** *42,* 253.
46. Hatschek, E. *Kolloid-Z.* **1911,** *8,* 34.
47. Sibree, J. O. *J. Chem. Soc. Faraday Trans.* **1930,** *26,* 26.
48. Sibree, J. O. *J. Chem. Soc. Faraday Trans.* **1931,** *27,* 161.
49. Eilers, H. *Kolloid-Z.* **1941,** *97,* 313.
50. Eilers, H. *Kolloid-Z.* **1943,** *102,* 154.
51. Mooney, M. *J. Colloid Sci.* **1951,** *6,* 162.
52. Barnea, E.; Mizrahi, J. *Chem. Eng. J.* **1973,** *5,* 171.
53. Krieger, I. M.; Dougherty, T. J. *Trans. Soc. Rheol.* **1959,** *3,* 137.
54. Frankel, N. A.; Acrivos, A. *Chem. Eng. Sci.* **1967,** *22,* 847.
55. Sherman, P. In *Encyclopedia of Emulsion Technology;* Becher, P., Ed.; Dekker: New York, 1983; Vol. I, pp 419–420.
56. Marjerrison, D. M.; Sayre, J. A. *J. Can. Pet. Tech.* **1988,** *27,* 68–72.
57. Yan, Y.; Pal, R.; Masliyah, J. *Chem. Eng. Sci.* **1991,** *46,* 985.
58. Yan, Y.; Pal, R.; Masliyah, J. *Can. J. Chem. Eng.* **1991,** in press.
59. Barnes, H. A. *J. Rheol.* **1989,** *33(2),* 329–355.
60. Goodwin, J. N. *Colloid and Interface Science;* Academic Press: New York, 1967; Vol. IV.
61. Hoffman, R. L. In *Science and Technology of Polymer Colloids;* Poehlein, G. W.; Ottewill, R. H.; Goodwin, J. W., Eds.; Martinus Nijhoff: Boston, MA, 1983; Vol. 2, p 570.
62. Mewis, J.; Metzner, A. B. *J. Fluid Mech.* **1974,** *62,* 593–600.
63. Clarke, B. *Trans. Inst. Chem. Eng.* **1967,** *45,* T251–T256.
64. Yan, Y.; Pal, R.; Masliyah, J. *Chem. Eng. Sci.* **1991,** *46,* 1823.

RECEIVED for review December 18, 1990. ACCEPTED revised manuscript May 14, 1991.

Fluid Dynamics of Oil–Water–Sand Systems

Hisham A. Nasr-El-Din

Petroleum Recovery Institute, 3512 33rd Street N.W., Calgary, Alberta, Canada T2L 2A6

Heavy-oil-in-water emulsions have different rheological behaviors for different emulsion qualities. With low oil volume fractions, these emulsions behave as Newtonian fluids. However, with oil volume fractions ≥0.5, they often behave as shear-thinning fluids. The friction loss and power requirements for pipeline transportation of heavy-oil emulsions depends on their rheological behavior. Various formulas for predicting the friction loss of the flow of heavy-oil-in-water emulsions in smooth pipes are discussed. Fine sand particles, which are usually produced with heavy oil, change the friction loss of the flow of these emulsions in pipelines. The effect of the fine particles depends on the solids concentration profile in the pipe. Various methods of measuring in situ solids concentration in pipelines are reviewed, including sampling and electrical conductivity probes.

As WORLD RESERVES OF CONVENTIONAL CRUDE OIL continue to decline, heavy oil and bitumen are becoming increasingly important sources of energy. In general, heavy crude oils and bitumen have viscosity ranges from a few hundred to several thousand centipoises. Because of their high viscosities, it is not feasible to transport them in conventional pipelines without reducing their viscosities. Three methods were introduced to reduce the viscosity of heavy oils and enable them to be transported in conventional pipelines: heating the oil during transportation, adding a diluent, and emulsifying the heavy oil in water. The first two methods are expensive at 1991 prices. However, the emulsification method has a potential application whenever an ample water supply is available.

Transport of viscous crude oils as concentrated oil-in-water emulsions

0065–2393/92/0231–0171 $12.75/0
© 1992 American Chemical Society

has been demonstrated on a large scale in an Indonesian pipeline and in California (1). A major disadvantage of this mode of transport is that it requires dewatering of the emulsion after transport. Consequently, the oil-in-water emulsions must be prepared with the highest possible oil volume fraction.

To stabilize emulsions, a surfactant, which increases the repulsive force between oil droplets, is used. Nonionic surfactants are the preferred type because they are effective in brines, are generally cheaper, and often form less viscous emulsions than do ionic surfactants. In addition, their emulsions are easier to break, and they do not introduce inorganic residues that might lead to refinery problems. They are chemically stable at oil reservoir temperatures and are noncorrosive and nontoxic. The surfactant type and concentration required for a particular situation can be determined by conducting laboratory tests. A typical concentration of 0.1 lb of surfactant per barrel of oil is used for emulsions containing about 50–70% oil (2).

The presence of natural organic acids in some crude oils, especially asphaltic crude oils, may eliminate the need for expensive surfactants. These acids react with strong alkali (usually NaOH) to form petroleum soaps. These soaps diffuse into the oil–water interface, decrease interfacial tension, and form emulsions. Many researchers have used dilute alkali solutions ($\simeq$0.1 wt% NaOH) to form stable oil-in-water emulsions containing up to 75% oil (1, 3).

Another aspect of the transportation of heavy-oil-in-water emulsions, especially for short-distance pipelines, is the presence of sand particles. Fine sand particles are usually produced with heavy oils. The presence of these particles will change the flow resistance and pumping requirements for heavy-oil-in-water emulsions. First, the dynamic viscosity of an emulsion will change in the presence of fine particles (4). Second, sand particles, because of their higher density, will flow in a distinct layer at the bottom of the pipe (5).

The objectives of this chapter are (1) to give a brief review of various formulas to predict friction losses for flow of oil-in-water emulsions in smooth pipes, and (2) to discuss various methods that measure in situ solids concentration in pipelines.

Predicting the Pressure Drop for Flow of Emulsions in Pipelines

A large body of literature is available on estimating friction loss for laminar and turbulent flow of Newtonian and non-Newtonian fluids in smooth pipes. For laminar flow past solid boundaries, surface roughness has no effect (at least for certain degrees of roughness) on the friction pressure drop of either Newtonian or non-Newtonian fluids. In turbulent flow, however, the nature of the flow is intimately associated with the surface roughness. Significant increases in friction loss in turbulent flow over rough surfaces have been reported 6).

Extensive studies (6) have been conducted to understand the effect of pipe roughness on friction loss in turbulent flow of Newtonian fluids in rough pipes. The phenomenon of turbulent flow with non-Newtonian fluids in rough pipes, however, has received very little attention (7).

Flow of Newtonian Fluids in Smooth Pipes.   Estimates of friction losses in laminar flow

$$\mathrm{Re} = \frac{\rho_f U_b D}{\mu_f} \leq 2100$$

[where Re is the Reynolds number, ρ_f is the fluid density (kg/m^3), U_b is the bulk (average) fluid velocity (m/s), D is the pipe inner diameter (m), and μ_f is fluid viscosity (Pa·s)] of Newtonian fluids in smooth pipes can be obtained from the Hagen–Poiseuille equation (6):

$$f = 16/\mathrm{Re} \tag{1}$$

The Fanning friction factor (f) is defined as

$$f = \frac{2\tau_w}{\rho_f U_b^2} \tag{2}$$

where τ_w is the shear stress at the wall of the pipe (Pa). The friction factor can be also expressed in terms of the pressure gradient along the pipe ($\Delta p/L$, where p is pressure and L is pipe length). For steady flow, a force balance yields

$$\frac{\pi}{4} D^2 \Delta p = \pi D L \tau_w \tag{3a}$$

or

$$\tau_w = \frac{D \Delta p}{4L} \tag{3b}$$

where D is the diameter and L is length of the pipe. Substituting equation 3b in equation 2 yields

$$f = \frac{D}{2\rho_f U_b^2} \left[\frac{\Delta p}{L} \right] \tag{4}$$

The shear rate at the wall (γ_w) for laminar flow in a pipe can be calculated as follows:

$$\gamma_w = -\frac{du}{dr}(r)\bigg|_{r=D/2} \tag{5}$$

where $u(r)$ is the local fluid velocity and r is the radial position. The velocity profile for fully developed steady flow of a Newtonian fluid flowing under laminar conditions in a pipe is

$$u(r) = \frac{(D/2)^2 \, \pi 45 \, r^2}{4\mu_f}\left[\frac{\Delta p}{L}\right] \tag{6}$$

Combining equations 1, 4, 5, and 6 gives the shear rate at the wall of the pipe:

$$\gamma_w = \frac{8U_b}{D} \tag{7}$$

Many useful correlations have been published to determine the friction factor for fully developed turbulent flow of Newtonian fluids in smooth pipes. One of the earliest correlations was given by Blasius (8) as follows:

$$f = 0.079/\text{Re}^{0.25} \tag{8}$$

Equation 8 is valid for $3000 \le \text{Re} \le 100{,}000$. Another commonly used correlation was given by Drew et al. (9):

$$f = 0.0014 + 0.125(\text{Re})^{-0.32} \tag{9}$$

Equation 9 is valid for $3000 \le \text{Re} \le 3{,}000{,}000$.

Flow of Power Law Fluids in Smooth Pipes.

Oil-in-water emulsions having oil volume fractions greater than 0.5 are often non-Newtonian shear-thinning fluids (3, 10–13). For such fluids, the shear stress (τ) and the shear rate (γ) can be related by the power law model:

$$\tau = k\gamma^n \tag{10}$$

For a Newtonian fluid, the power law index $n = 1$, and k is the fluid viscosity. Also, for shear-thinning (pseudoplastic) fluids, $n < 1$.

The friction losses for the flow of non-Newtonian pseudoplastic fluids under laminar flow conditions can be determined by using the method suggested by Metzner and Reed (14) as follows:

$$f = 16/\text{Re}' \tag{11}$$

Re', the Metzner–Reed modified Reynolds number, is defined as

$$\text{Re}' = \frac{D^{n'} U_b^{2-n'} \rho_f}{k'(8)^{n'-1}} \tag{12}$$

where n' and k' are the Metzner–Reed modified power law constants for pipe flow. These constants are related to the power law constants obtained with a viscometer as follows:

$$n' = n \tag{13a}$$

$$k' = k \left[\frac{1 + 3n}{4n} \right]^n \tag{13b}$$

The shear rate at the wall for the flow of power law fluids in smooth pipes under laminar conditions can be calculated as follows (*15*):

$$\gamma_w = \frac{8 U_b}{D} \left[\frac{1 + 3n}{4n} \right] \tag{14}$$

For $n = 1$ (Newtonian fluids), equation 14 reduces to equation 7.

Dodge and Metzner (*16*) presented an extensive theoretical and experimental study on the turbulent flow of non-Newtonian fluids in smooth pipes. They extended von Kármán's (*17*) work on turbulent flow friction factors to include the power law non-Newtonian fluids. The following implicit expression for the friction factor was derived in terms of the Metzner–Reed modified Reynolds number and the power law index:

$$\frac{1}{\sqrt{f}} = \frac{4}{n'^{0.75}} \log (\text{Re}' f^{(1-n')/2}) - \frac{0.4}{n'^{1.2}} \tag{15}$$

Dodge and Metzner (*16*) obtained excellent agreement between calculated (with equation 15) and experimental friction factors over values of n' from 0.36 to 1 and Re' from 2900 to 36,000.

The flow of oil-in-water emulsions in pipelines was examined by various researchers both in laminar and turbulent flow regimes (*3, 10, 18–20*). These studies showed that pressure drop predictions based on equations 11 and 15 are in some cases higher than the experimental measurements in both laminar and turbulent flow regimes. In the laminar flow regime, this difference was explained by Wyslouzil et al. (*3*) and Gillies and Shook (*5*) in terms of droplet migration away from the pipe wall as a result of high shear rates. However, in the turbulent flow regime, the viscoelastic properties of oil-in-water emulsions may be the cause for this difference (*10*).

Measuring the Solids Concentration in Pipelines

In practice, testing an emulsion for purposes of pipeline design requires a sample to be removed from a container or a pipeline. Although the testing is often straightforward, sampling is not, especially when an emulsion contains sand. Because the concentration and particle size distribution of the dispersed phase are so important, the rest of this review will deal with this aspect.

Sampling Methods. A number of methods have been used to measure solids concentration in pipelines. Reviews of these methods are given by Kao and Kazanskij (*21*) and recently by Baker and Hemp (*22*). In general, the principle of any of these methods is to find a specific property that is significantly different for the two phases, for example, electrical conductivity, dielectric constant, density, refractive index, or absorption of electromagnetic radiation. Solids concentration can be determined by measuring this property for the mixture, then using a calibration curve. Any of these methods will give inaccurate measurements if the values of the specific property of the two phases approach one another or if the solids concentration is very low.

Sampling is widely used in industry to measure in situ solids concentration, composition, and size distribution from fluid–solid systems (*23*). It is probably the only reliable method for use at low solids concentration. It is also used to calibrate and evaluate newly developed methods of measuring solids concentration (*24*). A number of methods of sampling differ primarily in the geometry of the sampling device. Figure 1 shows schematic diagrams of the most commonly used sampling methods.

Serious errors in measuring solids concentration arise as a result of improper sampling. The effectiveness of sampling devices is usually expressed as the ratio of the measured solids concentration, C, to the upstream local solids concentration, C_0. The concentration ratio (C/C_0) is also known as the aspiration coefficient (*25*), separation coefficient (*26*), or sampling efficiency (*27*).

Three main factors can cause the sampling efficiency of a sampling device to deviate from unity (i.e., ideal sampling):

1. particle inertia
2. particle bouncing
3. flow structure ahead of the sampler

In the following sections the effect of these parameters on the performance of various sampling devices will be discussed.

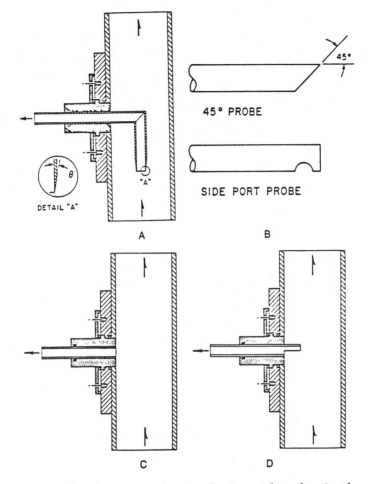

Figure 1. Sampling devices: A, L-shaped probe; B, straight probes; C, side-wall sampling; and D, side-wall sampling with a projection. (Reproduced with permission from reference 23. Copyright 1989 Gulf Publishing Company.)

Particle Inertia.   Particle inertia is a major source of sampling errors when the densities of the two phases are significantly different. Because particle inertia is different from that of an equivalent volume of fluid, particle motion does not follow the distorted fluid streamlines. Consequently, sample solids concentration and composition will be significantly different from those in the pipe. Sampling errors due to inertia depend on

- how the sampling device disturbs the flow field
- how the particles respond to this disturbance

Thin L-shaped probes are commonly used to measure solids concentration profile in slurry pipelines (28–33). However, serious sampling errors arise as a result of particle inertia. To illustrate the effect of particle inertia on the performance of L-shaped probes, consider the fluid streamlines ahead (upstream) of a sampling probe located at the center of a pipe, as shown in Figure 2. The probe has zero thickness, and its axis coincides with that of the pipe. The fluid ahead of the sampler contains particles of different sizes and densities. Figure 2A shows the fluid streamlines for sampling with a velocity equal to the upstream local velocity (isokinetic sampling). Of course, the probe does not disturb the flow field ahead of the sampler, and consequently, sample solids concentration and composition equal those upstream of the probe.

Sampling with a velocity different from the upstream local velocity

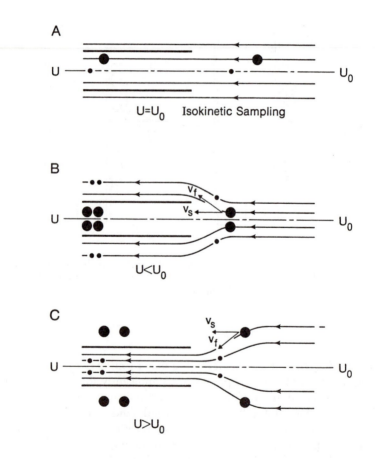

Figure 2. Isokinetic and anisokinetic sampling. (Reproduced with permission from reference 23. Copyright 1989 Gulf Publishing Company.)

(anisokinetic sampling) will distort the fluid streamlines ahead of the sampler. The distortion of the fluid streamlines depends on the ratio of the sampling velocity, U, to the upstream local velocity, U_0. If the velocity ratio (U/U_0) is less than unity, the fluid streamlines will diverge away from the probe, as shown in Figure 2B. Particles of low inertia will follow the fluid streamlines, whereas those of high inertia will move in straight lines like bullets. As a result, the sample obtained has a higher solids concentration, with more coarse and dense particles, than in the pipe.

An opposite trend occurs if the velocity ratio is higher than unity. In this case the fluid streamlines converge into the probe (Figure 2C), but the particles will respond according to their inertia; particles of low inertia will follow the streamlines into the probe, whereas those of high inertia will miss the probe. One ends up with a sample with a lower solids concentration, with more fine and light particles, than in the pipe.

The preceding discussion shows that the sampling efficiency for thin L-shaped probes is a function of two parameters: the deviation from the isokinetic conditions and the response of the particles to the deflection of the fluid streamlines upstream of the sampler. The deviation from the isokinetic conditions is a function of the velocity ratio (U/U_0), whereas the particle response is a function of the ratio of particle inertia to fluid drag. This ratio in a dimensionless form is known as the particle inertia parameter, the Stokes number, or the Barth number (K), defined as:

$$K = \frac{\rho_s d_s^2 U_0}{18 \mu_f R_{sm}} \tag{16}$$

where ρ_s is the solids density, d_s is the mean particle diameter, and R_{sm} is the sampler radius. The effect of particle inertia on sampling efficiency for thin L-shaped probes has been studied extensively in fluid–solid systems of low solids concentration. Reviews on the performance of thin L-shaped probes to sample from gas–solid systems were given by Fuchs (27), and recently by Stevens (34). Unlike gas–solid systems, few investigations have been conducted on sampling from liquid–solid systems. Rushton and Hillestad (28) measured solids concentration profiles in vertical and horizontal slurry pipelines by using different sampling techniques. For L-shaped probes, they found a linear relation between the inverse of the sampling velocity $(1/U)$ and the concentration ratio (C/C_b), where C_b is the average solids concentration over the pipe cross section. The slope of the line was found to be a function of the settling properties of the solids. Nasr-El-Din et al. (33) examined both theoretically and experimentally the performance of L-shaped probes when used to sample from slurry pipelines. Figures 3–5 show good agreement between their model and their experimental measurements for sand particles having a mean particle size, d_{50}, of 0.19 mm (fine sand), 0.45 mm (medium sand), and 0.91 mm (coarse sand), respectively.

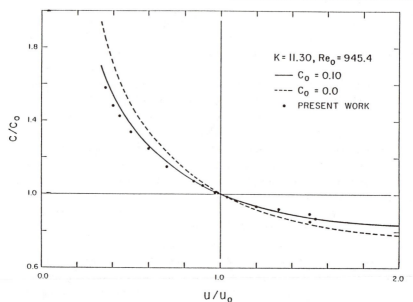

Figure 3. Predicted and observed sampling efficiencies for the fine sand. (Reproduced with permission from reference 33. Copyright 1984.)

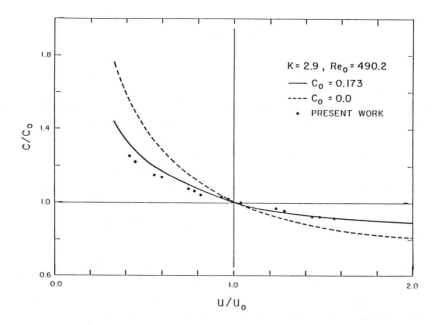

Figure 4. Predicted and observed sampling efficiencies for the medium sand. (Reproduced with permission from reference 33. Copyright 1984.)

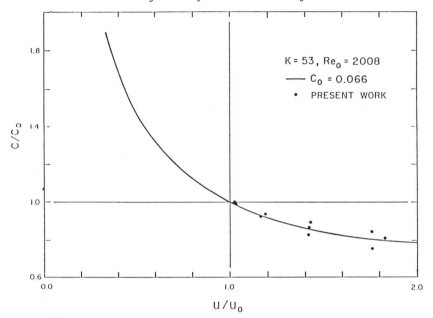

Figure 5. Predicted and observed sampling efficiencies for the coarse sand. (Reproduced with permission from reference 33. Copyright 1984.)

Another way to collect a sample from a pipeline or a container is by withdrawing the sample from an opening in the wall (*see* Figure 1C). This method of sampling, known as side-wall sampling, is widely used in industry, not only for slurry pipelines (28), but also for mixing vessels (35–37) and slurry heat exchangers (38). The advantage of this technique is its simplicity of operation, because it uses a small aperture in the wall of the pipe and does not disturb the flow with a probe. On the other hand, the main disadvantage is that the sampling efficiency is a strong function of particle inertia and the solids distribution upstream of the sampler.

Rushton (35) was the first to draw attention to the errors associated with wall sampling. Sharma and Das (37) mentioned that the mechanism of particle collection using an opening flush with the wall is different from the concept of isokinetic sampling. Moujaes (38) used wall sampling to measure solids concentration in upward vertical slurry flows. He found the sample concentration to be consistently lower than the true values in the pipe, especially with the coarse sand particles.

Torrest and Savage (39) studied collection of particles in small branches. The sampling transport efficiency, E, defined as the ratio of the solids flow rate in the branch to that in the main pipe, was found to be a function of particle settling velocity (V_t) and the upstream bulk velocity (U_b) as follows:

$$E = 158.7xQ \left[\frac{40(V_t + U_b) - 58.4}{1 - 125(V_t + U_b)} \right] \tag{17}$$

where Q is the branch flow rate (m^3/s) and $(V_t + U_b)$ is in meters per second. This correlation is valid for the range of $0.04 \leq (V_t + U_b) \leq 0.4$ m/s.

Nasr-El-Din and co-workers (40, 41) studied wall sampling from an upward vertical slurry flow. They found that this type of sampling caused serious errors in measuring solids concentration and particle size distribution. Figures 6–8 show that the sampling efficiency for side-wall sampling from a vertical pipeline is always less than unity and is dependent on particle size, upstream solids concentration, and sampler diameter, respectively. Figures 9 and 10 show that the sample mean particle diameter using side-wall sampling is smaller than that in the pipe, especially at low sampling velocity ratios.

The results discussed so far indicate that the sampling efficiency of a side-wall sampler from a vertical pipeline is always less than unity. One way to increase sample solids concentration is by using a side-wall sampler with a projection (see Figure 1D). Nasr-El-Din et al. (40) examined the performance of such sampling devices. They found that the projection increased the sample solids concentration. However, the variation of the sampling efficiency with the velocity ratio was different from that obtained with a

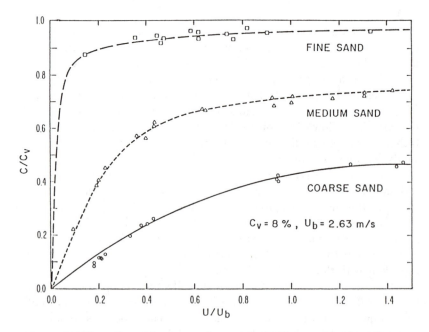

Figure 6. Effect of particle size on the sampling efficiency of side-wall sampling. (Reproduced with permission from reference 40. Copyright 1985.)

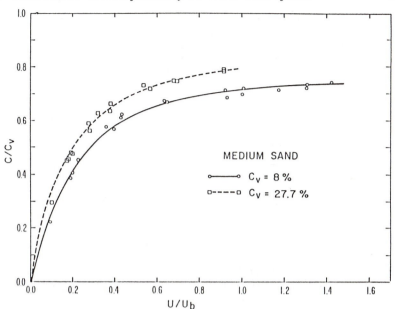

Figure 7. Effect of solids concentration on the sampling efficiency of side-wall sampling. (Reproduced with permission from reference 40. Copyright 1985.)

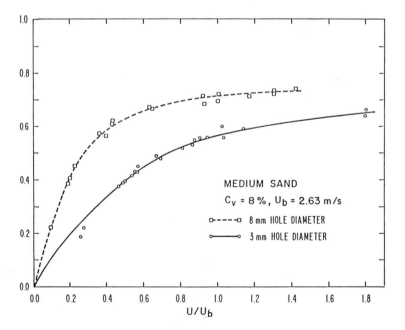

Figure 8. Effect of sampler diameter on the sampling efficiency of side-wall sampling. (Reproduced with permission from reference 40. Copyright 1985.)

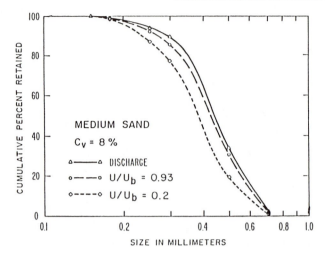

Figure 9. Effect of sampling velocity ratio on the sample particle size distribution using a side-wall sampler. (Reproduced with permission from reference 40. Copyright 1985.)

side-wall sampler without a projection. This difference occurs because the projection changes the flow pattern ahead of the sampler.

Unlike wall sampling from vertical slurry flows, the sampling efficiency of a side-wall sampler from a horizontal slurry flow may exceed unity in some cases. Nasr-El-Din et al. (42, 43) showed that the sampling efficiency for

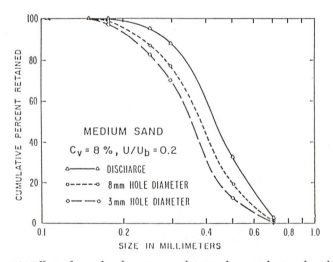

Figure 10. Effect of sampler diameter on the sample particle size distribution using a side-wall sampler. (Reproduced with permission from reference 40. Copyright 1985.)

wall sampling from a horizontal slurry pipeline is a strong function of the sampler orientation (upwards, sideways, and downwards). Sampling efficiencies greater than unity were observed only with the downwards orientation.

Particle Bouncing.   A second source of sampling errors occurs as a result of particle bouncing effects. A typical example of this effect is sampling particles of high inertia using thick (blunt) L-shaped probes. In this case, particles may hit the probe wall, lose some of their inertia, and enter the probe. Consequently, the sample solids concentration is higher than the upstream solids concentration, even when the sampling velocity equals the upstream local velocity, that is, when $U/U_0 = 1$.

The effect of particle bouncing on the sampling efficiency of thick L-shaped probes was first noted in gas–solid systems by Whitely and Reed (44). They found that sampling efficiency for thick L-shaped probes was higher than unity at $U/U_0 = 1$. To estimate the sampling efficiency due to particle bouncing at the isokinetic velocity, Belyaev and Levin (25) and Yoshida et al. (45) proposed an empirical equation. This equation can be written in a slightly different form as

$$C/C_0 = 1 + B(2T + T^2) \tag{18}$$

where T is the probe relative wall thickness and B is the fraction of particles that hit the nozzle edge and enter the probe.

To establish the performance of blunt probes when used to sample from liquid–solid systems, a set of L-shaped probes of different thicknesses was tested. Figure 11, from Nasr-El-Din and Shook (46), shows the effect of the probe relative wall thickness on the sampling efficiency for the medium sand at solids concentration of 10%. The sampling efficiency at $U/U_0 = 1$ is higher than unity, an observation found for sampling sand particles using thick probes. As the relative wall thickness is increased, C/C_0 at $U/U_0 = 1$ increases. Also, to obtain the correct concentration using these probes, samples should be taken at a velocity greater than the isokinetic one. This velocity was found to be a function of the solids concentration, the particle inertia parameter, and the probe relative wall thickness. The increase in the sample solids concentration at isokinetic conditions was much less than the corresponding values obtained from equation 18 with $B = 0.5$.

Figure 12 shows the sampling efficiency versus the velocity ratio for L-shaped probes having a tip angle (θ) of 18° and probe relative wall thicknesses of 0.4, 0.8, and 1.2. The fine sand at 6.3% discharge concentration and 2.63-m/s bulk velocity was used in these experiments. At this angle, the increase of C/C_0 at $U/U_0 = 1$ is eliminated. These results seem to confirm the explanation given previously about the bouncing effect and agree with the trend previously obtained by Whitely and Reed (44) in gas–solid systems.

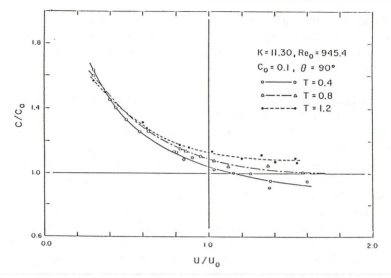

Figure 11. Effect of the probe relative wall thickness on the sampling efficiency. (Reproduced with permission from reference 46. Copyright 1985.)

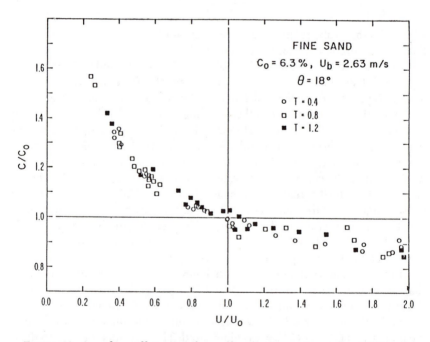

Figure 12. Sampling efficiencies for probes having a tip angle of 18° and various probe relative wall thicknesses. (Reproduced with permission from reference 46. Copyright 1985.)

Figure 13, from Nasr-El-Din et al. (47), shows the effect of the probe relative wall thickness on the sampling efficiency for polystyrene particles of 0.3-mm mean diameter. Samples were taken from the center of the pipe at a mean solids concentration of 37% and a bulk velocity of 3.4 m/s, with probes of relative wall thicknesses of 0.05, 0.5, 0.8, and 1.2. Unlike the results obtained with the sand particles, shown in Figure 11, the effect of the sampling velocity on the sampling efficiency is not significant. This result is reasonable because the polystyrene particles have a density of 1050 kg/m^3, which is very close to water. This finding implies that these particles can follow the fluid streamlines, and consequently the sampling efficiency for these particles is very close to unity, no matter what the sampling velocity.

Figure 13 illustrates that the sampling efficiency appears to be independent of the probe relative wall thickness at sampling velocities equal to and higher than the upstream local velocity. This observation contrasts with the results obtained with the sand particles shown in Figure 11. This difference can be explained as follows: In the presence of a blunt probe, the fluid streamlines deflect ahead of the probe nozzle even at a velocity ratio $U/U_0 = 1$, and the deflection increases as the probe relative wall thickness is increased. Particles of high inertia, such as coarse sand particles, are not significantly affected by fluid deflection, and strike the nozzle wall. Some of the particles bounce into the probe aperture and thereby cause higher sampling concentrations. Particles of low inertia, such as the polystyrene particles, follow the fluid streamlines to a greater extent and should not strike the nozzle wall as frequently. To account for particle rebound and inertia effects simultaneously, a modification was introduced by Nasr-El-

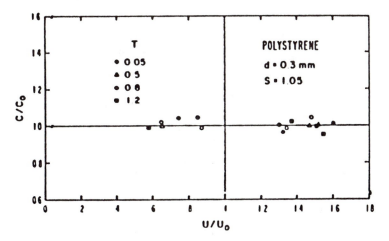

Figure 13. Effects of sampling velocity and probe relative wall thickness on the sampling efficiency for polystyrene particles. (Reproduced with permission from reference 47. Copyright 1986.)

Din and Shook (*46*). Figure 14 compares the calculated sampling efficiency for a thick probe having a probe relative wall thickness of 0.8, considering the inertial effect alone and with the particle bouncing effect, with the experimental measurements. Clearly, the agreement is much better when both effects are considered.

A second example of the particle bouncing effect is sampling using straight probes. Although L-shaped thick probes are more practical than thin probes, they will require a relatively large aperture in the wall of the pipe. Straight probes are robust, simple to construct, require a minimum size of aperture in the wall of the pipe, and can be withdrawn after sampling. The performance of two different straight probes, a side-port probe and a 45° probe (*see* Figure 1B) for measuring solids concentration of liquid–solid systems was examined (*46*). Figure 15 shows C/C_0 versus U/U_0 for the thin-walled L-shaped and the circular-port probes. For the circular-port probe, the sampling efficiency is higher than unity at the isokinetic velocity. Thus, to get the correct concentration, the velocity ratio would have to be greater than unity.

The increase in the sample concentration at the isokinetic conditions resembles that of thick L-shaped probes. Particles rebounding from the probe wall probably enter the probe and thus cause higher concentrations at sampling velocities equal to and greater than isokinetic.

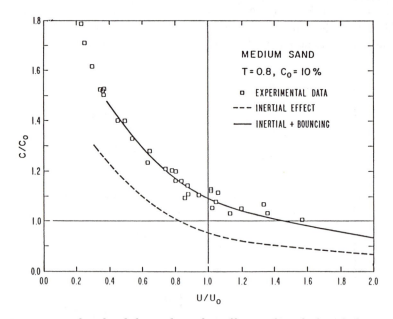

Figure 14. Predicted and observed sampling efficiency for a thick probe having a probe relative wall thickness of 0.8. (Reproduced with permission from reference 46. Copyright 1985.)

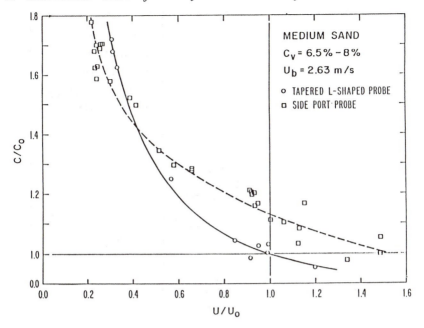

Figure 15. Sampling efficiency for circular and thin-walled L-shaped probes. (Reproduced with permission from reference 46. Copyright 1985.)

Figure 16 shows a comparison between the 45° and the thin-walled probes. The 45° probe gives higher sampling concentrations at the isokinetic velocity and above, and the difference increases as the sampling velocity is increased. For subisokinetic velocities, this probe gives lower sampling efficiencies than the thin L-shaped probe and side-port probe. This lower sampling efficiency results from the difference in the flow field ahead of the sampling point. The projected area for each probe was used in calculating the sampling velocity. Figures 15 and 16 show that the scatter for straight probes is greater than that obtained using the thin-walled L-shaped probes.

The Flow Structure Ahead of the Sampler.   A third source of sampling errors is not directly related to the geometry of the sampling device, but to the flow structure ahead of the sampler. Obviously, if the flow field ahead of the sampler is strongly three-dimensional, it will be very difficult to obtain a representative sample. To illustrate this point, consider the flow field downstream of a 90° elbow. Whenever a fluid flows along a curved pipe, a pressure gradient must occur across the pipe to balance the centrifugal force. The pressure is greatest at the wall farther from the center of curvature (pressure wall), and lowest at the nearer wall (suction wall). Because of inertia, the fluid in the core moves across the pipe from the suction wall

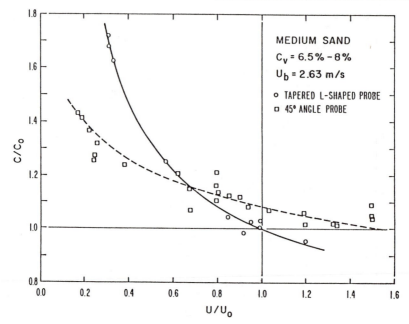

Figure 16. Sampling efficiency for 45° and thin-walled L-shaped probes. (Reproduced with permission from reference 46. Copyright 1985.)

toward the pressure wall and returns to the inner edge along the wall, as shown in Figure 17. A pair of symmetrical, counter-rotating vortices is formed as a result of the fluid inertia. This secondary flow is superimposed on the main stream, so the resultant flow consists of helical motion on each side of the plane of the bend passing through the axis of the pipe. The strength of the secondary flow depends, among other factors, on the flow Reynolds number and the curvature of the elbow. Flow in curved pipes has been studied extensively both experimentally and theoretically. A recent review on this work was given by Ito (48).

This type of flow affects sampling in two ways: First, because of the helical motion, it is very difficult to align the probe with the fluid velocity vector. Consequently, and because of the inertial effect, sample concentration will be always less than the upstream concentration (49). Second, the inertial effects on the elbow plane and the centrifugal force on a plane perpendicular to that of the elbow will produce a nonuniform solids distribution downstream of the elbow.

A few studies considered the solids distribution downstream of elbows. Ayukawa (50) and Toda et al. (51) observed an accumulation of coarse particles at the outer wall of vertical bends. Toda et al. (52) noted some changes in the solids distribution downstream of 90° bends. However, no concentration measurements were taken.

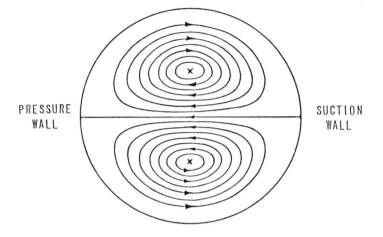

PRESSURE
WALL

SUCTION
WALL

Figure 17. Secondary flow pattern caused by flow through 90° elbows. (Reproduced with permission from reference 55. Copyright 1987.)

To overcome the effect of 90° bends, various methods have been suggested. Stairmand (53) recommended using a flow straightener in a reduced pipe section to get a uniform concentration profile. Davis (54) suggested that the secondary flow generated by bends could be eliminated by using straightening vanes. Fuchs (27) recommended taking measurements five pipe diameters downstream of the elbow. Nasr-El-Din and Shook (55) studied the effect of a 90° elbow on solids distribution downstream of a vertical elbow. They tested sand–water slurries of various solids concentrations and particle sizes. The slurry flows were turbulent, and the particle Stokes number based on the pipe diameter and bulk velocity varied from 0.5 to 3. The solids distribution downstream of a vertical elbow was found to be a function of the radius of curvature of the elbow, solids concentration, and particle size.

Figure 18 shows the solids concentration profile 22 pipe diameters downstream of a short-radius elbow. The concentration profile is symmetrical, and a minimum solids concentration appears at the center of the pipe. Also, the solids concentration gradually increases toward the pipe wall. This variation in concentration across the pipe is evidently a consequence of the centrifuging action of the secondary flow that is generated by the bend upstream. Figure 18 also shows that the concentration profiles are concentration dependent, and as the solids concentration is increased, the profiles become flatter. Other results (55) showed that these profiles are also functions of the particle size and the radius of curvature of the elbow.

To understand and follow the concentration variations, measurements were obtained just downstream of the elbow (1.5 pipe diameters). Figure 19 shows the effect of the inverse of the sampling velocity on the concentration

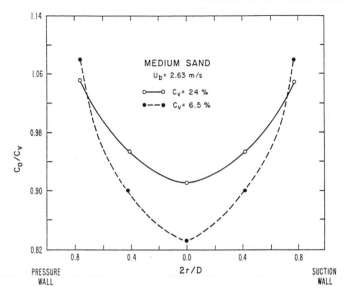

Figure 18. Concentration profiles 22 pipe diameters downstream of a short-radius elbow. (Reproduced with permission from reference 55. Copyright 1987.)

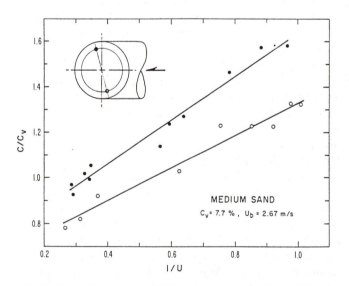

Figure 19. Effect of the inverse of the sampling velocity on C/C_v for sand particles at two positions 1.5 pipe diameters downstream of a short-radius elbow. (Reproduced with permission from reference 55. Copyright 1987.)

ratio C/C_v for the medium sand at two positions equidistant from the center of the pipe (C_v is the discharge solids concentration). The figure shows that the linear relation between C/C_v and $1/U$ holds, even in the region of strong secondary flow.

An attempt was made to measure the particle local velocity at 1.5 pipe diameters downstream of the elbow by using a particle velocity probe (56). However, the technique failed, presumably because the strong secondary flow prevented the velocity probe from being aligned with the velocity vector. For this reason, velocities obtained at 22 pipe diameters downstream of the elbow had to be used to estimate the concentrations at this level (1.5 pipe diameters). Figure 20 shows the estimated solids concentration normalized by the discharge concentration (C_0/C_v) for fine and medium sand particles 1.5 pipe diameters downstream of the elbow. Most of the relative concentrations are lower than unity, and consequently the mean concentration based on these measurements would be lower than the true value. Similar findings were obtained by Sansone (57) in gas–solid systems downstream of a 90° elbow. This phenomenon occurs because the velocity vector and the probe axis are not colinear, so that the concentration results are only of qualitative value.

Figure 20 also shows that the solids are more concentrated near the pipe wall, and a minimum solids concentration appears at the center of the pipe. The location of the maximum solids concentration depends on the particle

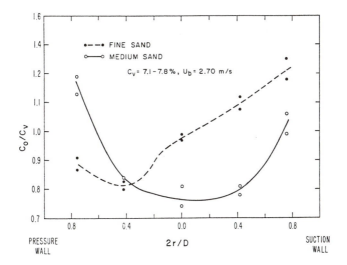

Figure 20. Effect of particle size on solids concentration profile downstream of a short-radius elbow. (Reproduced with permission from reference 55. Copyright 1987.)

size. For the medium sand, probably because of the higher inertia in the elbow plane, the particles are relatively concentrated at the pressure wall. For the fine sand, the secondary flow seems to play an important role, and the maximum concentration occurs at the suction wall.

To establish a uniform concentration profile downstream of a 90° elbow, straightening vanes 10 cm long were inserted just downstream of the short-radius elbow. Figure 21 shows the effect of these vanes on the concentration profile. Although the concentration becomes flatter, a distinct minimum at the center of the pipe still exists. These results imply that the solids are already distributed at the exit of the elbow, and the vanes merely increase the rate of diffusion of the particles.

The concentration profiles discussed so far were obtained in a vertical pipeline downstream of an elbow with a horizontal approach. Colwell and Shook (58) examined concentration profiles in a horizontal slurry pipeline downstream of a 90° elbow. According to their results, a length of at least 50 pipe diameters downstream of the elbow is needed to obtain fully developed concentration profiles.

Conductivity Methods.   The electrical conductivity of a mixture of two or more phases is an important property of the mixture. Many details

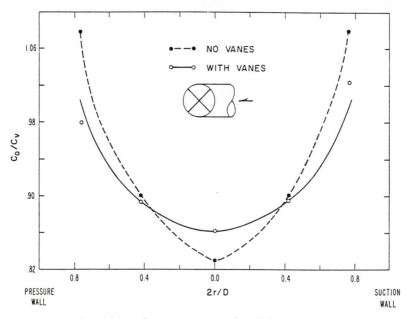

Figure 21. Effect of straightening vanes on the solids concentration profile 22 pipe diameters downstream of a short-radius elbow. (Reproduced with permission from reference 55. Copyright 1987.)

regarding the mixture's structure can be inferred from its electrical conductivity.

According to the nature of the dispersed phase in the mixture, uses of electrical conductivity can be divided into two major groups. In the first group, the dispersed phase (the solid particles in slurry systems or the oil droplets in oil-in-water emulsions) consists of loose particles dispersed in a continuous phase (matrix). The particles have a defined shape and size distribution, but the concentration of the dispersed phase is less than the corresponding maximum packing concentration. In this group, electrical conductivity is used to measure the concentration and the particle size distribution of the dispersed phase within the system. Typical examples of such systems are measuring concentration of the dispersed phase in mixing tanks (*59, 60*), in pipelines (*24, 61, 62*), and in three-phase fluidized beds (*63*).

In the second group, the solid-phase concentration is high, and solids particles are either loose but in contact, or consolidated. In this case, the solid phase is the matrix, while the liquid phase is the dispersed phase. In this group, electrical conductivity is used to measure the effective porosity of the porous medium (*64, 65*). Also, if two immiscible fluids, for example, oil and water, are present in a porous medium, the electrical conductivity can be employed to measure the relative saturations of the two fluids and to give an indication of the wettability of the porous medium (*66, 67*).

The electric conductivity methods are widely used in both categories because they are simple to operate and give quick response, accurate results, and a continuous reading; that is, they can be used as a measuring element in any control loop.

Besides these examples, knowing the relationship between the mixture effective conductivity and the porosity or the concentration of the dispersed phase is important. Such a relation can be used to predict other transport coefficients such as the diffusion coefficient, dielectric constant, and thermal conductivity. Of course, such relations are useful in many practical applications.

Although estimating the electrical conductivity of a mixture of two or more phases looks simple and straightforward, it is a very complicated problem, both theoretically and experimentally. This complexity explains the huge volume of work devoted to solving this problem since the pioneering work of Maxwell (*68*) and Lord Rayleigh (*69*).

Although this section deals with the conductivity of liquid–solid systems, similar treatment can be used in other similar two-phase systems such as gas–liquid dispersions (*70, 71*), oil-in-water emulsions (*72*), and foams (*73*).

Definition of the Mixture Effective Conductivity.   If a liquid–solid mixture is placed between two electrodes of different potential, the resulting potential difference will cause a current to flow from the electrode of higher potential to the electrode of lower potential. The current and potential gradient are related by the following diffusion-type equation:

$$I = \lambda_m \nabla \phi \tag{19}$$

where I and ϕ are the volume-average values of the current and the potential, respectively. The proportionality constant λ_m is the effective conductivity of the mixture. For a homogeneous and isotropic mixture, λ_m is a scalar quantity, whereas for a homogeneous and anisotropic mixture, λ_m is a symmetrical second-order tensor. This fact explains why most of the previous theoretical and experimental studies were devoted to random mixtures of monosized spheres. Of course, these mixtures are homogeneous and isotropic. Therefore, it is relatively easier to measure and/or to determine the electrical conductivity of the mixture.

In the following sections, various methods to determine the effective conductivity of two phases will be discussed, especially for homogeneous and isotropic mixtures.

Mathematical Description of Effective Conductivity.   In very general terms, the electrical conductivity of a mixture is a function of the electrical conductivity of its constituents, their relative amounts, and their distribution within the system. Models and expressions to predict the mixture effective conductivity can be divided according to the degree of complexity of the mixture into two major categories. In the first category, the mixture consists of particles of definite shape (e.g., spheres, spheroids, and ellipsoids) at low solids concentration. For these mixtures, describing the boundary conditions is straightforward. Also, the effect of the surrounding particles can be neglected. For this category, rigorous solutions are available for particles of simple geometrical shapes. A rigorous solution in this case means solving Laplace's equation for the potential and using appropriate boundary conditions.

The second category includes mixtures of high concentrations. Unlike dilute mixtures, particle–particle interactions cannot be neglected. Also, it is very difficult to describe the boundary conditions. Because the problem is basically a boundary value problem, no rigorous solutions are available for concentrated mixtures, except for ordered arrays. To overcome these problems, various approaches have been considered. In this review, the following cases will be discussed:

- approximate solutions based on Maxwell's theory
- empirical formulas

Effective Conductivity of Dilute Mixtures.   The simplest, best defined case, is a cluster of spherical particles dispersed in a liquid and located in a uniform electrical field. If the particles have the same conductivity as the liquid, the potential around the particles will not be distorted, and the mixture conductivity is equal to that of the liquid. If the particles have a lower conductivity, the streamlines will diverge away from the particles, and the mixture conductivity will be lower than that of the liquid. If the particles have a higher conductivity, the streamlines will converge into the particle, and the mixture conductivity will be higher than that of the liquid.

Maxwell (68) calculated the potential distribution for a single spherical particle immersed in a conducting medium and subjected to a uniform electrical field: He solved Laplace's equation within the two regions subject to continuity of potential, and continuity of the normal component of the current density, at the surface of the particle. Maxwell then extended his single-sphere solution to dilute mixtures and obtained the following expression for λ_m:

$$\lambda_m = \lambda_f \left[\frac{2\lambda_f + \lambda_s - 2(\lambda_f - \lambda_s)C}{2\lambda_f + \lambda_s + (\lambda_f - \lambda_s)C} \right] \tag{20}$$

where λ_f and λ_s are the electrical conductivities of the liquid and solid phases, respectively; and C is the volumetric concentration of the dispersed phase. The assumptions used to derive equation 20 are very important:

- The particles are spherical, of uniform size, and have the same electrical conductivity.

- The electrical field around any particle or droplet is not affected by the presence of other particles; that is, particle diameter is much smaller than the distance between the particles. Obviously this condition can be met only for very dilute mixtures.

- The effect of surface conductance is negligible.

- The mixture is homogeneous and isotropic.

Equation 20 indicates that

- Mixture conductivity does not follow the additivity rule, which is sometimes used as a simplifying assumption. This relation is not linear, except at extremely low concentrations.

- Equation 20 satisfies the following three limiting conditions:

  1. As $C \to 0$, $\lambda_m \to \lambda_f$.
  2. As $C \to 1$, $\lambda_m \to \lambda_s$.
  3. As $\lambda_s \to \lambda_f$, $\lambda_m \to \lambda_f$, for all concentrations.

The second condition can be obtained only with mixtures having an infinitely wide distribution. For monosized particles, C can not be greater than the maximum packing concentration (C_M). Furthermore, the third condition can be used to measure the solids conductivity by using solutions of known conductivities.

- The mixture conductivity is independent of particle size for monosized spheres. This condition is observed to be true in practice provided that the particle size is much smaller than the spacing between the two sensor electrodes.

- For a mixture of nonconducting spheres ($\lambda_s = 0$) in a conducting liquid, equation 20 reduces to

$$\lambda_m = \lambda_f \left[\frac{2(1 - C)}{2 + C} \right] \qquad (21)$$

- Equation 20 is not symmetrical with respect to λ_f and λ_s; that is, one has to know which phase is the continuous phase (matrix) and which phase is the dispersed.

Effective Conductivity of Concentrated Mixtures.   So far, we have considered dilute mixtures of random spheres (*68*). This case has defined boundaries and consequently, equation 19 has a rigorous solution. Unfortunately, a rigorous solution is not possible for random concentrated suspensions for which it is very difficult to describe the boundaries. Because of this difficulty, it was necessary to introduce more simplifying assumptions. In this section, the most important approaches are reviewed.

The first approach to estimate for concentrated suspensions was introduced by Bruggeman (*74*). He considered the electrical conductivity of spherical particles of a random size distribution. Basically, his derivation is an extension of Maxwell's theory. According to Bruggeman, a suspension of high solids concentration is formed by continuously adding the particles (dispersed phase) to the liquid (matrix). The addition process starts with the smallest particles; then, in each step larger particles are added. At any step, the suspension of smaller particles is treated as a continuum with a conductivity that can be calculated from Maxwell's equation. The conductivity of the suspension (after adding larger particles), can be determined by applying Maxwell's equation once more. This process is repeated to the desired concentration.

Regarding Bruggeman's assumptions, two points are important:

1. At each step, the suspension of smaller particles is not a continuum.

2. The suspension must have an infinite range of particle sizes. This situation is seldom encountered in practice.

Using these assumptions and applying Maxwell's equation, Bruggeman derived the following implicit equation for λ_m:

$$\lambda_m - \lambda_s)(\lambda_m/\lambda_f)^{-0.33} = (1 - C)(\lambda_f - \lambda_s) \tag{22}$$

For nonconducting solids in a conducting liquid, equation 22 gives

$$\lambda_m = \lambda_f(1 - C)^{1.5} \tag{23}$$

De La Rue and Tobias (75) measured the conductivities of random suspensions of spheres, cylinders, and sand particles in aqueous solutions of zinc bromide of approximately the same densities as the particles. They found the suspension conductivity could be calculated from the following expression:

$$\lambda_m = \lambda_f(1 - C)^x \tag{24}$$

where $x = 1.5$ for a solids concentration in the range 0.45–0.75. This equation is similar to that of Bruggeman (74). Equation 24 is usually written by petroleum engineers in terms of the formation factor (F), where F is the reciprocal of λ_m/λ_f.

Begovich and Watson (63) found experimentally that the mixture conductivity in a liquid–solid fluidized bed is proportional to the liquid holdup. Their equation can be written as

$$\lambda_m = \lambda_f(1 - C) \tag{25}$$

Still another empirical expression was given by Machon et al. (60):

$$\lambda_m = \lambda_f(1 - aC) \tag{26}$$

where a is a constant to be determined experimentally. Machon et al. found this constant by measuring the conductivity of a bed of nonmoving particles. The bed solids concentration was in the range 0.6–0.65. Equation 26 is linear, and according to this equation, $\lambda_m = \lambda_f$ for $C = 0$; this result is similar to equation 20. However, at $C = 1$, equation 26 does not agree with Maxwell's prediction unless $a = 1$. This observation and the fact that it has no theoretical justification suggest that equation 26 should be used with caution.

A comparison of these expressions is given in Table I. This table shows the increase in the mixture resistance due to the presence of nonconducting particles or droplets, $R_m - R_l$, divided by the fluid resistance as a function of the dispersed-phase concentration. R_m and R_l are the mixture and fluid resistances, respectively. Maxwell's (68) and Bruggeman's (74) relations give

Table I. Comparison Between Various Expressions
for $(R_m - R_l)/R_l$

C (%)	Ref. 63	Ref. 68	Ref. 74
10	0.111	0.167	0.171
20	0.25	0.375	0.398
30	0.429	0.643	0.708
40	0.667	1.0	1.52
50	1.0	1.5	1.829
60	1.5	2.05	2.953
70	2.33	3.5	5.086
80	4.0	6.0	10.18
90	9.0	13.5	30.623
100	infinite	infinite	infinite

very similar results at low solids concentrations. However, at higher solids concentrations, Bruggeman's relation gives higher values. Begovich and Watson's (63) relation predicts lower values for all concentrations, and the deviation from the other two relations increases as the concentration is increased.

Conductivity Probe for Local Solids Concentration Measurements. On the basis of the preceding discussion, local solids concentration can be determined by measuring the mixture conductivity, then using a calibration curve, for example, equation 20. However, using this method to measure solids concentration or dispersed-phase concentration is not an easy task. In the following sections, the development a new conductivity probe will be summarized (24). Also, various problems encountered with conductivity methods will be discussed.

Description. The probe, shown in Figure 22, has an L-shaped configuration. It is constructed from 3/16-in. stainless-steel tubing. To minimize the effect of flow disturbances, the probe terminates with a conical stainless steel tip, and the approach length to the sensor electrodes is 10 probe diameters. The two field electrodes are flush with the surface of the tubing and completely insulated from each other. The field electrode of larger area is grounded to the pipeline. The field electrode circuit consists of a function generator, a ballast resistance, and an ammeter. The two sensor electrodes are also flush with the surface of the tubing, 1 mm apart, and are located between the field electrodes. The sensor electrodes are constructed from 28-gauge platinum. They are also completely insulated from each other and from the field electrodes. The sensor electrodes are connected to a voltmeter from which a time-average reading can be obtained.

The probe has two unique features:

1. The field electrodes are mounted on the probe itself and not on the pipe wall, as commonly used (59). This feature is very

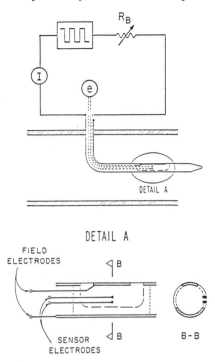

Figure 22. Conductivity probe for local solids concentration measurement. (Reproduced with permission from reference 24. Copyright 1987.)

important because it eliminates the need for higher voltages for measurements in large pipes. It also allows the study of the solids distribution within the pipe.

2. The potential is sensed for a small region (1-mm diameter) in the applied field. This means that resistivity and solids concentration can be measured over a small volume in space.

Operation. The operation of the probe relies on the variation of the slurry resistivity as the solids concentration changes. To understand the probe's principle, assume the probe is surrounded by a conducting liquid such as tap water; then if a potential is applied across the field electrodes (of the order of 5 V), a small current flows from one field electrode to another. The value of this current, for a fixed probe geometry and applied signal, depends on the total resistance of the medium surrounding the field electrodes. If nonconducting particles (e.g., sand particles) are added to this fluid, then the resistivity of the mixture will increase. As the solids concentration is increased, the mixture resistivity increases, and consequently the field circuit current diminishes.

One way of measuring the solids concentration, similar to that used by previous workers, is to relate the field circuit current change to the solids concentration. This method has a serious disadvantage because the field circuit current depends on both the slurry resistivity and the polarization resistance developed on the surfaces of the field electrodes. This polarization resistance is velocity dependent. This method yields calibration curves that are functions of velocity (59).

To avoid this problem, the total current was not used to measure the solids concentration. Instead the voltage was measured across the two sensor electrodes located between the field electrodes, as shown in Figure 22. Because the impedance of the sensor circuit is virtually infinite, practically no current flows into the sensors. Consequently, no polarization occurs on their surfaces. Thus, the calibration curve obtained should be independent of velocity, as shown in Table II.

To minimize the effect of polarization on the surfaces of the field electrodes and to facilitate a constant-current operation, a ballast resistance can be used in the field electrode circuit (24). Also, to eliminate fluid electrolysis, a square wave of 1 kHz and 5-V amplitude can be employed.

Calibration.   Various methods were used to calibrate the probe. These studies were conducted to find the relation between voltage (e) and solids concentration and to compare the experimental measurements with the predictions of the Maxwell and Bruggeman relationships. It was also necessary to establish an efficient probe calibration procedure.

**Table II. Normalized Sensor Voltages
as a Function of Position
Measured over a Velocity Range of 0 to 4 m/s**

Position (Y)	e(0,Y)/ e(0,0.5)	Standard Deviation
0.1	1.03	0.009
1.15	1.02	0.005
0.2	1.0	0.005
0.25	0.99	0.007
0.3	0.99	0.005
0.35	0.99	0.006
0.4	0.99	0.005
0.45	1.0	0.005
0.5	1.0	0
0.55	1.0	0.007
0.6	0.99	0.007
0.65	0.99	0.007
0.7	0.99	0.007
0.75	0.99	0.008
0.8	1.0	0.007
0.85	1.02	0.007
0.9	1.04	0.01

The first test was conducted with the conductivity probe mounted in the pipeline, as shown in Figure 23. Polystyrene particles of 0.3-mm mean diameter were used in these slurries. These particles were chosen because of their tendency to give a uniform concentration profile across the pipe. Concentrations were obtained by isokinetic sampling at the center of the pipe over a temperature range of 8 to 25 °C. Figure 24 shows the results; the effect of temperature in tap water is shown for comparison. At a fixed temperature, increasing solids concentration causes the sensor voltage to increase. This result is reasonable because polystyrene particles are nonconducting and their presence increases slurry resistivity.

The curves obtained at various concentrations are almost parallel. This observation means that the rate of change of voltage with respect to temperature is independent of the solids concentration. By cross-plotting the results shown in Figure 24, a set of calibration curves can be prepared with temperature as a parameter. When such curves were prepared, they indicated that the value of e at $C = 0$, obtained by extrapolation, was lower than the corresponding value obtained for tap water at the same temperature. A review of the procedure of this experiment indicated that the only possible reason for this difference was the fact that a small amount of a wetting agent (an anionic surfactant) was added with the solids to increase the wettability of the polystyrene particles.

To check this effect, the loop was operated with tap water, and measurements were taken at various surfactant concentrations (C_d). Figure 25 shows the results obtained. At a given temperature, as the surfactant concentration is increased, the voltage decreases. This decrease is reasonable the surfac-

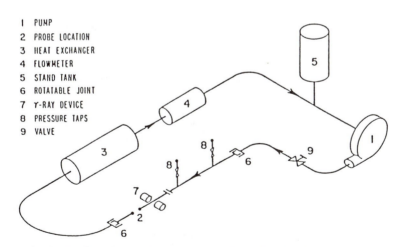

1 PUMP
2 PROBE LOCATION
3 HEAT EXCHANGER
4 FLOWMETER
5 STAND TANK
6 ROTATABLE JOINT
7 γ-RAY DEVICE
8 PRESSURE TAPS
9 VALVE

Figure 23. Test loop for conductivity probe. (Reproduced with permission from reference 24. Copyright 1987.)

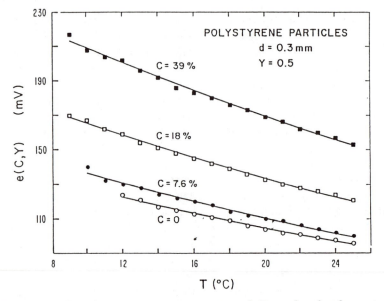

Figure 24. *Effect of temperature on sensor potential. (Reproduced with permission from reference 24. Copyright 1987.)*

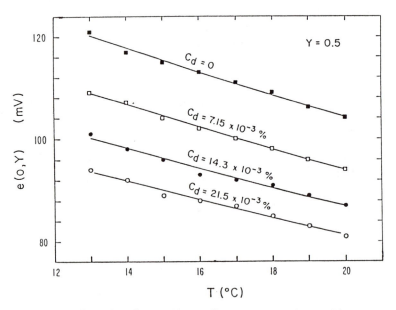

Figure 25. *Effect of surfactant (anionic) concentration (C_d, wt%) on sensor potential. (Reproduced with permission from reference 24. Copyright 1987.)*

tant contains sodium salts of organic acids. Its presence would decrease the fluid resistivity and consequently the sensor voltage. These results show

1. A calibration curve based on these measurements is not acceptable because of poor control on the amount of surfactant.
2. The voltage measured is strongly affected by any small change in the chemical composition of the conducting liquid.

The results shown in Figures 24 and 25 indicate that using e as a measure for solids concentration is not appropriate because it is strongly dependent on temperature, chemical composition, and position. Therefore, the following function was used in calibrating the probe:

$$\frac{e(C,Y) - e(0,Y)}{e(0,Y)}$$

This function will correct the sensor voltage at any concentration (C) for temperature, chemical composition, and position (Y) in the pipe.

Sedimentation was the second method tested to calibrate the probe. Polystyrene particles of 0.3-mm mean diameter were again used, without the wetting agent. These particles were chosen because of their very low settling velocities, which allowed sufficient time for voltage readings. Tests were done in a 5-cm acrylic pipe with both tap water and a glycol solution of the same density as the particles.

Figure 26 shows the voltage measurements, expressed in terms of the function just defined, as a function of solids concentration in the two fluids. The figure indicates good agreement between the experimental measurements and Maxwell's relation. Although Maxwell's relation was supposed to be valid for low concentrations, it actually agrees very well with all the experimental results. This observation agrees with previous work (76–78). Furthermore, changing the solution conductivity had no effect on the measurements. This result demonstrates that by using the sensor output function, the effects of all variables on the probe output except C should be isolated.

The second step was to examine the effect of particle size on the calibration curve. This step was not possible by sedimentation, because coarser particles have higher settling velocities. Therefore, a liquid–solid fluidized bed was used. A fluidization column was constructed with a 5-cm acrylic pipe. Weighed quantities of solids were used, and solids concentration was varied by changing the liquid flow rate. Measurements for these experiments included voltage, bed height, and temperature. To allow a precise determination of concentration from bed height, narrow sizes of particles were used.

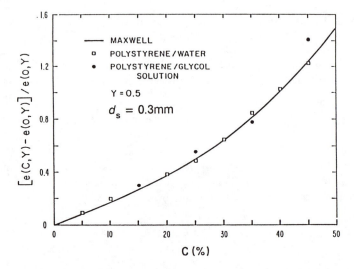

Figure 26. Probe calibration in sedimentation tests. Maxwell's relation is shown for comparison. (Reproduced with permission from reference 24. Copyright 1987.)

Figure 27 shows the experimental data obtained for a sand fraction of 0.6-mm mean diameter and narrow size distribution. Measurements in this experiment were taken at the pipe center. Maxwell's and Bruggeman's relations are also shown for comparison. Good agreement between the experimental measurements and both relations is observed at low concentrations. However, at high solids concentrations, the experimental data exceed Maxwell's predictions, but show good agreement with Bruggeman's equation.

Figures 26 and 27 show that particle shape has an effect on the results. Good agreement between the experimental results for spherical particles (0.3-mm polystyrene particles) and Maxwell's equation is observed. For irregular sand particles, Maxwell's equation underpredicts the experimental results, especially at high concentrations. The scatter in the experimental data is higher for sand particles than for polystyrene particles. The scatter increases for sand particles as concentration is increased. This observation implies that it is difficult to reproduce the same particle packing for irregular sand particles.

Figure 28 shows measurements for the same sand fraction at dimensionless radial positions ($R = 2r/D$) of 0.0, 0.7, and 0.8, where r is the radial position measured from the pipe center. The effect of position on $[e(C,R) - e(0,R)]/e(0,R)$ is significant. Results obtained at the other positions show the same deviation from Maxwell's relation at higher concentrations.

Because measurements refer to a small volume in space, the effect of particle size is of interest. This effect was examined through two sets of

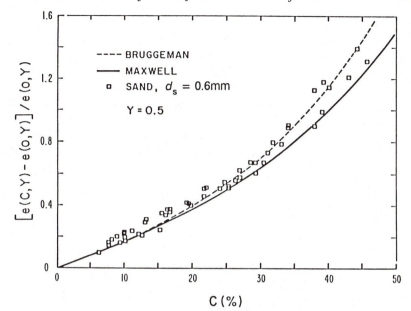

Figure 27. Probe calibration in fluidization tests. Maxwell's and Bruggeman's relations are shown for comparison. (Reproduced with permission from reference 24. Copyright 1987.)

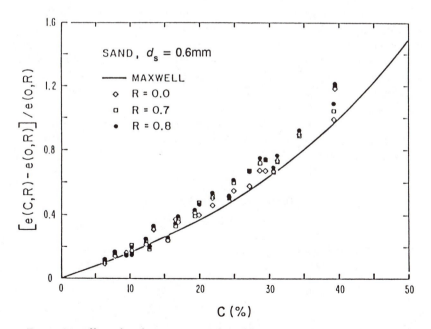

Figure 28. Effect of probe position on the calibration curve for sand particles. (Reproduced with permission from reference 24. Copyright 1987.)

experiments. In the first test, sand fractions of 0.6- and 1.5-mm mean diameter were used. These sand fractions were obtained from the medium and coarse sands (*see* Table III), respectively. In the second, glass beads of 1.5-, 2.8-, and 5.5-mm mean diameter and spherical shape were used.

Figure 29 shows the results obtained for the sand particles. The coarser particles show the same deviation at high concentration from Maxwell's equation as observed previously for the 0.6-mm sand. The figure also shows that measurements for the coarser particles are slightly lower than those of the 0.6-mm sand. The difference is generally small and could probably be neglected, especially if the probe were used to measure the concentration of particles of a wide size distribution.

Figure 30 shows the results obtained for the glass beads. For particles of 1.5-mm mean diameter, good agreement with Maxwell's equation was found up to 30% solids concentration. For higher concentrations, the experimental data are significantly lower than Maxwell's equation. For particles of 2.8-mm diameter, the experimental results are slightly lower than Maxwell's predictions up to 15%, then start to deviate significantly. For 5.5-mm diameter particles, the experimental results are significantly lower than Maxwell's for all concentrations.

Figures 29 and 30 show that particle size has no significant effect on the probe calibration curve for particles of diameter comparable to the sensor electrode spacing and smaller. For coarser particles, particle size has an effect on the calibration curve. As particle size is increased, the sensor voltage decreases, and it becomes lower than Maxwell's predictions. These results indicate that there is a limitation on the use of a probe with a fixed geometry for measuring solids concentration. This limitation would have to be considered when selecting sensor electrode spacings. The effect is probably due to packing, because as the particle diameter is increased, the mean concentration at the probe surface falls.

The next step was to test the probe performance in the pipeline in comparison with accepted methods for measuring solids concentrations: isokinetic sampling and γ-ray absorption methods.

Table III. Particle Properties

Particles	Mean Diameter (mm)	Density (g/cm³)	Shape
Glass beads	1.5	3.0	spherical
Glass beads	2.8	2.5	spherical
Glass beads	5.5	2.3	spherical
Polystyrene	0.3	1.05	spherical
Polystyrene	1.4	1.06	irregular
Fine sand	0.19	2.65	irregular
Medium sand	0.45	2.65	irregular
Coarse sand	0.9	2.65	irregular

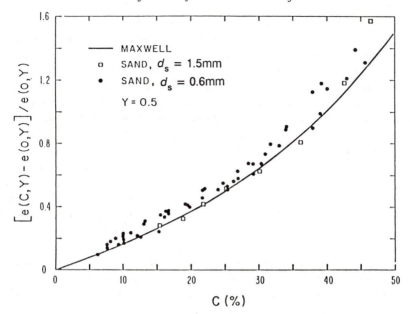

Figure 29. Effect of particle size on the calibration curve for sand particles. (Reproduced with permission from reference 24. Copyright 1987.)

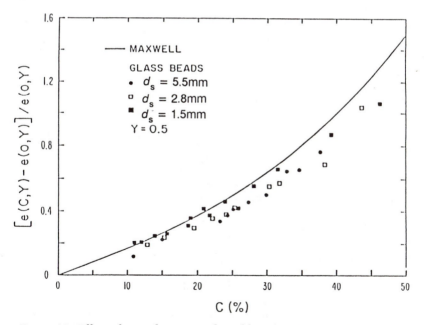

Figure 30. Effect of particle size on the calibration curve for glass beads. (Reproduced with permission from reference 24. Copyright 1987.)

Figure 31 shows the resistivity measured by the conductivity probe and the local concentration measured by the isokinetic sampling method. Measurements were made for slurries of polystyrene particles of 1.4-mm mean diameter at the pipe center and at radial positions of $R = 0.8$. Good agreement with calibration results is seen in these tests. Because the particles used in these experiments were large, no samples could be withdrawn from the center of the pipe at concentrations higher than 35%. Also, no voltage measurements could be taken closer to the pipe wall because particles tended to be trapped between the probe and the wall.

Figure 32 shows a typical sensor voltage profile in the vertical plane. Sand of 0.45-mm mean diameter and 10% solids concentration was used in this experiment, at a bulk velocity of 2 m/s and a temperature of 22 °C. This profile shows that the voltage is high at the bottom of the pipe where most of the solids are expected to be at these operating conditions. The voltage decreases as Y increases, that is, as the solids concentration decreases.

Figure 32 also shows the variation of the ballast resistance with position. The change of resistance across the pipe is much less than that of the sensor voltage. This difference demonstrates the response of the sensor electrodes to a smaller spatial region than the field electrodes.

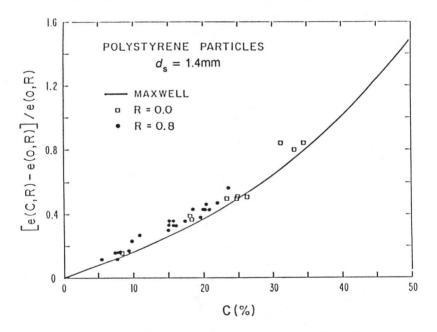

Figure 31. Pipe flow calibration curve using isokinetic sampling. (Reproduced with permission from reference 24. Copyright 1987.)

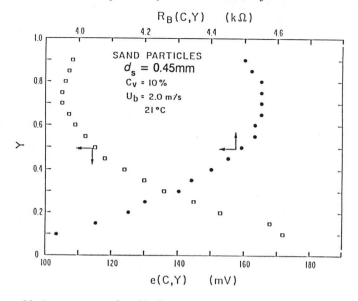

Figure 32. Sensor potential and ballast resistance variations using sand particles. (Reproduced with permission from reference 24. Copyright 1987.)

The profiles shown in Figure 32 are typical for sand–water flows, where because of gravity, more particles are found near the bottom of the pipe. Solids concentrations were obtained from measuring sensor voltages such as those shown in Figure 32 and from the calibration curve. This calibration curve was obtained by using a least-squares fit of the experimental data of Figure 27.

A second test for the conductivity probe was conducted with the γ-ray method. Two scans were conducted simultaneously on sand slurries, with the conductivity probe and with the γ-ray method. The γ-ray values were obtained with a collimated beam of 1-mm diameter.

Figure 33 shows the concentration profile for sand particles of 0.19-mm mean diameter and a bulk velocity of 2 m/s. Good agreement between the γ-ray method and probe measurements was obtained. The figure shows some scatter at the top of the pipe where the solids concentration is very low and both methods are subject to error. These results are indirect evidence of a constant concentration across the pipe under the conditions of these measurements.

Figure 34 shows another comparison for the same sand, but at a bulk velocity of 3.5 m/s. Again, good agreement was obtained. This result is important because it confirms the results shown in Table II in that the calibration curve is independent of velocity.

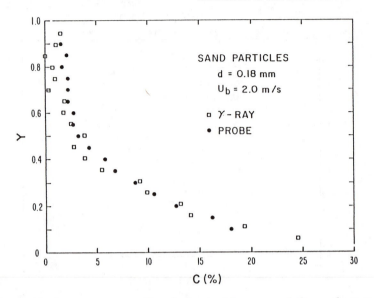

Figure 33. Concentration profiles obtained with γ-ray and the conductivity probe (low velocity). (Reproduced with permission from reference 24. Copyright 1987.)

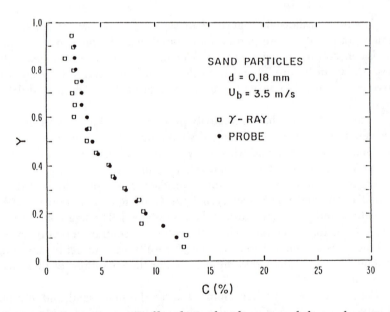

Figure 34. Concentration profiles obtained with γ-ray and the conductivity probe (high velocity). (Reproduced with permission from reference 24. Copyright 1987.)

Concluding Remarks

Heavy-oil-in-water emulsions having oil volume fractions greater than 0.5 are often non-Newtonian shear-thinning fluids. The friction loss for the flow of such fluids in smooth pipelines can be predicted by using the procedures developed by Metzner and Reed (*14*) and Dodge and Metzner (*16*).

Another aspect of the transportation of heavy-oil-in-water emulsions, especially for short-distance pipelines, is the presence of sand particles. In situ solids concentration and emulsion quality can be measured with various sampling devices. However, serious errors in measuring both parameters arise from improper sampling.

Conductivity probes can be employed to measure in situ solids concentration or oil-in-water quality with Maxwell's equation. However, the effects of temperature and ionic strength on the conductivity of the continuous phase should be isolated before proper measurements can be taken.

List of Symbols

B	fraction of particles that enter a probe
c	local solids concentration, volume fraction
C	solids concentration in the sampler, volume fraction
C_b	average solids concentration over the pipe cross section, volume fraction
C_d	surfactant concentration, wt%
C_0	local solids concentration upstream of the sampler, volume fraction
C_M	particle maximum packing concentration, volume fraction
C_v	discharge solids concentration, volume fraction
d_s	mean particle diameter, m
D	pipe inner diameter, m
D_{sm}	sampler diameter, m
e	voltage, V
E	sampling transport efficiency
f	Fanning friction factor
F	formation factor
I	volume-average current
k	power law constant, $Pa \cdot s^n$
k'	Metzner–Reed modified power law constant, $Pa \cdot s^n$
K	particle inertia parameter based on probe radius (also called Stokes number or Barth number)
L	pipe length, m
n	power law index

n' Metzner–Reed modified power law index
p pressure, Pa
Q branch flow rate, m³/s
r radial position measured from the center of the pipe, m
R dimensionless radial position, $2r/D$
R_b $\rho_f U_b d_s / \mu_f$
R_l fluid resistance, ohm
R_m mixture resistance, ohm
R_{sm} sampler radius, m
Re   Reynolds number, $\rho U_b D / \mu$
Re_0 Reynolds number, $\rho_f U_0 d_s / \mu_f$
Re'   Metzner–Reed modified Reynolds number
S ρ_s / ρ_f
t probe wall thickness, m
T probe relative wall thickness, t/R_{sm}
$u(r)$ local velocity, m/s
U sampling velocity, m/s
U_b bulk velocity, m/s
U_0 upstream local velocity, m/s
V_t particle settling velocity, m/s
y vertical position measured from the bottom of the pipe, m
Y dimensionless vertical position, y/D

Greek

γ shear rate, s⁻¹
γ_w shear rate at the pipe wall, s⁻¹
θ probe tip angle
λ_f fluid conductivity, S
λ_s solids conductivity, S
λ_m mixture conductivity, S
μ_f fluid viscosity, Pa·s
ρ_f fluid density, kg/m³
ρ_s solids density, kg/m³
τ shear stress, Pa
τ_w shear stress at the pipe wall, Pa
ϕ volume-average potential, V

Acknowledgment

The assistance of B. Fraser in typing this manuscript is greatly appreciated.

References

1. Plegue, T. H.; Frank, S. G.; Fruman, D. H.; Zakin, J. L. *Soc. Pet. Eng. Prod. Eng.* **1989**, *3*, 181.
2. Marsden, S. S.; Raghavan, R. *J. Inst. Petrol.* **1973**, *59*, 273.
3. Wyslouzil, B. E.; Kessick, M. A.; Masliyah, J. H. *Can. J. Chem. Eng.* **1987**, *65*, 353.
4. Pal, R.; Masliyah, J. H. *Can. J. Chem. Eng.* **1990**, *68*, 25.
5. Gillies, R. G.; Shook, C. A. Presented at the Quarterly Meeting of the Canadian Heavy Oil Association, Calgary, Alberta, Canada, November 14, 1989.
6. Gerhart, P. M.; Gross, R. J. *Fundamentals of Fluid Mechanics;* Addison-Wesley: Reading, MA, 1985.
7. Shah, S. N. *SPE Prod. Eng.* **1990**, *4*, 151.
8. Blasius, H. *Forsch. Ingenieurwes.* **1913**, 131.
9. Drew, T. B.; Koo, E. C.; McAdams, W. H. *Trans. Am. Inst. Chem. Eng.* **1932**, *28*, 56.
10. Zakin, J. L.; Pinaire, R.; Borgmeyer, M. E. *Trans. ASME* **1979**, *101*, 100.
11. Mao, M. L.; Marsden, S. S. *J. Can. Petrol. Technol.* **1977**, *16*, 54–59.
12. Flock, D. L.; Steinborn, R. Presented at the 33rd Annual Petroleum Society Technological Meeting, Calgary, Alberta, Canada, June, 1982; CIM paper 82-33-60.
13. Pilehvari, A.; Saadevandi, B.; Halvaci, M.; Clark, P. E. *Proceedings of the 3rd ASME International Symposium on Liquid–Solid Flows;* American Society of Mechanical Engineers: Chicago, IL, 1988.
14. Metzner, A. B.; Reed, J. C. *AIChE J.* **1955**, *1*, 434.
15. *The Flow of Complex Mixtures in Pipes;* Govier, G. W.; Aziz, K., Eds.; Van Nostrand Reinhold: New York, 1972.
16. Dodge, D. W.; Metzner, A. B. *AIChE J.* **1959**, *5*, 189.
17. von Kármán, T. National Advisory Committee for Aeronautics Technical Memo, No. 611, 1931, Washington, DC.
18. Pal, R.; Bhattacharya, S. N.; Rhodes, E. *Can. J. Chem. Eng.* **1986**, *64*, 3.
19. Pal, R. *Chem. Eng. J.* **1990**, *43*, 53.
20. Pal, R.; Rhodes, E. Presented at the 36th Annual Technology Meeting of the Canadian Institute of Mining and Metallurgy, Edmonton, Alberta, Canada, June 1985; CIM paper 85-36-12.
21. Kao, D. T.; Kazanskij, I. *Proceedings of the 4th International Technical Conference on Slurry Transportation;* Slurry Transport Association (now Coal and Slurry Technology Association): Washington, DC, 1979, p 102.
22. Baker, R. C.; Hemp, J. *Fluid Eng. Ser.* (British Hydrodynamics Research Association) **1981**, *8*, 3.
23. Nasr-El-Din, H. In *Encyclopedia of Environmental Control Technology;* Cheremisinoff, N. P., Ed.; Gulf Publishing Company: Houston, TX, 1989; Vol. 3, pp 389–422.
24. Nasr-El-Din, H.; Shook, C. A.; Colwell, J. *Int. J. Multiphase Flow* **1987**, *13*, 365.
25. Belyaev, S. P.; Levin, L. M. *J. Aerosol Sci.* **1972**, *3*, 127.
26. Rehakova, M.; Novosad, Z. *Coll. Czech. Chem. Commun.* **1971**, *36*, 3004.
27. Fuchs, N. A. *J. Atmos. Environ.* **1975**, *9*, 698.
28. Rushton, J. H.; Hillestad, J. Y. Paper No. 52–64, American Petroleum Institute, Washington, DC, 1964, p 517.
29. Karabelas, A. J. *AIChE J.* **1977**, *23*, 426.

30. Hayashi, H.; Sampei, T.; Oda, S.; Ohtomo, S. *Proceedings of the 7th International Conference on Hydraulic Transport of Solids in Pipes;* British Hydrodynamics Research Association: Cranfield, United Kingdom, 1980; Paper D2, p 149.
31. Iinoya, K. *Kagaku Kogaku Ronbunshu* **1970,** *34,* 69.
32. Akers, R. J.; Stenhouse, J. I. T. *Proc. Inst. Mech. Eng.* **1976,** 45.
33. Nasr-El-Din, H.; Shook, C. A.; Esmail, M. N. *Can. J. Chem. Eng.* **1984,** *62,* 179.
34. Stevens, D. C. *J. Aerosol Sci.* **1986,** *17,* 729.
35. Rushton, J. H. *AIChE Symp. Ser. No. 10;* American Institute of Chemical Engineers: New York, 1965, p 3.
36. Barresi, A.; Baldi, G. *Chem. Eng. Sci.* **1987,** *42,* 2949.
37. Sharma, R. N.; Das, H. C. L. *Coll. Czech. Chem. Commun.* **1980,** *45,* 3293.
38. Moujaes, S. F. *Can. J. Chem. Eng.* **1984,** *62,* 62.
39. Torrest, R. S.; Savage, R. W. *Can. J. Chem. Eng.* **1975,** *53,* 699.
40. Nasr-El-Din, H.; Shook, C. A.; Esmail, M. N. *Can. J. Chem. Eng.* **1985,** *63,* 746.
41. Nasr-El-Din, H.; Shook, *Int. J. Multiphase Flow* **1986,** *12,* 427.
42. Nasr-El-Din, H.; Afacan, A.; Masliyah, J. H. *Chem. Eng. Commun.* **1989a,** *82,* 203.
43. Nasr-El-Din, H.; Afacan, A.; Masliyah, J. H. *Int. J. Multiphase Flow* **1989b,** *15,* 659.
44. Whitely, A. B.; Reed, L. E. *J. Inst. Fuel* **1959,** *32,* 316.
45. Yoshida, H.; Yamashita, K.; Masuda, H.; Iinoya, K. *J. Chem. Eng. Jpn.* **1978,** *11,* 48.
46. Nasr-El-Din, H.; Shook, C. A. *J. Pipelines* **1985,** *5,* 113.
47. Nasr-El-Din, H.; Shook, C. A.; Colwell, J. *Hydrotransport* **1986,** *10,* 191.
48. Ito, H. *JSME Int. J.* **1987,** *30,* 543.
49. Lundgren, D. A.; Durham, M. D.; Mason, K. W. *Am. Ind. Hyg. Assoc.* **1978,** *39,* 640.
50. Ayukawa, K. *Bull. JSME* **1969,** *12,* 1388.
51. Toda, M.; Ishikawa, T.; Sait, S.; Maeda, S. *J. Chem. Eng. Jpn.* **1973,** *6,* 140.
52. Toda, M.; Komori, N.; Sait, S.; Maeda, S. *J. Chem. Eng. Jpn.* **1972,** *5,* 4.
53. Stairmand, C. J. *Trans. Inst. Chem. Eng.* **1951,** *29,* 15.
54. Davies, R. E. *Int. J. Air Water Pollut.* **1964,** *8,* 177.
55. Nasr-El-Din, H.; Shook, C. A. *J. Pipelines* **1987,** *6,* 239.
56. Brown, N. P.; Shook, C. A.; Peters, J.; Eyre, D. *Can. J. Chem. Eng.* **1983,** *61,* 597.
57. Sansone, E. *Am. Ind. Hyg. Assoc.* **1969,** *30,* 487.
58. Colwell, J. M.; Shook, C. A. *Can. J. Chem. Eng.* **1988,** *66,* 714.
59. Lee, K. T.; Beck, M. S.; McKeown, K. J. *Meas. Control* **1974,** *7,* 341.
60. Machon, V.; Fort, I.; Skrivanek, J. *Proceedings of the 4th European Conference on Mixing;* BHRA Fluid Engineering Centre: Cranfield, United Kingdom, 1982; p 289.
61. Ong, K. H.; Beck, M. S. *Meas. Control* **1975,** *8,* 453.
62. Pal, R.; Rhodes, E. *Proceedings of the 3rd Multi-Phase Flow and Heat Transfer Symposium;* Clean Energy Research Institute: Coral Gables, FL, 1983.
63. Begovich, J. M.; Watson, J. S. *AIChE J.* **1978,** *24,* 351.
64. Perez-Rosales, C. *J. Pet. Technol.* **1976,** *28,* 819.
65. Perez-Rosales, C. *Soc. Pet. Eng. J.* **1982,** *22,* 531.
66. Sweeney, S. A.; Jennings, H. Y., Jr. *J. Phys. Chem.* **1960,** *64,* 551.
67. Keller, G. V. *Oil Gas J.* **1953,** 62.
68. *A Treatise on Electricity and Magnetism,* 3rd ed.; Maxwell, J. C., Ed.; Dover: New York, 1954; Vol. 1, Article 314.

69. Lord Rayleigh *Phil. Mag.* **1892,** *34,* 481.
70. Pearce, C. A. R. *Br. J. Appl. Phys.* **1955,** *6,* 113.
71. Clark, N. O. *J. Phys. Chem.* **1945,** *49,* 93.
72. Lorentz, H. A. *Ann. Phys. Chem.* **1880,** *9,* 641.
73. Lorenz, L. *ibid.* **1880,** *11,* 70.
74. Bruggeman, D. A. G. *Ann. Physik.* **1935,** *24,* 636.
75. De La Rue, R. M.; Tobias, C. W. *J. Electrochem. Soc.* **1959,** *106,* 827.
76. Merilo, M.; Dechene, R. L.; Cichowlas, W. M. *J. Heat Transfer* **1977,** *99,* 330.
77. Neale, G. H.; Nader, W. K. *AIChE J.* **1973,** *19,* 112.
78. Turner, J. C. R. *Chem. Eng. Sci.* **1976,** *31,* 487.

RECEIVED for review December 18, 1990. ACCEPTED revised manuscript May 23, 1991.

6

Flow of Emulsions in Porous Media

Sunil L. Kokal, Brij B. Maini, and Roy Woo

Petroleum Recovery Institute, 3512 33rd Street N.W., Calgary, Alberta, Canada T2L 2A6

A comprehensive review of the important factors that affect the flow of emulsions in porous media is presented with particular emphasis on petroleum emulsions. The nature, characteristics, and properties of porous media are discussed. Darcy's law for the flow of a single fluid through a homogeneous porous medium is introduced and then extended for multiphase flow. The concepts of relative permeability and wettability and their influence on fluid flow are discussed. The flow of oil-in-water (O/W) and water-in-oil (W/O) emulsions in porous media and the mechanisms involved are presented. The effects of emulsion characteristics, porous medium characteristics, and the flow velocity are examined. Finally, the mathematical models of emulsion flow in porous media are also reviewed.

THE FLOW OF EMULSIONS IN POROUS MEDIA is encountered in the production of oil from underground reservoirs containing oil, water, and gas. Emulsions may form naturally during simultaneous flow of oil and water in porous rock formations, or they may be promoted by injection of external chemicals. In emulsion flooding for heavy-oil recovery (*see* Chapter 7), externally generated emulsions are injected into the reservoir. Emulsion flow through a porous medium may also be encountered in the chemical process industry in fixed-bed catalytic reactors involving two immiscible liquids. The physics of such flows is very complex because it involves flow of a complex and unstable fluid in an extremely complex geometry. To develop a working knowledge for solving problems involving the flow of emulsions in porous media, one must possess a knowledge of the nature and properties of emulsions, an understanding of the characteristics of porous media, and a working knowledge of the basic mechanisms involved in the flow of simpler fluids in porous media. Because the other chapters of this book provide an in-depth coverage

0065–2393/92/0231–0219 $012.00/0
© 1992 American Chemical Society

of the nature and properties of emulsions, we will not discuss the bulk behavior of emulsions.

Properties of Porous Media

In this section we will define some of the terms used to characterize a porous medium and briefly discuss those properties of porous materials that may have relevance to the flow of emulsions.

Porous Medium. A porous medium is simply a solid containing holes or void spaces. However, a metal block with a few holes drilled through it is not a porous medium, at least not the kind of porous medium that concerns us. The porous medium is a solid containing a large number of voids dispersed throughout in either a regular or random manner, provided that these voids occur frequently enough so that even a small subvolume (small compared to the bulk dimensions of the solid) will contain some voids.

Porosity. The fraction of total volume occupied by the voids is called the porosity of the porous medium. A distinction can be made between the pores that are interconnected and the pores that are totally isolated. The absolute or total porosity is the fraction of bulk volume occupied by all voids, connected or not. The effective porosity is the fraction of bulk volume occupied by interconnected pores.

The porous media can also be classified by the type of porosity involved, depending on the size and shape of voids. In sandstones and unconsolidated sands, the voids are between the adjoining sand grains, and this type of porosity is called intergranular. Carbonate rocks are generally more complex and may contain more than one type of porosity. The small voids between the crystals of calcite or dolomite constitute intercrystalline porosity. Often carbonate rocks are naturally fractured. The void volume formed by fractures constitutes the fracture porosity. Carbonate rocks sometimes contain relatively large holes, called vugs, and these constitute the vugular porosity. At the extreme end of the pore size scale, some carbonate formations may contain very large channels and cavities (several meters in size), which constitute the cavernous porosity.

Permeability. The permeability of a porous medium is a measure of the ease with which a fluid can flow through it. In other words, it is a measure of the fluid conductivity of the medium that determines the flow rate of a given fluid for a given pressure gradient.

The equation that defines permeability was discovered by Darcy (1) and is called Darcy's law. For linear, horizontal, isothermal flow of a fluid, the equation is

$$q = \frac{kA}{\mu} (dP / dL) \tag{1}$$

where q is the flow rate (mL/s), k is permeability (darcies), A is cross-sectional area of the medium (cm^2), μ is viscosity of the fluid (mPa·s), and dP/dL is the pressure gradient (atm/cm).

Pore Size Distribution. The pore size distribution is a measure of the average size of the pores and the variability of pore sizes. It is usually determined by mercury porosimetry. This technique is based on a simple conceptual model of the pores that treats the pores as capillary tubes. The pressure required to force mercury into a pore (assuming that the pore behaves like a circular capillary) can be related to the radius of the pore by

$$P_c = \frac{2\sigma \cos (\theta)}{r} \tag{2}$$

where P_c is the capillary pressure required to force mercury into the porous medium, r is the radius of the pore being invaded, σ is the surface tension of mercury, and θ is the contact angle. By measuring the volume of mercury entering into a sample of the porous medium as a function of the applied capillary pressure, the pore size distribution can be determined.

The size measured by capillary porosimetry is that of the entrance to the void space being invaded. The actual size of the void space can be much larger than the entrance size. However, an improved technique using very sensitive pressure transducers and computerized data acquisition and analysis has been developed to measure the size distribution of both the pore entrances (throats) and pore bodies (2).

Specific Surface Area. The specific surface area is defined as the area of internal surfaces per unit volume (or weight) of the porous material. The specific surface area of porous materials is very high. Petroleum reservoir rocks typically possess specific surface area in the range of 150 to 3000 cm^2/cm^3. The high internal surface area is responsible for many interesting characteristics of porous media. It plays an important role in processes involving adsorption of material from fluids flowing through the porous medium. It is also an important parameter in deep-bed filtration, ion exchange, and processes involving a chemical reaction between the solid matrix and a flowing fluid. The specific surface area also has a direct influence on the permeability of the medium (3).

Chemical Composition. The chemical composition of the porous medium can be very important in processes involving exchange of material between the solid grains and the flowing fluid. Such would be the case when the material forming the porous solid dissolves in the fluid or a chemical reaction occurs between the solid and the fluid. However, in this study we will assume that the solid is totally inert; that is, the flowing emulsion does not change the physical or chemical characteristics of the solid, and the solid does not change the chemical composition of the fluids flowing through it.

Flow of a Single Fluid: Darcy's Law

A fluid's motion is a function of the properties of the fluid, the medium through which it is flowing, and the external forces imposed on it. For one-dimensional steady laminar flow of a single fluid through a homogeneous porous medium, the relationship between the flow rate and the applied external forces is provided by Darcy's law:

$$q = -\frac{kA \left(\Delta P / \Delta L + \rho g \sin (\theta)\right)}{\mu} \tag{3}$$

where ρ is density and g is acceleration due to gravity.

Darcy's law simply says that the flow rate is proportional to the permeability of the medium, the cross-sectional area, and the sum of pressure gradient and the gradient of hydrostatic head along the direction of flow; and that the flow rate is inversely proportional to the viscosity of the liquid.

For flow in more than one direction, a more general form of equation 3 is required:

$$v_x = -(k_x/\mu) \, (\partial P/\partial x) \tag{4}$$

$$v_y = -(k_y/\mu) \, (\partial P/\partial y) \tag{5}$$

$$v_z = -(k_z/\mu) \, (\partial P/\partial z + \rho g) \tag{6}$$

where v_x, v_y, and v_z are the superficial velocities in the x, y, and z directions, respectively. The z direction is parallel to vertical. Also, different permeabilities (k_x, k_y, and k_z) are used in different directions to recognize the fact that porous media often exhibit different permeabilities in different directions.

In applying equations 4–6, the directional permeabilities are treated as point functions, that is, as a property of a point in the medium. The point value of the directional permeability can be visualized as a statistical average of the fluid conductance in the given direction of all pores contained in a small volume surrounding the point in question. This small volume must be visualized as being small compared to the size of the medium but large compared to the size of the individual flow channels.

Equations 4–6 combined with the law of conservation of mass are sufficient to derive equations for the flow of a single fluid in systems of complex geometry. Several excellent books and review articles are available on this topic (3–7).

Multiphase Flow in Porous Media

When two or more immiscible fluids are flowing simultaneously through a porous medium, the flow process becomes more complex. We will use the simpler case of two-phase flow to review the basic mechanisms involved in such processes.

The presence of two mobile phases means that each fluid can interact with both the porous medium and the other immiscible fluid. If the porous medium is visualized as a collection of interconnected flow paths, only a fraction of the total flow paths become available to a given fluid, the rest being occupied by the other fluid. This condition necessitates the introduction of fluid saturation as an important parameter. The saturation of a fluid phase is defined as the fraction of total void space occupied by that fluid. For two-phase systems, the sum of the two fluid saturations is equal to unity, because any void space not occupied by one fluid must be occupied by the other fluid.

By analogy to single-phase flow, under steady-state conditions, the flow rate of each fluid should be directly proportional to the applied pressure gradient and the cross-sectional area of the medium and inversely proportional to the fluid viscosity. Therefore, an equation analogous to equation 1 can be written for each fluid:

$$q_i = \frac{k_i A}{\mu_i} \left(\Delta P_i / \Delta L \right) \qquad\qquad i = 1,2 \qquad\qquad (7)$$

where the subscript i refers to a specific fluid. Using P_i in place of P is necessary to account for the local pressure discontinuity existing at the interface between the two fluids.

Capillary Pressure. Because of the small size of pores, the fluid–fluid interfaces within the porous medium are highly curved, and the pressure difference across the interface can be substantial. This local pressure difference across the fluid–fluid interface is called capillary pressure. In general, one of the two fluids preferentially wets the solid and is called the wetting fluid. The capillary pressure is usually defined as the pressure in the nonwetting fluid minus the pressure in the wetting fluid.

$$P_c(S_w) = P_{nw} - P_w \tag{8}$$

where the subscripts nw and w refer to the nonwetting phase and the wetting phase, respectively; and S_w is the saturation of the wetting phase.

Wettability. The wettability of the porous medium refers to its preference for one or the other fluid in becoming wet. It is defined as the "tendency of one fluid to spread on or adhere to a solid surface in the presence of other immiscible fluids" (7). In a rock–oil–brine system, it is a measure of the preference that the rock has for either the oil or the water. A water-wet rock is preferentially wetted by the water phase, and similarly for an oil-wet system, the rock primarily makes contact with the oil phase.

Relative Permeability. A comparison of equation 7 with equation 1 also shows that for the two-phase system we have used k_i, the effective permeability for the fluid, in place of the absolute permeability k, which is a property of the porous medium alone. This effective permeability term, k_i, depends on the absolute permeability, the type of fluid involved, and the saturation of this fluid.

The contribution of the absolute permeability is isolated by modifying equation 1 as follows:

$$q_i = \frac{kk_{ri}A}{\mu_i}(\Delta P_i/\Delta L) \qquad\qquad i = 1,2 \tag{9}$$

The term k_{ri} in equation 9 is called relative permeability of the fluid i. It represents the fraction by which the fluid conductivity of the porous medium must be modified to account for the presence of the other fluid. The presence of the other fluid implies that some of the flow paths would not be available to this fluid, thus the term k_{ri} in equation 9 must always be less than, or at most equal to, 1. Furthermore, when more of the fluid i is present in the medium, it will occupy more of the available channels, and hence its effective permeability will be higher. Therefore, the relative permeability term is expected to increase with increasing saturation of this fluid.

A porous medium in general will have flow channels of many different sizes; consequently, the relative permeability of a given fluid will depend not only on what fraction of the available pore space it occupies but also on what types of flow channels it occupies. If the fluid occupies smaller channels, its relative permeability will be smaller. Therefore, the distribution of the fluids is an important factor in determining relative permeability.

If only one fluid occupies any given channel, the fluid conductivity of this channel would remain unchanged. Therefore, if the two fluids were to occupy totally separate channels, the sum of relative permeabilities of the two fluids would be unity. However, both fluids can occupy different parts of the same flow channel. In this situation, the flow of one fluid might be impeded by the presence of the other fluid. Therefore, the sum of the relative permeabilities of the two fluids is often less than unity.

Clearly, the relative permeability of a fluid in a given porous medium is determined by its distribution within the pore space. This distribution depends on several factors, including the relative amounts of each fluid present, the past history of the system (i.e., how the fluids were introduced into the system), and the balance of various forces acting on the fluids. Under static condition, these forces include the gravitational or buoyancy forces resulting from the density differences and the capillary forces arising from the interfacial tensions or surface energies of the fluids. When fluids are in motion, the viscous drag forces and inertial effects may also play a role in determining the fluid distribution within a porous medium. Therefore, the factors that can affect relative permeability of a given fluid include the following:

1. wetting preference of the solid in relation to other fluid present in the porous medium

2. pore geometry and pore size distribution of the medium

3. saturation of the fluid

4. saturation history of the porous medium

5. densities of different fluids present

6. viscosities of different fluids present

7. interfacial tension between the two fluids

8. relative velocity of the fluids

In most situations of practical interest in petroleum reservoir engineering, the local distribution of fluids within the porous rock is dominated by capillary forces. Therefore, a reasonable assumption is that capillary equilibrium exists between the two fluids even under dynamic conditions. Under these conditions, the local distribution of fluids and relative permeability do not depend on the last four factors listed. A large volume of published experimental research has shown that the relative permeability is not a strong function of the fluid viscosities and interfacial tension, provided that these variables remain within their usual range. Experimental evidence also suggests that the relative permeability does not change significantly with the magnitude of relative velocity and the density difference between the two fluids.

At a given fluid saturation in a given porous medium, the wetting preference of the solid for one of the two fluids present determines the fluid distribution within the porous medium, and consequently, it also determines the relative permeability behavior. In the context of reservoir engineering, either water or oil may preferentially wet the reservoir rock. However, because of the complex geometry of the internal surfaces of natural porous media, the relative permeabilities for a given fluids–rock system cannot be predicted from a knowledge of the system wettability and pore geometry. Therefore, experimentally measured values obtained in representative samples of the porous medium must be used. Typical experimentally determined oil–water relative permeability characteristics for water-wet and oil-wet rocks are shown in Figures 1 and 2, respectively.

The relative permeability, for either of the two fluids, becomes zero at nonzero saturations of the respective fluids. In other words, a significant saturation of either fluid can become immobile and may not be displaced by the other fluid. Although both wetting and nonwetting fluids can become trapped, their distribution within the porous medium is very different. In strongly wetted systems, a thin layer of the wetting fluid always covers the solid surfaces. Therefore, the wetting fluid always remains continuous even at the irreducible saturation. It becomes immobile because of the strong interaction between the solid and the wetting fluid and the inability of the nonwetting fluid to enter very fine pores and crevices where bulk of the irreducible saturation exists.

Because of the presence of this thin layer of wetting fluid between the nonwetting fluid and the solid, the nonwetting fluid is always surrounded by the wetting fluid. A natural consequence of this condition is that parts of the

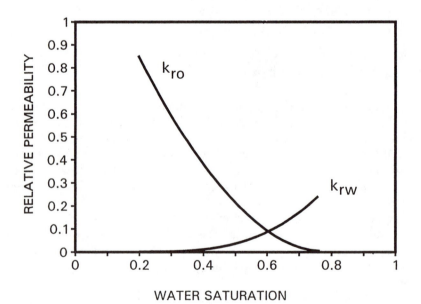

Figure 1. Relative permeabilities for a water-wet system. (k_{ro} is the relative oil permeability; k_{rw} is the relative water permeability.)

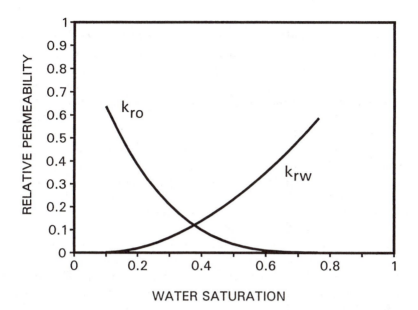

Figure 2. Relative permeabilities for an oil-wet system. (k_{ro} is the relative oil permeability; k_{rw} is the relative water permeability.)

nonwetting fluid can become isolated and may stop flowing. At the terminal saturation where the relative permeability of the nonwetting phase becomes zero, its distribution within the porous medium is discontinuous. The trapped saturation of nonwetting fluid exists as isolated blobs completely surrounded by the wetting phase. These blobs can be large enough to fill several tens of pores or they may be small and fully contained within individual pores. They remain trapped because the pressure gradient resulting from the flow of the wetting phase is not sufficient to overcome the capillary resistance encountered in forcing the blob through narrow pore throats.

Equations 8 and 9 combined with the law of conservation of mass are sufficient for mathematical description of one-dimensional two-phase flow. The only additional information needed is the functional relationship between relative permeability and fluid saturation for both fluids. For flow in more than one dimension, a generalized form of equation 9 is used. Collins (5), Richardson (6), and Craig (7) present more information on this subject.

Flow of Oil-in-Water Emulsions in Porous Media

The flow of oil-in-water (O/W) emulsions in porous media is a more complex process because of the complex nature of the emulsion itself in addition to the complexities of the porous medium. A major issue to be considered is whether to treat the emulsion as a homogeneous fluid. If the emulsion droplets are very small compared to the geometry of the flow channels, a reasonable approach is to use a continuum model of the material and totally ignore the microscopic structural details. Equivalent homogeneous properties can then be assigned to the fluid. Unfortunately, in most cases of practical interest, the emulsion droplet sizes are not much smaller than pore sizes. Therefore, treating the emulsions as a pseudo-single-phase fluid would be objectionable in most cases.

To develop an understanding of the emulsion flow in porous media, it is useful to consider differences and similarities between the flow of an O/W emulsion and simultaneous flow of oil and water in a porous medium. As discussed in the preceding section, in simultaneous flow of oil and water, both fluid phases are likely to occupy continuous, and to a large extent, separate networks of flow channels. Assuming the porous medium to be water-wet, the oil phase becomes discontinuous only at the residual saturation of oil, where the oil ceases to flow. Even at its residual saturation, the oil may remain continuous on a scale much larger than pores. In the flow of an O/W emulsion, the oil exists as tiny dispersed droplets that are comparable in size to pore sizes. Therefore, the oil and water are much more likely to occupy the same flow channels. Consequently, at the same water saturation the relative permeabilities to water and oil are likely to be quite different in emulsion flow. In normal flow of oil and water, oil droplets or ganglia become trapped in the porous medium by the process of snap-off of oil filament at pore throats (8). In the flow of an O/W emulsion, an oil droplet is likely to become trapped by the mechanism of straining capture at a pore throat smaller than the drop.

Permeability Reduction by Flow of Oil-in-Water Emulsion.
McAuliffe (9) proposed the concept of permeability reduction by emulsion flow in a porous medium. Consider a single droplet of an oil emulsion entering a pore throat smaller than itself, as shown in Figure 3. The radius of curvature of the leading edge is smaller than the radius of curvature of the trailing edge of the droplet in the pore throat, and consequently the capillary pressure is greater at the front of the droplet than at its back. A certain pressure is then required to force the droplet through the constriction. This Laplacian differential pressure required to move the droplet through the pore throat is given by

$$\delta P = 2\sigma \left[\frac{1}{r_1} - \frac{1}{r_2} \right] \tag{10}$$

where σ is the interfacial tension at the oil–water interface and r_1 and r_2 are the radii of curvature at the trailing and leading edges of the drop, respec-

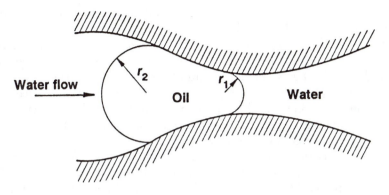

Figure 3. Emulsion blockage mechanism. (r_1 and r_2 are the radii of curvature at the trailing and leading edges of the drop, respectively.) (Reproduced with permission from reference 9. Copyright 1973 Society of Petroleum Engineers.)

tively. If the actual pressure differential across the pore is less than that predicted by equation 10, the emulsion droplet plugs the pore throat. This consideration is the fundamental basis for McAuliffe's (9) theory for permeability reduction in the flow of emulsions in porous media.

For an emulsion to be an effective blocking agent, the oil droplet size should be slightly larger than the pore throat size. Emulsion droplets have a range of sizes, as do pore throats, hence a small amount of emulsified oil can be very effective in restricting flow. However, pore throat constrictions should not be excessively large. For adverse mobility ratios, viscous instabilities develop inside the reservoir, and water starts to finger through. When an O/W emulsion is injected (or formed), a greater amount of emulsion enters the more permeable zones. As the emulsion restricts the flow of fluids in the more permeable zones, the displacing water (and emulsion) begins to flow into the less permeable zones and thus effectively improves the sweep efficiency.

When the pressure drop across the pore throat is larger than that predicted by equation 10, the oil droplet undergoes distortions to pass through the pore constriction. The droplets pass undistorted through those pore constrictions that have diameters larger than the droplet diameter. Thus, when the droplet diameter is much smaller than pore throat size (a microemulsion, for example), the rock "sees" the fluid as a homogeneous fluid.

This concept of emulsion flow in porous media, however, is not without contradictions (10, 11). Blockage of the pore throats by oil droplets necessitates an increase of differential pressure as given by equation 10. This feature implies that the interfacial tension σ and/or the droplet radius be increased. However, for emulsification to occur, interfacial tension must be decreased. Therefore, to maintain emulsion stability (low σ) and provide effective blockage (high σ), the interfacial tension has to be minimized to some optimum value and the droplet size maximized to some diameter larger than the pore throat diameter.

Soo and Radke (11) confirmed that the transient permeability reduction observed by McAuliffe (9) mainly arises from the retention of drops in pores, which they termed as straining capture of the oil droplets. They also observed that droplets smaller than pore throats were captured in crevices or pockets and sometimes on the surface of the porous medium. They concluded, on the basis of their experiments in sand packs and visual glass micromodel observations, that stable O/W emulsions do not flow in the porous medium as a continuum viscous liquid, nor do they flow by squeezing through pore constrictions, but rather by the capture of the oil droplets with subsequent permeability reduction. They used deep-bed filtration principles (12, 13) to model this phenomenon, which is discussed in detail later in this chapter.

Effect of Emulsion Characteristics.   The flow of emulsions in porous media is affected by a large number of variables. This section describes the properties of emulsions, such as stability, quality, droplet size distribution, oil viscosity, water–oil interfacial properties, and their effect on its flow in porous media.

Emulsion Stability.   An emulsion is a thermodynamically unstable system and has a natural tendency to separate into two phases. This tendency is due to the fact that when one phase is dispersed in another, the interfacial area increases and leads to an increase of the free energy of the system. Any oil–water system tends to minimize this free energy by reducing the interfacial area and by inducing coalescence of the dispersed oil droplets. However, apparent emulsion stability may be attained for a certain time period by using stabilizing agents or emulsifiers. These agents are either added or could be naturally occurring in the oil reservoir, and they suppress the mechanisms (flocculation, coalescence, creaming, phase inversion, etc.) that cause emulsion breakdown.

The stability of emulsions was discussed in Chapters 1 and 2 and will be discussed here very briefly in relation to their flow in porous media. The DLVO (Derjaguin, Landau, Verwey, and Overbeek) theory of colloid stability is often used to describe the short-range interaction between droplets causing flocculation and coalescence. According to this theory, the total potential energy of interaction is the sum of the London–van der Waals attractive energy and electrical double-layer repulsive energy between particles. The total interaction energy as a function of distance of separation is shown in Figure 4. Droplet interactions can occur as a result of hydrodynamic effects, such as mixing and flow in porous media, and nonhydrodynamic effects, such as diffusion and surface phenomena. In these situations, the destabilization of the emulsion due to flocculation,

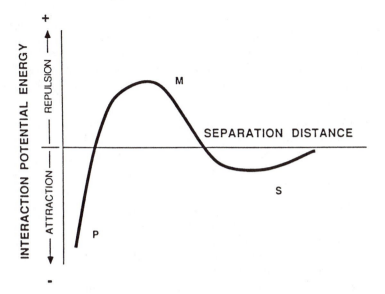

Figure 4. DLVO theory of colloid stability. (Reproduced with permission from reference 10. Copyright 1988 Canadian Institute of Mining, Metallurgy, and Petroleum.)

coalescence, and creaming is determined by the shape of the interaction potential energy curve in Figure 4. This curve is determined by the surface potential, thickness of the electrical double layer, ionic strength, droplet size, Hamakar constant, and a number of other factors (*10, 14*). In Figure 4 the primary minimum, P, represents the potential energy of two droplets in close proximity due to repulsion as a result of the overlap of electron shells. The maximum, M, represents the repulsive energy barrier to coalescence, and the secondary minimum, S, represents the separation distance where droplets will flocculate into aggregates but may be dispersed.

The presence of surfactants, either natural or added, promotes emulsion stability by the reduction of interfacial tension and the formation of highly rigid films on the surface of the droplets. This reduction of interfacial tension can increase the maximum, M, in Figure 4 significantly through charge stabilization or steric stabilization (*15*). Because the nature and shape of the interaction energy curve determine the stability of O/W (and other types) of emulsions, any process, parameter, or phenomenon that affects the shape of this curve will ultimately control emulsion stability.

Some of the parameters that affect the stability (*16*) are the following:

- **Temperature.** Temperature can affect emulsion stability in a number of ways. Temperature affects interfacial tension, which generally decreases with increasing temperature (*17,*

18). A lower interfacial tension will lead to a more stable emulsion. Temperature affects physical properties of oil, water, interfacial films, and surfactant solubilities in the oil and water phases, which can all affect emulsion stability. Further, the rheology of the emulsion itself is affected significantly by temperature.

- **Pressure.** Reservoir pressure has a less significant effect on emulsion stability than temperature. Interfacial tension decreases as the pressure of the system increases. Pressure effects probably have an indirect effect on emulsion stability because of the dependence of physical properties on pressure.

- **Surface-active agents.** Surface-active agents such as emulsifiers and surfactants play a very significant role in the stability of emulsions. They greatly extend the time of coalescence, and thus they stabilize the emulsions. Mechanisms by which the surface-active agents stabilize the emulsion are discussed in detail by Becher (*19*) and Coskuner (*14*). They form mechanically strong films at the oil–water interface that act as barriers to coalescence. The emulsion droplets are sterically stabilized by the asphaltene and resin fractions of the crude oil, and these can reduce interfacial tension in some systems even at very low concentrations (*17, 20*). In situ emulsifiers are formed from the asphaltic and resinous materials found in crude oils combined with ions in the brine and insoluble dispersed fines that exist in the oil–brine system. Certain oil-soluble organic acids such as naphthenic, fatty, and aromatic acids contribute to emulsification (*21*).

The interfacial films formed by different crude oils have different characteristics. The physical characteristics of the films are a function of the crude-oil type and gas content, the composition and pH of water, the temperature, the presence of nonionic polar molecules in the water, the extent to which the adsorbed film is compressed, and the contact time allowed for adsorption and concentration of polar molecules in the oil phase (*14, 22, 23*). The rheological properties of the adsorbed emulsifier film have an important effect on the stability of emulsions.

Very few studies have focused on the stability of O/W emulsions in porous media. Sarbar et al. (*24*) conducted a study to determine the effect of chemical additives on the stability of O/W emulsion flow through porous media. They injected 1, 5, and 10% O/W emulsions in sand packs with varying pH and surfactant concentrations and found that there was an optimal value of the surfactant concentration at which emulsions were the most stable. Addition of sodium chloride to the aqueous phase had a detrimental effect on the stability of the emulsion. For their system they found that there was an optimum value of pH (10) at which the emulsions were the most stable. Unstable emulsions flow in a markedly different manner from stable emulsions. Unstable emulsions have relatively large interfacial tensions resulting in large oil droplets that were observed to be deposited at the inlet of the core. This oil bank grows until it stalls and re-emulsification of the oil takes place. This re-emulsified oil is carried to another position, and another oil bank begins to form. This mechanism was not observed for the stable emulsion because there was less oil breaking from the emulsion to become available for the banking phenomenon.

Emulsion Quality. The quality of an emulsion is defined as the volume fraction (or percent) of the dispersed phase in the emulsion. The quality of emulsions strongly affects their rheology. Several studies have been reported for the relationship of isothermal shear stress to shear rate for emulsions of different qualities. O/W emulsions having qualities less than 0.5 (or 50%) exhibit Newtonian behavior, and those having higher qualities exhibit non-Newtonian behavior (9, 16, 25).

Figure 5 (25) shows the effect of emulsion quality on the pressure drop in porous media as a function of the flow rate. For emulsion of quality up to 40%, the relationship between pressure drop and flow rate is linear, an effect showing that Darcy's law describes the steady-state, laminar flow of a Newtonian macroemulsion through porous media. For emulsion qualities higher than 40%, the relationship between pressure drop and flow rate is not linear, a result indicating non-Newtonian behavior. This kind of behavior is typical of non-Newtonian pseudo-plastic fluids. Figure 6 shows a log–log plot of pressure drop versus flow rate for a 40 and 60% macroemulsion flowing through porous media. The 40% emulsion has a linear relationship on the plot with a slope of 1, and thus shows Newtonian behavior. The 60% macroemulsion, on the other hand, gives a slope of less than 1 and thus shows non-Newtonian, pseudo-plastic behavior.

Similar results were also obtained by Uzoigwe and Marsden (26). They conducted flow tests in capillary tubes and in porous media using different qualities of O/W emulsions and observed Newtonian behavior up to 50% qualities and non-Newtonian behavior at higher qualities, as shown in Figures 7 and 8. For the high-quality emulsions, Newtonian behavior was exhibited at low shear rates and shear stresses. Transition to non-Newtonian behavior occurred after some critical shear rates that were found to depend on emulsion quality. The higher the quality, the lower the value of this critical shear rate. The transition shear rate values reported suggest an exponential dependence on emulsion quality. After the transition to non-Newtonian behavior, these high-quality emulsions exhibited pseudo-plastic shear-thinning behavior.

Uzoigwe and Marsden (26) explained this behavior using particle–particle interactions that are caused by forces of attraction or repulsion between them. For dilute, low-quality emulsions, the repulsion forces are quite high

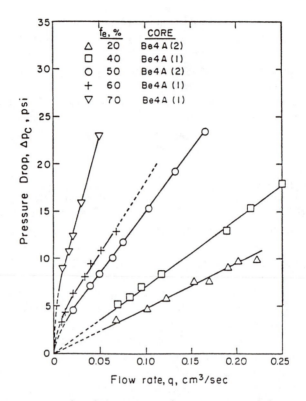

Figure 5. Cartesian plot of Newtonian and non-Newtonian behavior of flow of O/W macroemulsions through porous media. (f_e is emulsion quality.) (Reproduced with permission from reference 25. Copyright 1979 Society of Petroleum Engineers.)

relative to the attractive forces, and consequently minimum aggregation takes place. However, in concentrated or high-quality emulsions, the repulsive forces are reduced, and the attractive forces will lead to the formation of aggregates causing coalescence and flocculation. At low shear rates these aggregates rotate like single particles, and the viscosity is high. As shear is increased, the aggregates break down, and the viscosity decreases. The breakup of aggregates results in the observed pseudo-plastic behavior. On the other hand, for stable low-quality emulsions, the particles are far apart, and the net response of the emulsions to shear is similar to that of the Newtonian nature of the continuous phase.

Figure 9 from Uzoigwe and Marsden (26) shows a plot of apparent viscosity versus emulsion quality. This graph shows that apparent viscosities increase sharply with quality particularly at low shear rates. It also shows the Newtonian behavior at low qualities and non-Newtonian behavior at high qualities.

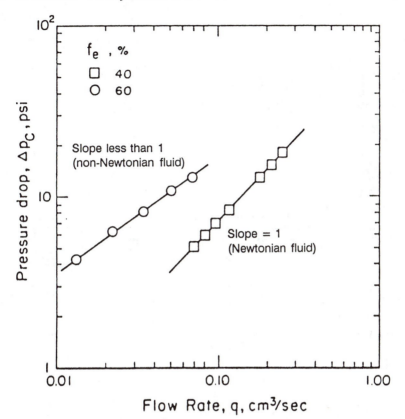

Figure 6. A log–log plot of Newtonian and non-Newtonian behavior of flow of macroemulsions through porous media. (f_e is emulsion quality.) (Reproduced with permission from reference 25. Copyright 1979 Society of Petroleum Engineers.)

When using stable, dilute Newtonian emulsions through porous media, the flowing permeability, k_f, must be used in Darcy's law to describe its behavior instead of the initial or conventional permeability. When plugging due to the flow of Newtonian macroemulsions occurs, only the permeability of the porous medium should be adjusted. Emulsion rheology with respect to Newtonian and non-Newtonian behavior will be reviewed under the section "Mathematical Models of Emulsion Flow in Porous Media".

Average Droplet Size and Droplet Size Distribution. All practical emulsions show some form of droplet size distribution with an average value representing this size distribution. The average droplet size and the droplet size distribution affect the rheology of emulsion (discussed in Chapter 4). Droplet size in relation to pore throat size affects the flow of fluids in porous media, as discussed previously (Figure 3).

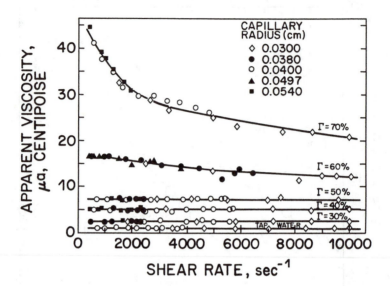

Figure 7. Apparent viscosity vs. shear rate at various emulsion qualities (Γ). (Reproduced with permission from reference 26. Copyright 1970 Society of Petroleum Engineers.)

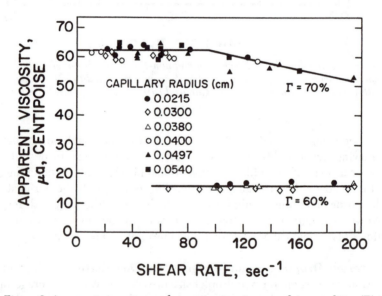

Figure 8. Apparent viscosity vs. shear rate at various emulsion qualities (Γ) in the low-shear apparatus. (Reproduced with permission from reference 26. Copyright 1970 Society of Petroleum Engineers.)

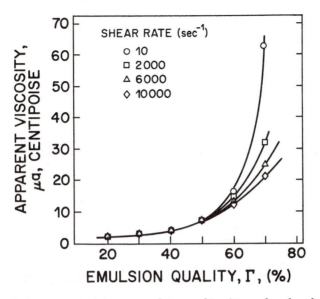

Figure 9. Apparent viscosity vs. emulsion quality. (Reproduced with permission from reference 26. Copyright 1970 Society of Petroleum Engineers.)

Droplet size depends on a number of factors such as the type of oil, brine composition, interfacial properties of the oil–water system, surface-active agents present (added or naturally occurring), flow velocity, and nature of porous material. For the study of O/W emulsions, McAuliffe (9) varied emulsion droplet sizes and size distributions by increasing the sodium hydroxide concentration in the aqueous phase, as shown in Figure 10. Higher NaOH concentration neutralizes more of the surface-active acids in the crude oil and produces an emulsion that has droplets of smaller diameters and is also more stable. Emulsion droplet size distribution can also be varied by varying the concentration of a surfactant added to the crude oil, as shown in Figure 11.

Soo and Radke (11) also studied the effect of average droplet size of emulsion on the flow behavior in porous media. The droplet size distribution of the emulsions that were prepared with surfactants and NaOH in a blender are shown in Figure 12. These droplet size distributions were found to be log-normal distributions. Others (9, 27) have also observed that the size of emulsion droplets was log-normally distributed. Soo and Radke (11) conducted experiments with emulsions having different average mean diameter in fine Ottawa water-wet sand packs. Their results of the reduced permeability, k/k_0, and reduced effluent volume concentration as a function of the pore volume of oil (in the emulsion) injected are shown in Figure 13. All emulsions were of 0.5% quality, and the initial permeability, k_0, was 1170 mD (millidarcies). The lines in the figure represent results of flow theory (12, 13) based on deep-bed filtration principles.

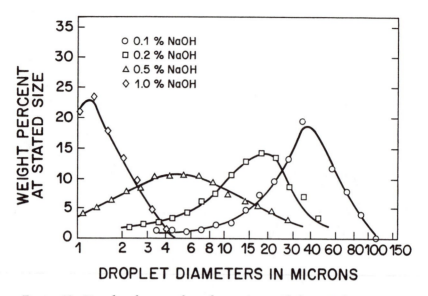

Figure 10. Size distribution of emulsions prepared from Midway–Sunset 26 C crude-oil and NaOH solutions. Oil content of emulsions was 70%. (Reproduced with permission from reference 9. Copyright 1973 Society of Petroleum Engineers.)

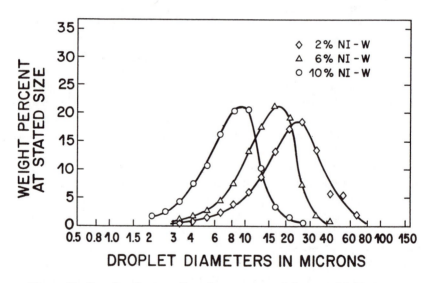

Figure 11. Size distribution of emulsions prepared from Richfield–Kraemer crude oil, injection water, and Chevron dispersant NI-W. Oil content of emulsion was 60%. (Reproduced with permission from reference 9. Copyright 1973 Society of Petroleum Engineers.)

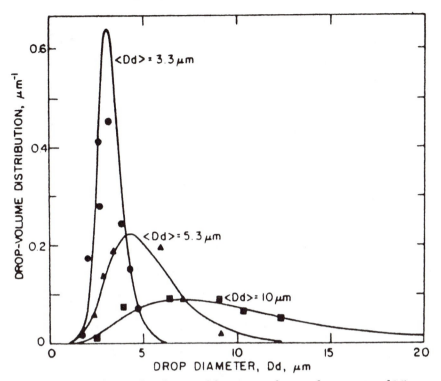

Figure 12. Droplet size distribution of the O/W emulsion with a viscosity of 1.5 mPa·s. (Reproduced from reference 11. Copyright 1984 American Chemical Society.)

Figure 13 shows that the core permeability drops over many injected pore volumes and ultimately levels off. The effluent profile (dashed lines) shows that the oil droplets appear after some time, and their concentration rises slowly and levels off at the inlet value of 0.5%. The time at which the permeability reduction stops is about the same time at which the oil droplets approach their inlet concentration.

Permeability reductions were also observed by McAuliffe (9), and his results are shown in Figure 14. He used a Boise sandstone core with an initial permeability of 1600 mD and injected a 0.5% O/W emulsion having average oil-droplet sizes of 1 and 12 μm. The small-diameter emulsion reduced the permeability from 1600 to 900 mD after 10 pore volumes of the injected emulsion; the 12-μm emulsion was much more effective in reducing the core permeability. After 10 pore volumes had been injected, the permeability was reduced to 30 mD, almost a 50-fold reduction.

Soo and Radke (*11*) also compared the effluent and inlet droplet size distributions. Their results for the 4.5-μm average droplet size emulsion

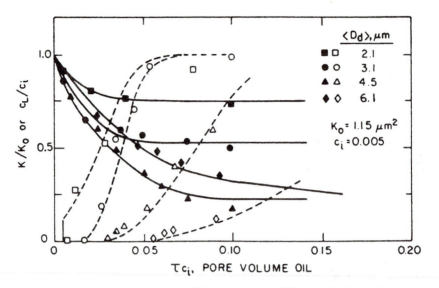

Figure 13. Experimental permeability reduction (filled symbols) and break-through concentration histories (open symbols) for varying droplet size in the 1170-mD core. Oil viscosity was 1.5 mPa·s; c_i is reduced effluent volume concentration, and D_d is droplet diameter. (Reproduced from reference 11. Copyright 1984 American Chemical Society.)

after 7 and 15 pore volumes are shown in Figure 15. Droplets eluting early were generally smaller than the injected distribution. However, with time, the effluent droplets shifted to larger diameters and ultimately matched the inlet droplet size distribution. This transient droplet size shift was not observed for the 2.1-μm emulsion and was less pronounced for the 3.1-μm emulsion. Figure 15 demonstrates little, if any, droplet coalescence in the porous medium. Similar results were also obtained by Alvarado and Marsden (27).

Figure 16 shows the results when 20 pore volumes of an emulsion having a 3.1-μm mean droplet size is injected into an 1170-mD sand pack and is followed by several pore volumes of water (11). After emulsion injection, a permeability reduction of about 50% is observed. With water injection, the effluent concentration drops to 0 after one pore volume, whereas the permeability is unaltered. For this dilute emulsion, the droplets are captured in the porous medium, and this capture leads to blocking of the flow paths. Figure 16 shows that once the droplets are captured, they do not re-enter the flow stream, velocity being constant. Soo and Radke (11) proposed the following physical interpretation for the results of Figure 15. Initially oil droplets are preferentially captured in the small-size pores, and as injection proceeds, more and more of the small pores become blocked. This blockage leads to a flow diversion toward larger size pores, and the rate

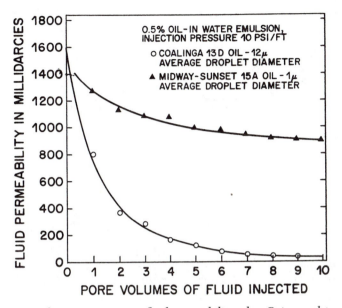

Figure 14. Changes in apparent fluid permeability when Boise sandstone was flooded with an emulsion of two different droplet sizes. (Reproduced with permission from reference 9. Copyright 1973 Society of Petroleum Engineers.)

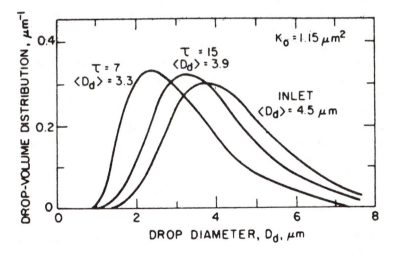

Figure 15. Comparison between inlet and effluent droplet size distribution after injection of 7 and 15 pore volumes. (Reproduced from reference 11. Copyright 1984 American Chemical Society.)

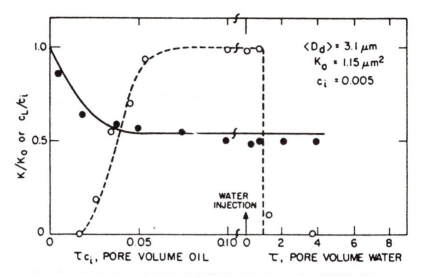

Figure 16. Experimental permeability reduction (filled symbols) and break-through concentration histories (open symbols) of a 20-pore-volume pulse followed by a pH 10 solution. (Reproduced from reference 11. Copyright 1984 American Chemical Society.)

of capture of the oil droplets decreases. Eventually, when the droplet capture sites are filled, capture ceases and steady state is reached. This interpretation is based on the concept of deep-bed filtration.

The results of Figure 13 suggest that as the droplet size increases, the emulsion retention increases. The large droplets have a higher capture probability and fill up more of the pores faster, a result that explains why they elute later than the smaller droplets. Emulsions with small droplet size diameters elute with essentially the inlet size distributions. Two factors control permeability reduction: the total volume of droplets retained and the effectiveness of these droplets in restricting flow. For a given porous medium, a critical mean droplet size of the emulsion controls permeability reduction. Below this value, retention of oil in porous media is dominant, and above the critical mean droplet size, their obstruction ability is pronounced. This situation explains the trends shown in Figure 13 for the effect of droplet size on permeability reduction. These conclusions are valid for stable, very dilute O/W emulsions and are based on a few experiments.

Oil Viscosity. The viscosity of O/W emulsion is generally dominated by the viscosity of the continuous phase, that is, water. As discussed earlier, emulsions of qualities up to 50% are Newtonian in behavior, whereas emulsions with higher qualities are generally non-Newtonian in nature. Table I from McAuliffe (9) shows the viscosities of O/W emulsions prepared from different crude oils varying in quality. The viscosities of O/W mixtures containing up to 50% oil are less than 20 times that of water, even though the

**Table I. Viscosities (at 75 °F) of O/W Emulsions
Prepared from Different Crude Oils**

Percent Oil in Emulsion	Midway–Sunset 15 A	Casmalia	Boscan
80	1200	1200–5000	—[a]
70	185	200	800
60	28	—	55
50	—	12	14
40	6	—	—
20	1.6	—	—
10	1.1	—	—

NOTE: All values are given in centipoises. The viscosities of the oils are as follows: Midway–Sunset 15 A, 3600 cP; Casmalia, 1,000,000 cP; and Boscan, 180,000 cP.
[a]The dash indicates not measured.
SOURCE: Reproduced with permission from reference 9. Copyright 1973 Society of Petroleum Engineers.

oil viscosities range from 3600 to 1,000,000 mPa·s at room conditions. For higher qualities, the O/W emulsions exhibit non-Newtonian flow behavior, and the viscosity (apparent) is dependent upon the shear rate. For very high qualities, the apparent viscosities of the emulsion (water-in-oil, W/O) can be substantially high (even higher than that of oil itself) at very low shear rates. Most of these high-quality emulsions exhibit shear-thinning or pseudo-plastic behavior (Figures 5–8).

Soo and Radke (*11*) demonstrated the role of viscosity of the oil phase on emulsion flow behavior. They compared results for injecting emulsions having similar droplet sizes (3.1 and 3.4 μm at 0.5% quality) with oil-phase viscosities of 1.5 and 23 mPa·s into a sand pack with a permeability of 1170 mD. The results are shown in Figure 17 for permeability reductions and effluent profiles as a function of pore volumes injected. Neither the permeability reduction nor the effluent concentrations changed significantly, whereas the oil viscosity increased by almost an order of magnitude. This result is not surprising, considering that the viscosity of emulsions in both cases should be very close to that of the water phase that makes up 99.5% of the emulsion. According to the authors, droplet sizes, pore sizes (discussed later), and surface chemistry are the most important factors for flow of emulsion in porous media, and viscosity of the oil phase does not affect this process. Although this condition is true for the dilute, stable emulsions that they studied, the viscosity of the oil phase will definitely affect the flow of higher quality emulsions in porous media. This aspect is discussed in more detail in the section "Mathematical Models of Emulsion Flow in Porous Media".

Effect of Porous Medium Characteristics.   The characteristics of porous media such as average pore size and wettability play an important role in the flow of O/W emulsion through porous media.

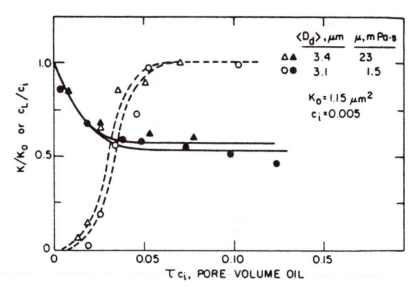

Figure 17. Experimental permeability reduction (filled symbols) and break-through concentration histories (open symbols) for two viscosities of oil phase with a 3-μm droplet size. (Reproduced from reference 11. Copyright 1984 American Chemical Society.)

Average Pore Size and Pore Size Distribution.   According to the concept of emulsion flow in porous media discussed earlier, capture of oil droplets in a porous medium is closely related to the pore size and the pore size distribution. Soo and Radke (*11*) studied the effect of two different sand packs with varying pore size distributions on emulsion flow. The average pore size and the pore size distribution of the two sand packs with permeabilities of 1170 and 580 mD are shown in Figure 18. These had average pore size diameters of 29.5 and 17.3 μm, respectively, and their porosities were 0.34 and 0.31, respectively. Figure 19 shows the effect of flowing a 0.5% O/W emulsion with a mean droplet size of 3.3 μm through these two sand packs. The emulsion breakthrough in the low-permeability sand pack occurred later than in the higher permeability sand pack and also resulted in a permeability reduction.

This observation was also made by McAuliffe (*9*), who injected a 0.5% O/W emulsion (3.8-μm average droplet size) into a Boise sandstone core (1170 mD) and Alhambra core (520 mD). The results are shown in Figure 20 and depict that the permeability of the Alhambra core was reduced more rapidly earlier during the injection period than that of the Boise core. The percentage reduction in permeability after 10 pore volumes of emulsion injection, however, was the same for the two cores. After 10 pore volumes of the O/W emulsion, distilled water was injected into the two cores. Distilled water, however, does not remove the oil droplets that are captured in the

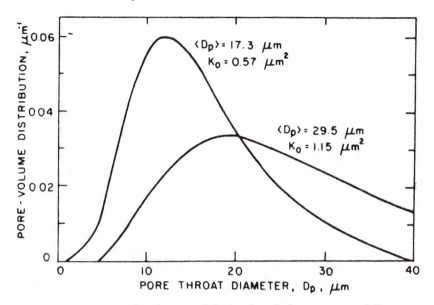

Figure 18. Pore size distribution of the sand-packed core at two different permeabilities. (Reproduced from reference 11. Copyright 1984 American Chemical Society.)

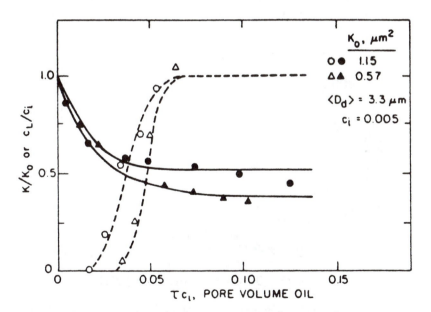

Figure 19. Experimental permeability reduction (filled symbols) and break-through concentration histories (open symbols) for two initial permeabilities with a 3.3-μm droplet size emulsion. (Reproduced from reference 11. Copyright 1984 American Chemical Society.)

pore throats and consequently does not improve the permeability reduction. Injection of brine into the core in the later part of the floods in Figure 20 failed to destroy the flow restrictions caused by the emulsion.

Wettability. Wettability of the porous medium controls the flow, location, and distribution of fluids inside a reservoir (7, 28). It directly affects capillary pressure, relative permeability, secondary and tertiary recovery performances, irreducible water saturations, residual oil saturations, and other properties.

The wettability of reservoir rocks can be altered by the adsorption of polar compounds or the deposition of organic material such as asphaltenes in the crude oil. Wettability alteration is determined by the interaction of oil constituents, mineral surface, and brine chemistry including ionic composition and pH. Any extraneous substance such as artificial surfactants that changes the mineral surface will change the wettability of the rock and consequently the flow of fluids inside the reservoir.

Clearly, wettability will affect the flow of O/W emulsions in porous media. Many surface-active compounds (which are normally needed for stable emulsions) will alter wettability, which will then affect the flow of the oil and water phases inside the reservoir. No studies have addressed the effect of wettability on the flow of emulsions in porous media. However, some effects of wettability appear to be obvious from simple intuitive reasoning. The nature of interactions between the internal surfaces of the

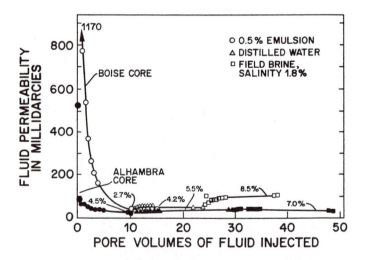

Figure 20. Reduction in water permeability by emulsion injection and residual effect of emulsion. Percentages at arrows compare fluid permeability at that point with original water permeabilities. (Reproduced with permission from reference 9. Copyright 1973 Society of Petroleum Engineers.)

porous medium and emulsion droplets would be affected by the wettability of the surface. If the surface is strongly water-wet, it can be perceived as being covered by a thin layer of strongly held water. Therefore, a direct collision between the solid surface and an emulsion droplet will require removal of this thin water layer and would be an unlikely event. Moreover, when an emulsion droplet does come in direct contact with the preferentially water-wet surface, it would be easily displaced from the surface by the surrounding water. On the other hand, if the surface was preferentially oil-wet, an oil droplet reaching the solid surface could easily adhere to the surface and may become trapped. Eventually, the oil-wet solid surface will become fully coated with oil, and this oil layer will grow in thickness by capturing more droplets until it becomes mobile. Thus the surface wettability may have a direct influence on the droplet capture mechanism.

Effect of Flow Velocity. The velocity of fluids in the reservoir plays an important role during the flow of emulsions in porous media. Velocity influences the shear rate, and for high-quality non-Newtonian emulsions, it directly affects their rheology. Its magnitude determines the capture and breakup rates of oil droplets through the rock. Soo and Radke (29) investigated the role of velocity and its effect on the flow of dilute, stable O/W emulsions through porous media. They used deep-bed filtration principles to describe its effect. According to this theory, the two mechanisms by which oil droplets are captured in the pores and thereby cause restriction to flow are straining capture and interception capture.

Straining capture refers to the condition that results when a particle lodges in a pore throat of size smaller than its own. The rate of capture is directly proportional to the velocity. Re-entrainment of strained droplets occurs either by squeezing of droplets through pore constrictions due to locally high pressures or by breakup of the oil droplets.

Interception capture refers to the situation where drops are captured on the surface of the rock, in vugs, and in recirculation eddies. Re-entrainment of the captured droplets can occur when a repulsive hydrodynamic force on the droplet is much larger than the van der Waals electrostatic attraction between the droplet and the rock surface. Droplet breakup will occur when interfacial tension is low and hydrodynamic forces are high.

These capture mechanisms are dependent on the critical re-entrainment velocity, which is the velocity at which droplets first break up from the rock surface and constrictions and are re-entrained into the flow stream. This critical velocity is a function of surface properties of the system and the droplet-size to pore-size ratio. Small double-layer repulsive forces and small droplet-size to pore-size ratios lead to large critical velocities. Soo and Radke (29) found the effect of velocity on emulsion flow in porous media to be dependent on the capillary number ($\mu v/\phi\sigma$, where μ is fluid viscosity, v is velocity, ϕ is porosity, and σ is surface tension). At low capillary numbers with low velocities and small droplet-size to pore-size ratios, droplets are captured in crevices and constrictions. They can also be captured by physical forces on the surface of the rock. For a larger droplet-size to pore-size ratio, droplets are captured in the pores. When velocity is close to the re-entrainment velocity, these captured droplets can squeeze through the pore constrictions and be re-entrained in the flowing stream. This re-entrainment leads to shear-thinning flows. Droplet breakup occurs at high capillary numbers (≥ 1).

Flow of Water-in-Oil Emulsions in Porous Media

The flow of W/O emulsions in porous media has received little attention in the literature. Only recently, with the recognition that it may be an important factor in controlling the oil–water ratio in production from steam-stimulated wells (*30*), the interest in this topic has increased. Theoretically, the flow of W/O emulsions in oil-wet media should be similar to the flow of O/W emulsions in water-wet porous media. Therefore, most of what was discussed in the preceding section can also be applied to the flow of W/O emulsions. By the same reasoning, the flow of W/O emulsions in water-wet media should be similar to the flow of O/W emulsions in oil-wet media. Unfortunately, not much information is available about flow of O/W emulsions in oil-wet media. What follows is mostly based on our own experimental work, which is still in its initial stage and has not been published.

Figure 21 shows a schematic diagram of the experimental equipment being used in our investigation of the flow behavior of W/O emulsions. Unconsolidated clean sand is held in a flow cell made from a high-strength plastic that is transparent to microwaves. Attenuation of a beam of millimeter microwaves passing through the cell is used to monitor the saturation of free water. Microwaves are absorbed by free water and are relatively unaffected by sand, the cell walls, emulsified water, and oil, and therefore the amount of microwave attenuation is directly proportional to the amount of free water in the core at the point of examination (*31–34*). As of mid-1991, only a single point on the cell is being monitored, but the apparatus will be modified so that the entire core may be scanned along its long axis.

Effect of Emulsion Characteristics.   As discussed in Chapter 4, the rheology of emulsions is affected by several factors, including the dispersed-phase volume fraction, droplet size distribution, viscosity of the continuous and dispersed phases, and the nature and amount of emulsifying surfactant present. All of these parameters would be expected to have some effect on flow behavior of the emulsion in porous media. However, the relationship between bulk rheological properties of an emulsion and its flow behavior in porous media is feeble at best because, in most cases, the volume

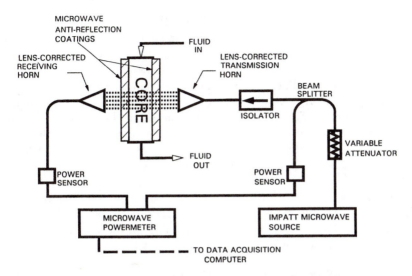

Figure 21. Schematic diagram of microwave apparatus for investigating W/O emulsions.

fraction of the dispersed phase and the droplet size distribution of a W/O emulsion flowing through a water-wet sand changes as a result of the flow. Part of the emulsified water drops out of the emulsion and flows as a bulk phase. The amount of free water present in the interior of the porous medium (away from the inlet) appears to be independent of the injected droplet size distribution but depends on the flow velocity, nature and amount of emulsifier present, the volume fraction of dispersed phase in the injected emulsion, and the viscosity of the oil. It has not been possible to measure the flowing droplet size distribution. However, the droplet size distribution present in the emulsion coming out of the production end of the flow cell appears to be independent of the injected droplet size distribution and varies strongly with the flow velocity. In fact, there was no significant difference in the droplet size distributions of the emulsion produced from the cell by injecting oil and water as separate bulk phases (which mixed only in the porous medium) and that produced by injecting an emulsion containing very fine droplets, provided all other factors remained the same.

Effect of Porous Medium Characteristics. Several characteristics of the porous medium, including the average pore size, pore size distribution, and the wettability of the porous medium, can influence the flow of W/O emulsions. Very little information is available on these issues. The role of wettability is intuitively obvious. A water-wet medium would be more conducive to capture of the water droplets at pore walls. This capture facilitates formation of a free-water phase. An oil-wet medium, on the other hand, would be a hindrance to formation of the free-water phase by droplet capture. A free-water phase can still form in an oil-wet medium during flow of a W/O emulsion by the mechanism of phase inversion when a change in the thermodynamic conditions is involved. However, in this discussion, the thermodynamic conditions are assumed to remain unchanged.

That the average droplet size of a W/O emulsion flowing through a porous medium under steady-state conditions should be related to the average pore size of the medium is intuitively apparent. However, the actual relationship, which depends on many other factors, such as the flow velocity, pressure gradient, pore size distribution, and oil–water interfacial properties, is not known. Our results show that for a given emulsion and porous medium, the average droplet size of the emulsion resulting from flow through the porous medium becomes smaller as the flow velocity is increased. No information is currently available on the effect of pore size distribution on the flow of W/O emulsions.

Effect of Flow Velocity. The flow velocity determines the shear rate and the pressure gradient. Therefore, the magnitude of a viscous force acting on a water droplet is directly related to flow velocity. This viscous force determines whether droplets can pass through pore throats smaller than themselves. It is also a factor in breakup of droplets into smaller droplets.

Simultaneous Flow of Emulsified and Bulk Dispersed Phase

As already noted, during the flow of a W/O emulsion in water-wet porous media, part of the water flows as a free phase; that is, it is continuous over macroscopic distances. Similarly, a part of the oil from an O/W emulsion flowing through an oil-wet medium is likely to flow as a bulk phase. Even in situations where the porous medium is not preferentially wetted by the dispersed phase, simultaneous bulk-phase flow of the dispersed phase may be involved in some situations of practical interest. Secondary recovery of oil from a water-wet reservoir by injection of a dilute O/W emulsion is an example. Schmidt et al. (35) studied this process and developed a simplified mathematical model to describe the flow behavior. Their theory is based on the assumption that, during simultaneous flow of bulk oil phase and a stable O/W emulsion in a water-wet porous medium, the bulk oil phase flows through a network of larger pores. Thus the emulsion flow is analogous to the single-phase situation, except that the larger pores are not available to the emulsion. This situation precludes any possibility of material exchange between the bulk oil phase and the emulsion.

In more realistic situations there is a certain probability of the emulsion droplets coalescing with the bulk oil phase or a part of the bulk oil becoming emulsified. The physics of such complex flow conditions is not well understood at present. The starting point of describing such a flow would be to treat it as a normal two-phase flow and use the concept of relative permeability and a model for the rheological properties of the emulsion phase. To account for the material exchange between the bulk phase and the emulsion phase, some form of droplet population balance model will be needed.

Mathematical Models of Emulsion Flow in Porous Media

The flow of emulsions in porous media is very complex, and to model it mathematically has been a challenge. It requires an understanding of the emulsion formation, its behavior, and its rheology inside the reservoir. Factors that affect the flow of emulsions through porous media were discussed earlier in this chapter, and the available mathematical models will be reviewed here.

Presently, three theories describe the flow of emulsions in porous media (*12*):

1. bulk viscosity or homogeneous models (*16, 25, 26*)

2. the droplet retardation model (*36, 37*)

3. the filtration model (*12, 13*)

Homogeneous Models.   The basic assumption in these models is that the emulsion is a continuum, single-phase liquid; that is, its microscopic features are unimportant in describing the physical properties or bulk flow characteristics. It ignores interactions between the droplets in the emulsions and the rock surface. The emulsion is considered to be a single-phase homogeneous fluid, and its flow in a porous medium is modeled by using well-documented concepts of Newtonian and non-Newtonian fluid flow in porous media (*26, 38*).

As discussed earlier, the rheology of emulsions depends on a number of factors, primary among which is the quality. Emulsions with qualities of less than 50% (oil) are considered Newtonian, whereas those having higher qualities exhibit non-Newtonian behavior.

Newtonian Emulsions.   For Newtonian emulsions, the viscosity is independent of the shear rate, and the simple Darcy's law is used for the flow of these emulsions in porous media. The viscosity of the emulsion, however, depends on several factors, such as the quality of the emulsion, viscosities of the dispersed and continuous phases, properties of the interfacial film, oil–water surface properties, droplet size, and droplet size distributions (16, 39). A number of correlations can be used to determine the viscosity of Newtonian emulsions (14, 16, 39) for use in Darcy's equation. The major drawback of this simple theory is that no permeability reduction is predicted, and that emulsion droplets must elute after one pore volume of the fluid injection. Experimental evidence (9, 11, 25) suggests that a permeability reduction takes place after injecting several pore volumes of dilute, stable emulsion into the porous medium. Also, in some cases the droplets in the effluent stream appear after several pore volumes of the injected emulsion.

Non-Newtonian Emulsions. Emulsions with qualities greater than 50% (oil) exhibit non-Newtonian flow behavior. For these non-Newtonian emulsions, the isothermal viscosity is a function of the shear rate. Darcy's law, which is applicable to the flow of Newtonian fluids in porous media, is no longer valid and has to be modified for describing non-Newtonian emulsion transport. Oil-field emulsions generally exhibit rheological behavior like that of pseudo-plastic fluids. Pseudo-plastic fluids show Newtonian behavior at low and very high shear rates and a shear-thinning behavior (apparent viscosity decreasing with shear rate) at intermediate shear rates (Figures 7–9). The rheological models for non-Newtonian fluids were discussed in Chapter 4. Many of the mathematical models that were developed for non-Newtonian polymer solutions (38) are used to describe emulsion flow.

Methods for predicting non-Newtonian flow in porous media can be divided into four categories (26, 39):

1. coupling of a particular model for a porous medium with a specific rheological model for the emulsion (26, 40)

2. generalized scale-up methods with a modified Darcy's law without a rheological model for the emulsion (41, 42)

3. dimensional analysis methods (43)

4. other methods based on viscoelastic and shear-thinning models (38)

The method that has received the most attention belongs to the first category. Specifically, the particular model is the capillary tube model for porous medium and the power law model for the emulsion (16). The shear-stress (τ) rate relationship for a power law fluid is given by

$$\tau = K\gamma^n \tag{11}$$

where K is the consistency factor, γ is the shear rate, and n is the behavior index. Both K and n can be obtained from capillary viscometry data. They are dependent on a number of factors such as the temperature, pressure, composition, surface-active components, droplet size, and quality (16). The popular capillary tube–power law model incorporates the following assumptions (16, 26, 38):

1. The emulsion is homogeneous on a macroscopic scale. However, its rheology parameters may be functions of other factors discussed.

2. The porous medium is represented by a tortuous capillary bundle model or by a capillary tube that has an equivalent hydraulic radius (44).

3. The average pore velocity is related to the flow velocity through the Dupuit–Forchheimer equations (44).

4. Permeability of the porous medium is not affected by the flow of emulsion through it. Alvarado and Marsden (25) and Ali and Abou-Kassem (16) account for permeability reduction that is observed by using the flowing permeability as a parameter.

5. The rheological behavior of the flowing fluid is independent of the geometry of the porous medium. Experimental evidence (25, 26, 38) suggest that the rheograms in both the viscometer and porous medium are parallel but do not coincide on a log–log plot; that is, the rheological parameter n is the same, but K may be different.

6. Flow is laminar, and viscoelastic effects are absent.

7. The rheological behavior of the flowing fluid is represented by a model that is valid in the range of shear rates to be encountered inside the porous medium.

The modified Darcy's law for non-Newtonian fluid through porous medium is

$$\overline{V}_c = \frac{k}{\mu_{\text{eff}}} \frac{\Delta P_c}{L_c} \tag{12}$$

where $\overline{V}_c$ is velocity through the capillary, L_c is the effective length of the capillary, and μ_{eff} is the effective viscosity of non-Newtonian fluid and is given by

$$\mu_{\text{eff}} = f\left[\frac{(\Delta P_c}{L_c}, k, \phi, K, n\right] \tag{13}$$

where ϕ is the porosity, or

$$\mu_{eff} = A\Psi \tag{14}$$

where Ψ is a functional form. The function A is specific to each model and is given in Table II. All the models reduce to the original Darcy's law for Newtonian fluids when $n = 1$, $K = \mu$, $\beta = 8$, and $F=1$ in Table II.

The homogeneous non-Newtonian capillary tube–power law model has a number of limitations. The models assume a power law relationship for the emulsion, and any deviations from this rheological behavior will lead to errors. The power law constants n and K are obtained by using viscometry, and their validity in porous media is questionable. No transient permeability reduction (assumption 4) is predicted, even though experimental evidence suggests otherwise. This model is seen to have validity only for high-quality emulsions that approach steady state quickly and have small droplet-size to pore-size ratios.

Droplet Retardation Model.   This model is based on the concept of droplet retardation during passage through pore constrictions encountered in a porous medium. When an emulsion droplet enters a pore constriction having a pore throat smaller than its own diameter (Figure 3), it deforms and squeezes through. During this process it experiences a capillary resistance force and as a result moves at a slower speed than the continuous phase and thereby causes a reduction in permeability. Steady state is reached when the emulsion breaks through the porous medium.

The mathematical formulation of this model was made by Devereux (36) based on the classical Buckley–Leverett theory for two-phase flow in porous media (49) and equations developed by Scheidegger (44). They solve a set of eight equations:

- Darcy equations for the oil and water phases
- continuity equations for the oil and water phases
- equations of state of the form $\rho_i = \rho_i(P_i)$ for the oil and water phases
- a statement of constant pore volume
- a definition of capillary resistance due to the retardation of oil droplets through pore constrictions.

In the original Buckley–Leverett theory, gravitational, compressibility and capillarity are ignored. Devereux (36) presents the solution for the case of constant pressure, and the constant-velocity case was derived by Soo and Radke (12). The model requires a knowledge of the capillary retarding force per unit volume of the porous medium, and the relative permeabilities of the oil droplets in the emulsion and the continuous water phase. These relative permeabilities are assumed to be functions of the oil saturation in the porous medium. These must be determined before the model can be used.

Table II. The Form of Function A for All Models

Model	Year	Porous Medium[a]	Parameters	Function A	Reported Error (%)
Christopher and Middleman (*45*)	1965	UC	—[b]	$(1/12)[9 + (3/n)]$	18
Gregory and Grisky (*46*)	1967	UC	—	1	15
Yu, Wein, and Bailie (*47*)	1968	UC	—	$\dfrac{(1/12)[9 + (3/n)]}{3^{\frac{1-n}{n}}\,\phi^{\frac{n+1}{n}}}$	11
Uzoigwe and Marsden (*26*)	1970	UC	C''	$\left[\dfrac{150}{C''}\right]^{n+1/2n}$	14–18
Kemblowski and Mertl (*48*)	1974	UC	—	$(1/12)[9 + (3/n)]$	—
Alvarado and Marsden (*25*)	1979	C	$F,\ \beta$	$4\left[\dfrac{2}{\beta}\right]^{\frac{1}{n}}\dfrac{\left[\dfrac{24}{25}\right]^{\frac{1-n}{2n}}}{F^{\frac{n+1}{2n}}}$	4
Ali and Abou-Kassem (*16*)	1990	UC + C	γ	$2\left[\dfrac{2}{\gamma}\right]^{\frac{n+1}{n}}\left[\dfrac{24}{25}\right]^{\frac{1-n}{2n}}$	2

NOTE: Symbols are defined in the List of Symbols.
[a]C is consolidated porous media; UC is unconsolidated porous media.
[b]The dash means no parameter.
SOURCE: Reproduced with permission from reference 16. Copyright 1990 Canadian Institute of Mining, Metallurgy, and Petroleum.

The model correctly describes the permeability reduction as a function of pore volume injected and takes into account the effect of emulsion droplet saturation and droplet-size to pore-size ratios. The main drawbacks of this theory are that the permeability reduction is caused as long as the emulsion is flowing and that the initial permeability is restored once the emulsion injection is followed by water alone. In other words, the emulsion droplets all pass through the porous medium, and none of them is captured inside. However, experimental evidence (9, 11) suggests that the permeability reduction cannot be restored after subsequent water injection (Figure 16).

Filtration Model.   A model based on deep-bed filtration principles was proposed by Soo and Radke (12), who suggested that the emulsion droplets are not only retarded, but they are also captured in the pore constrictions. These droplets are captured in the porous medium by two types of capture mechanisms: straining and interception. These were discussed earlier and are shown schematically in Figure 22. Straining capture occurs when an emulsion droplet gets trapped in a pore constriction of size smaller than its own diameter. Emulsion droplets can also attach themselves onto the rock surface and pore walls due to van der Waals, electrical, gravitational, and hydrodynamic forces. This mode of capture is denoted as interception. Capture of emulsion droplets reduces the effective pore diameter, diverts flow to the larger pores, and thereby effectively reduces permeability.

This mechanism is similar to that of a deep-bed filtration process with some differences (12). In the filtration process the particle-size to pore-size ratio is small, and the particles are mostly captured on the media surface. Thus interceptive capture dominates, and this capture does not alter the flow distribution in the porous medium. Permeability reduction is not significant and is ignored. On the other hand, the emulsion droplet size is generally of the same order of the pore size, and the droplets are captured both by straining and interception. This capture blocks pores and results in flow redistribution and a reduced permeability.

Re-entrainment of liquid droplets that are captured can also occur as a result of squeezing when the local pressure drop is increased to overcome the capillary resistance force. The shape of the liquid droplets depends on the wettability of the rock. On the basis of this physical picture, Soo and Radke (12) proposed a model to describe the flow of dilute, stable emulsion flow in a porous medium. The flow redistribution phenomenon and permeability reduction are included in the model. Both low and high interfacial tension were considered.

Soo and Radke (12) found that emulsion flow in a porous medium is characterized by three parameters: a filter coefficient, an interpore flow redistribution factor, and a local flow restriction factor. The filter coefficient

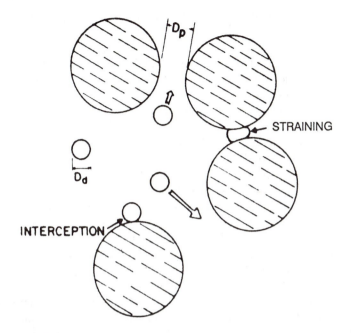

Figure 22. The two types of capture mechanisms of emulsion flow in porous media. (Reproduced with permission from reference 12. Copyright 1986 Pergamon Press PLC.)

controls the sharpness of the emulsion front, the flow redistribution parameter describes the redistribution phenomenon as well as the steady-state retention of emulsion droplets, and the flow restriction parameter addresses the effectiveness of the retained droplets in reducing permeability. A comparison between the homogeneous viscosity model, the droplet retardation model, and the filtration model was also made.

In Situ Emulsification in Porous Media

So far we have looked at the flow of emulsions in porous media; in this section we discuss some aspects of in situ emulsification in porous media that have received little attention. Some evidence suggests strongly that emulsions can be produced in the reservoir rock itself. A discussion on the formation of oil-field emulsions was given by Berkman and Egloff (50). They concluded that emulsions could be formed within the porous rock near the well bore where the velocity gradients (i.e., shear rates) were very high. Emulsions could also be formed as a result of mechanical agitation, for example, gas movement, fluid flow through constrictions, pumps, pipes, and chokes. Emulsions within the porous medium are formed as a result of

- the presence of surface-active agents, either native or externally added as, for example, during chemical flooding (discussed in Chapter 7)
- shear by the movement of fluids through pore throats

Theoretical aspects of emulsion formation in porous media were addressed by Raghavan and Marsden (51–53). They considered the stability of immiscible liquids in porous media under the action of viscous and surface forces and concluded that interfacial tension and viscosity ratio of the immiscible liquids played a dominant role in the emulsification of these liquids in porous media. A mechanism was proposed whereby the disruption of the bulk interface between the two liquids led to the initial formation of the dispersed phase. The analysis is based on the classical Raleigh–Taylor and Kelvin–Helmholz instabilities.

Evidence of in situ emulsification in porous media was recently provided by Vittortos (30) and Chen et al. (54). Field samples were taken with a high-pressure, low-rate sampler to preserve the bottom-hole emulsion produced during a cyclic steam stimulation project. Their data suggest that during the production of oil and water in the field, part of the produced water flows as a single phase, and part of the water will flow as an emulsion with the oil. This conclusion is in contrast to the traditional picture of the flow of water and oil in which the two phases are considered to be flowing separately. The implication of emulsion flow in porous media is that the existing two-phase relative permeability concept commonly used to describe immiscible displacement should be modified to allow for the mixing of the two phases to flow as an emulsion.

Concluding Remarks

A critical review of emulsion flow in porous media has been presented. An attempt has been made to identify the various factors that affect the flow of O/W and W/O emulsions in the reservoir. The present methods of investigation are only the beginning of an effort to try to develop an understanding of the transport behavior of emulsions in porous media. The work toward this end has been difficult because of the complex nature of emulsions themselves and their flow in a complex medium. Presently there are only qualitative descriptions and hypotheses available as to the mechanisms involved. A comprehensive model that would describe the transport phenomenon of emulsions in porous media should take into account emulsion and porous medium characteristics, hydrodynamics, as well as the complex fluid–rock interactions. To implement such a study will require a number of experimental and theoretical measurements. These experiments will consist of the following:

1. the effect of emulsion characteristics, for example, stability, quality, droplet size distribution, and rheology, on its flow in porous media

2. the effect of porous medium characteristics, for example, average pore size, pore size distribution, wettability, porosity, permeability, specific surface area, and chemical composition

3. emulsion–rock interactions, simultaneous flow of emulsion, and bulk dispersed phases

4. theoretical analysis taking into account all of the aforementioned

These experiments will help in a more fundamental understanding of the flow of emulsions in porous media.

List of Symbols

A	cross-sectional area of medium
C''	parameter (Table II)
dP/dL	pressure gradient
f_e	emulsion quality
F	parameter, $F = k_f/k$ (Table II)
k	permeability
K	consistency index
ΔL	length
L_c	effective length of the capillary
n	behavior index
P	pressure
ΔP	pressure drop
q	flow rate
r	radius of pore
r_1	radius of curvature at the trailing edge of the drop
r_2	radius of curvature at the leading edge of the drop
S	saturation
v	velocity
V_c	velocity through the capillary
x, y, z	directions

Greek
β	parameter (Table II)
γ	shear rate

Γ	emulsion quality
θ	contact angle
μ	viscosity of fluid
ρ	density
σ	surface tension
τ	shear stress
φ	porosity

Subscripts
0	initial
1	trailing edge index
2	leading edge index
c	capillary
eff	effective
f	flowing
i	specific fluid index
nw	nonwetting
r	relative
w	wetting
x	x direction
y	y direction
z	z direction

References

1. Craft, B. C.; Hawkins, M. F. *Applied Petroleum Reservoir Engineering;* Prentice Hall: Englewood Cliffs, NJ, 1959; p 259.
2. Yuan, H. H. In *Proceedings of the 64th Annual Technical Conference and Exhibition of the Society of Petroleum Engineers;* Society of Petroleum Engineers: Richardson, TX, 1989; paper no. 19617.
3. Bear, J. C. *Dynamics of Fluids in Porous Media;* American Elsevier: New York, 1972.
4. Nielsen, R. F. In *Petroleum Production Handbook;* Frick, T. C., Ed.; Society of Petroleum Engineers and American Institute of Mining, Metallurgical, and Petroleum Engineers: Dallas, TX, 1962.
5. Collins, R. E. *Flow of Fluids Through Porous Materials;* Reinhold: New York, 1961.
6. Richardson, J. G. In *Handbook of Fluid Dynamics;* Streeter, V. L., Ed.; McGraw-Hill: New York, 1961.
7. Craig, F. F. *The Reservoir Engineering Aspect of Waterflooding;* Society of Petroleum Engineers: Richardson, TX, 1971.
8. Mohanty, K. K.; Davis, H. T.; Scriven, L. E. *SPE Reservoir Eng.* **1987,** 113–128.
9. McAuliffe, C. D. *J. Pet. Technol.* **1973,** 727.
10. Decker, R. M.; Flock, D. L. *J. Can. Pet. Technol.* **1988,** 69.
11. Soo, H.; Radke, C. J. *Ind. Eng. Chem. Fundam.* **1984,** 23, 342–347.
12. Soo, H.; Radke, C. J. *Chem. Eng. Sci.* **1986,** 41(2), 263–272.
13. Soo, H.; Williams, M. C.; Radke, C. J. *Chem. Eng. Sci.* **1986,** 41(2), 273–281.

14. Coskuner, G. "Oil Field Emulsions"; Report No. 1988-18; Petroleum Recovery Institute: Calgary, Alberta, Canada, 1988.
15. Tadros, T. F.; Vincent, B. In *Encyclopedia of Emulsion Technology;* Becher, P., Ed.; Dekker: New York, 1983.
16. Ali, F. S. M.; Abou-Kassem, J. H. In *Proceedings of the Petroleum Society of the Canadian Institute of Mining and Metallurgy and the Society of Petroleum Engineering;* Society of Petroleum Engineers: Richardson, TX, 1990; paper no. CIM/SPE 90-97.
17. Egbogah, E. O., M.Sc. Thesis, University of Alberta, Edmonton, Alberta, Canada, 1975.
18. Hassan, M. E.; Nielsen, R. F.; Calhoun, J. C. *Am. Inst. Mech. Eng. Trans.* **1953**, *198*, 299–306.
19. Becher, P. *Emulsions: Theory and Practice;* Reinhold: New York, 1965.
20. Strassner, J. E. *J. Pet. Technol.* **1968**, 303–312.
21. Peak, E.; Hodgson, A. E. *J. Am. Oil Chem. Soc.* **1966**, *43*, 215–222.
22. Kimbler, O. K.; Reed, R. L.; Silberberg, I. H. *Soc. Pet. Eng. J.* **1966**, 153.
23. Jones, T. J.; Neustadter, E. L.; Wittingham, K. P. *J. Can. Pet. Technol.* **1978**, 100.
24. Sarbar, M.; Livesey, D. B.; Wee, W.; Flock, D. L. In *Proceedings of the Annual Technical Meeting of the Petroleum Society of the Canadian Institute of Mining and Metallurgy;* Canadian Institute of Mining and Metallurgy: Calgary, Alberta, Canada, 1987, paper no. CIM 87-35-25.
25. Alvarado, D. A.; Marsden, S. S. *Soc. Pet. Eng. J.* **1979**, 369–377.
26. Uzoigwe, A. C.; Marsden, S. S. In *Proceedings of the 45th Annual Fall Meeting of the Society of Petroleum Engineers;* Society of Petroleum Engineers: Richardson, TX, 1970, paper no. 3004.
27. Alvarado, D. A.; Marsden, S. S. In *Proceedings of the 2nd International Symposium on Heavy Crude Oils and Tar Sands;* Maracaibo, Venezuela, 1987, paper no. 275.
28. Anderson, W. G. *J. Pet. Technol.* **1986**, 1125–1144.
29. Soo, H.; Radke, C. J. *J. Coll. Interface Sci.* **1984**, *102(2)*, 462–476.
30. Vittortos, E. In *Proceedings of the Joint International Technical Meeting of the Petroleum Society of the Canadian Institute of Mining and Metallurgy and the Society of Petroleum Engineers;* Society of Petroleum Engineers: Richardson, TX, 1990, paper no. CIM/SPE 90-107.
31. Aggarwal, S.; Jonston, R. *IEEE Trans. Instrum. Meas.* **1985**, *IM-34(1)*, 21–25.
32. Hoekstra, P.; Cappillino, P. "U.S. Army Cold Regions Research and Engineering Laboratory Report", July 1971.
33. Thomas, C.; Perl, J. P.; Wasan, D. T. *J. Coll. Interface Sci.* **1990**, *139(1)*, 1–13.
34. Sarma, H., Ph.D. Thesis, University of Alberta, Edmonton, Alberta, Canada, 1988.
35. Schmidt, D. P.; Soo, H.; Radke, C. J. *Soc. Pet. Eng. J.* **1984**, 351–360.
36. Devereux, O. F. *Chem. Eng. J.* **1974**, *7*,121–128.
37. Devereux, O. F. *Chem. Eng. J.* **1974**, *6*, 129–136.
38. Savins, J. G. *Ind. Eng. Chem.* **1967**, *61*, 18–47.
39. Sherman, J. G. In *Encyclopedia of Emulsion Technology;* Becher, P., Ed.; Dekker: New York, 1983.
40. Sadowski, T. J.; Bird, R. B. *Soc. Rheol. Trans.* **1965**, *9(2)*, 243–250.
41. McKinley, R. M.; Jahns, H. O.; Harris, H. W.; Greenkorn, R. A. *AIChE J.* **1966**, *12(1)*, 17–20.
42. Kozicki, W.; Hsu, C. J.; Tiu, C. *Chem. Eng. Sci.* **1967**, 487–502.
43. Slattery, J. C. *AIChE J.* **1967**, *13(6)*, 1066–1071.
44. Scheidegger, A. E. *The Physics of Flow Through Porous Media*, 3rd ed.; University of Toronto: Toronto, Ontario, Canada, 1974.

45. Christopher, R. H.; Middleman, S. *Ind. Eng. Chem. Fundam.* **1965,** *4(4),* 422–426.
46. Gregory, D. R.; Grisky, R. G. *AIChE J.* **1967,** *13(1),* 122–125.
47. Yu, Y. H.; Wen, C. Y.; Bailie, R. C. *Can. Chem. Eng. J.* **1968,** *46,* 149–154.
48. Kemblowski, Z.; Mertl, J. *Chem. Eng. Sci.* **1974,** *29,* 213–223.
49. Bukley, W. E.; Leverett, M. C. *Trans. Am. Inst. Min. Met. Eng.* **1942,** *146,* 107–116.
50. Berkman, S.; Egloff, G. *Emulsions and Foams;* Reinhold: New York, 1941.
51. Raghavan, R.; Marsden, S. S. *Soc. Pet. Eng. J.* **1971,** 153–161.
52. Raghavan, R.; Marsden, S. S. *Q. J. Mech. Appl. Math.* **1973,** *26(2),* 205–216.
53. Raghavan, R.; Marsden, S. S. *J. Fluid Mech.* **1971,** *48(1),* 143–159.
54. Chen, T.; Chakrabarty, T.; Cullen, M. P.; Thomas, R. R.; Sieben, M. C. In *Proceedings of the Joint CIM/AOSTRA Annual Meeting;* Canadian Institute of Mining and Metallurgy: Calgary, Alberta, Canada, 1991, paper no. CIM/AOSTRA 91-78.

RECEIVED for review December 18, 1990. ACCEPTED revised manuscript May 8, 1991.

Emulsions in Enhanced Oil Recovery

Kevin C. Taylor and Blaine F. Hawkins

Petroleum Recovery Institute, 3512 33rd Street N.W., Calgary, Alberta, Canada T2L 2A6

Micellar–polymer flooding and alkali–surfactant–polymer (ASP) flooding are discussed in terms of emulsion behavior and interfacial properties. Oil entrapment mechanisms are reviewed, followed by the role of capillary number in oil mobilization. Principles of micellar–polymer flooding such as phase behavior, solubilization parameter, salinity requirement diagrams, and process design are used to introduce the ASP process. The improvements in "classical" alkaline flooding that have resulted in the ASP process are discussed. The ASP process is then further examined by discussion of surfactant mixing rules, phase behavior, and dynamic interfacial tension.

E MULSIONS ARE OF GREAT IMPORTANCE in enhanced oil recovery (EOR) techniques. In some cases, emulsions may be an unwelcome consequence of the process, but in other cases, the use of emulsions is critical and fundamental to the oil recovery process. In general, processes that rely on the injection of surfactants or surfactant-forming materials into a reservoir rely heavily on emulsion technology. Micellar–polymer flooding and alkali–surfactant–polymer flooding are two examples in which emulsion technology specific to the process has evolved to meet special needs. In these processes it is necessary to understand the behavior of an emulsion as it is injected into or formed in a reservoir, as it travels through that reservoir over a period of weeks or months, and as it flows out of the reservoir through a producing well. This chapter discusses the basics required for an appreciation of these processes.

Throughout this chapter, microemulsions will be treated as a type of emulsion, even though there are fundamental differences between the two. Microemulsions are thermodynamically stable and will not segregate with

0065–2393/92/0231–0263 $08.75/0

time. In contrast, emulsions will eventually separate with time and owe their stability to processes that delay the approach to equilibrium.

The progression from micellar–polymer flooding to alkali–surfactant–polymer (ASP) flooding has special significance in this chapter. Micellar–polymer flooding is technically well-developed, relatively well-understood, and has undergone numerous technically successful field trials. However, this process is inherently expensive because of the relatively large surfactant concentrations that must be injected into the reservoir. Alkali–surfactant–polymer flooding is a much newer technology, is more complex, and is not technically well-developed. Many of the lessons learned from micellar–polymer flooding can be applied to the ASP process. Alkali–surfactant–polymer flooding is inherently much less expensive than the micellar–polymer process, primarily because the surfactant concentration is significantly lower. Field trials are in progress, although many of the details remain confidential. This technology is at the stage that the micellar–polymer process was in during the early 1970s. As more is learned, this process may come into much more widespread use.

Oil Entrapment and Mobilization in Porous Media

Oil Entrapment Mechanisms.   Enhanced oil recovery processes depend in large part on the elimination or reduction of capillary forces. Capillary forces are the strongest that occur under typical reservoir conditions, and are most responsible for oil entrapment. Viscous forces, which act to displace oil, are composed of the applied pressure gradient, gravity, density differences between phases, and viscosity ratio. In a permeable medium, capillary forces result when the pores constrain the oil–water interface to a high degree of curvature. From the Laplace equation, the capillary pressure P_c in a capillary tube can be derived:

$$P_c \equiv P_o - P_w = \frac{2\,\sigma\,\cos\theta}{r} \tag{1}$$

where P is the pressure in a fluid phase (subscripts o and w for oil or water, respectively), σ is the interfacial tension, r is the radius of curvature of the interface, and θ is the contact angle, which is the angle that the oil–water interface makes with the solid surface, as measured through the water phase. When the contact angle is zero, the surface is said to be highly water-wet. If the contact angle is 180°, then oil completely wets the surface and it is oil-wet. The preference of the surface for oil or water is its wettability. In crude-oil reservoirs, wettability varies from relatively oil-wet in carbonate reservoirs to relatively water-wet in sandstone reservoirs, with many exceptions.

Although the capillary tube is a simple representation of fluid in a pore, several valid comments can be made. The capillary pressure increases as pore diameter decreases, or as interfacial tension (IFT) or contact angle increases. Capillary forces are therefore influenced by pore geometry, interfacial tension, and surface wettability.

When an aqueous phase flows through a porous medium containing oil, some oil will be readily displaced, but capillary forces will act to trap oil in some of the pore spaces. No matter how much aqueous phase flows through the material, a certain amount of oil, called residual oil, will remain trapped. The residual oil saturation is generally expressed as a percentage of the original oil in place (%OOIP), and can be greater than 40%. This oil is the target of many enhanced oil recovery techniques.

Two models help to explain the mechanisms by which oil is entrapped in porous media: the pore doublet and the snap-off models.

The pore doublet model was introduced by Moore and Slobod in 1956 (*1*), and has been critically reviewed by Chatzis and Dullien (*2*). In this model, two capillaries of different diameters are connected at the inlet and outlet ends to create an idealized model of a pore structure. Initially, the doublet is full of oil, and water with viscosity the same as that of the oil flows in from one end (Figure 1a). As the water continues to flow, three outcomes are possible:

1. Oil could be completely displaced from both capillaries.

2. Some oil can remain in the smaller capillary.

3. Some oil can remain in the larger capillary.

For the first outcome to occur, viscous and capillary forces would have to be equal, which is unlikely under realistic reservoir conditions. More likely, viscous forces will be small relative to capillary forces under normal flow conditions in a reservoir. In this case, under water-wet conditions, capillary pressure is much greater in the capillary with a small diameter, so the water moves more rapidly, trapping the oil in the larger capillary (Figure 1b). Qualitatively, the model predicts that the nonwetting phase will be trapped in large pores, while the wetting phase will be trapped in small cracks and crevices. As capillary forces are lowered, a decrease in trapping will result in the water-wet case, because the relative velocity of the water in the small capillary will decrease relative to that in the large capillary. The pore doublet model is a simple one that tends to greatly overestimate residual oil saturation in porous media. However, it does show how a nonwetting phase can become trapped in water-wet porous media.

The snap-off model has been detailed with experimental work by Chatzis et al. (*4*). In this model, oil initially fills a series of connected pore

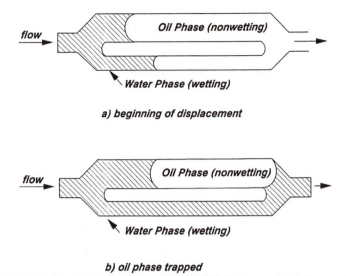

Figure 1. Pore doublet model. (Reproduced with permission from reference 3. Copyright 1989 Prentice Hall.)

bodies (Figure 2). The surface of each pore is water-wet, however, and is coated with a thin film of water. Capillary pressure varies along the flow path, being highest at the pore throat constrictions. As water flows into one end of the pore series, some oil will be displaced. But if the capillary pressure becomes high enough, the oil phase will snap off into globules within the pores in the flow path. Snap-off begins to occur as the ratio of pore diameter to pore throat diameter (aspect ratio) exceeds a critical value. In porous media the pore structure is much more complex than in the simple snap-off model, displaying a range of pore sizes and aspect ratios. In Berea sandstone, about 80% of the trapped nonwetting phase occurs in the snap-off geometry, and the remaining approximately 20% occurs in pore doublets or combinations of the two (4). The nonwetting residual oil becomes trapped in larger pores in globules that can be several pore diameters long.

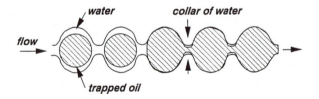

Figure 2. Pore snap-off model. (Reproduced with permission from reference 3. Copyright 1983 Society of Petroleum Engineers.)

Capillary Number in Oil Mobilization. The capillary number N_c is a dimensionless ratio of viscous to capillary forces; it provides a measure of how strongly trapped residual oil is within a given porous medium (5). Various definitions have been used for capillary number, but the following equation is common:

$$N_c = \frac{\upsilon\mu}{\sigma} \qquad (2)$$

where υ is Darcy velocity, μ is the viscosity of the displacing phase, and σ is the interfacial tension between the displaced and displacing phases. The Darcy velocity is expressed in units of distance over time and is obtained by dividing the flow rate into a porous media by the cross-sectional area through which flow occurs. Figure 3 shows several capillary number curves obtained from the literature for water-wet Berea sandstone. The shape of the curve is affected by wettability and pore size distribution. For the oil phase, mobilization of residual oil usually begins at a capillary number of about 10^{-5} (the critical capillary number), and complete oil recovery occurs at high values of N_c of about 10^{-2}. For a variety of water-wet sandstones, Chatzis and Morrow (6) found a critical capillary number of 10^{-5} and complete recovery of oil at N_c of 10^{-3}.

The capillary number for flow in a typical oil reservoir undergoing water-flooding can easily be calculated. At a flow rate of about 0.26 m/day (3 × 10^{-6} m/s), an oil–water interfacial tension of 30 mN/m, and water viscosity

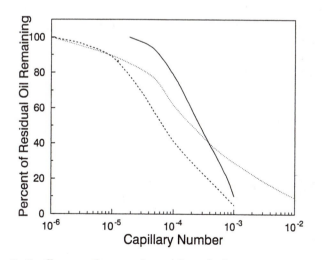

Figure 3. Capillary number correlations from the literature. Key: . . ., Gupta and Trushenski (34); - - -, Taber (5); and —, Chatzis and Morrow (6).

of 1 mPa·s, the capillary number is 1×10^{-7}. This number is 2 orders of magnitude below the critical value required to begin mobilization of residual oil. To recover all oil, the capillary number must be increased by a factor of 10^4. This increase could be done by decreasing the IFT to 3×10^{-3} mN/m; such a decrease is possible in some surfactant–oil systems. An increase in viscosity by 4 orders of magnitude is not feasible because of well-bore injectivity problems, although an increase of 1 order of magnitude combined with a decrease in IFT by 3 orders of magnitude would suffice. An increase of the flow rate by several orders of magnitude is impractical because the resulting pressure would fracture the oil-bearing formation. Also, injection wells usually do not have significant unused injection capacity. So, to greatly increase capillary number, interfacial tension can be decreased by a large amount, and displacing phase viscosity can be increased moderately.

Interfacial Parameters Important in Enhanced Oil Recovery

Interfacial Tension. As already seen, interfacial tension is of fundamental importance in determining the capillary forces acting on trapped oil within porous media. Interfacial tension arises from an imbalance in the forces of attraction between molecules in the bulk phase and molecules at the interface. It can be defined as the isothermal reversible work of formation of unit area of interface in a system of constant composition (7).

If interfacial tension between two phases becomes zero, then the two phases become miscible. This result is the ultimate aim of many types of EOR: to make oil–water interfacial tension equal to 0, so that a displacing fluid can miscibly displace oil trapped in the porous medium. In practice, it is difficult to make interfacial tension approach 0 for liquids of such different characteristics as oil and water.

Very low values of IFT are usually measured with the spinning drop method made practical by Cayias et al. (8), because this method is the most versatile method for taking measurements at values of 0.1 mN/m and lower. Classical methods such as the capillary rise, du Nouy ring, Wilhelmy plate, and drop weight (and volume) methods are generally used for interfacial tension measurements of greater than 1 mN/m. The spinning drop method also has the advantage of being able to measure dynamic changes in IFT. The only other potentially useful method to date is laser light scattering, which is largely unproven for crude-oil–water interfacial tension measurement (9).

In the spinning drop method, an oil droplet of approximately 1 μL is placed in a glass capillary (2 mm i.d.) full of aqueous solution. The glass tube is sealed and spun horizontally around its axis at 5000 to 10,000 rpm. Centrifugal force will cause the lighter phase (usually the oil) to form an elongated drop in the center of the glass capillary tube. The shape of this drop is the result of an equilibrium between rotational forces acting to elongate the drop and contraction by interfacial tension acting to minimize surface area. Providing that the drop is quite long relative to its width, interfacial tension can be calculated with Vonnegut's formula (*10*):

$$\sigma = \frac{\Delta\rho\omega^2 R^3}{4} \qquad (3)$$

where $\Delta\rho$ is the density difference between the two phases, ω is the angular velocity, and R is half of the drop width. When the drop is not long relative to its width, a more complicated treatment is required (*8*).

Interfacial tension is generally at a minimum when the interfacial concentration of adsorbed surfactant material is at a maximum. Possibly, maximum adsorption at the interface occurs when the surfactant is equally soluble in the oil and aqueous phases. Shah et al. (*11*) showed that the IFT between an oil and an aqueous surfactant-containing phase is at a minimum when the partition coefficient is unity. This condition occurs when the surfactant is equally soluble in both phases.

Interfacial Viscosity. In a clean system in which two pure liquids produce an interface, the viscosity of the interface should be the same as the bulk solution viscosity. However, surfactant or impurity adsorption at an interface can cause a resistance to flow to occur that can be measured as the interfacial shear viscosity. This viscosity is defined as the ratio between the shear stress and the shear rate in the plane of the interface (*12*). Methods used to make these measurements include a viscous traction surface viscometer (*12*), droplet–droplet coalescence (*13*), the rotating ring viscometer (*14*), and surface laser light scattering (*9*).

Low interfacial viscosity is desirable in enhanced oil recovery operations, so that displaced oil globules may readily coalesce into an oil bank. Emulsion stability decreases as interfacial viscosity decreases, and this condition increases the ease with which an oil bank can be formed. Wasan et al. (*15*) found a qualitative correlation between coalescence rates and interfacial viscosities for crude oil.

In extreme cases, material can adsorb at an interface to create a film. Interfacial film formation can occur in crude-oil systems and has been reported by Blair (*16*), and by Reisberg and Doscher (*17*). Film formation is relatively common with crude oils and can effectively stabilize emulsions by preventing droplet coalescence even with high values of interfacial tension.

Surface Charge at Interfaces. The interface in a crude-oil–water system usually carries a net charge, which can be caused by the adsorption of surface-active ions. These surfactants may be carboxylic acids that originate from the oil or synthetic surfactants added to the system. Thermal motion of the counterions in the aqueous phase results in a diffused double layer adjacent to the interface with an excess of counterions. These counterions screen the attraction between ions adsorbed at the interface and counterions that are more distant. This screening causes counterion concentration to decrease exponentially with distance from the interface, and a net charge results.

Information about the surface charge of droplets in an emulsion can be obtained from electrophoresis. In this process, a solution containing small oil droplets is placed between oppositely charged electrodes. By measuring the velocity of a particle under a known field gradient and dividing velocity by field gradient, the electrophoretic mobility is obtained. In general, the absolute value of electrophoretic mobility is at a maximum when IFT is at a minimum in surfactant–crude-oil systems (9, 18) as well as in acidic crude-oil–sodium hydroxide systems (19). Thus, surface charge is at a maximum when surface-active ions are present at maximum concentration at the interface.

Changes in surface charge can result in changes in the way that oil droplets react with their surroundings. A negative surface charge will reduce the attraction of oil droplets to negatively charged sand surfaces, for instance. Chiang et al. (18) found that oil recovery in sand packs and Berea sandstone was at a maximum when the absolute value of surface charge was at a maximum. They found that surface charge was at a maximum and interfacial viscosity at a minimum between 3.5 wt% sodium chloride solution and Seeligson crude oil. In sand packs, recovery improved from 65% OOIP to 73% OOIP when increasing the salinity of the displacing phase from 0 to 3.5%. In Berea sandstone, recovery improved from 48 to 58% OOIP. Improved oil recovery could not be accounted for by either the alteration of wettability or the small increase in capillary number.

Surface Wettability.   As mentioned earlier, wettability affects capillary pressure and thus the entrapment and displacement of oil in porous media. The importance of wettability on the displacement of oil by water or brine solutions has long been known. In 1956, after several thousand flooding experiments in a variety of porous media, Moore and Slobod (1) came to the conclusion that wettability is the single most important factor affecting water-flood recovery efficiency. Later (20, 21), intermediate wettability was shown to be unfavorable for enhanced oil recovery processes, because the capillary number required for oil mobilization is higher than occurs with water-wet porous media, and less oil is available to be recovered. Water-flood oil recovery from weakly water-wet cores was higher than that obtained with strongly water-wet Berea sandstone. Surface wettability is particularly important in enhanced oil recovery, because the fluids used can change the surface wettability during the course of the recovery process.

This change in wettability in turn can change relative permeability and flow characteristics of both oil and water in porous media.

Principles of Micellar–Polymer Flooding

Micellar–polymer flooding relies on the injection of a surfactant solution to lower interfacial tension to ultralow levels, on the order of 10^{-3} mN/m. The resulting increase in capillary number allows the recovery of residual oil from porous media. The term "micellar" is used because the concentrations of injected surfactant solutions are always above their critical micelle concentration. That is, they are always above the concentration at which micelles form.

Microemulsions.   The structure of microemulsion systems has been reviewed (22). Both bicontinuous and droplet-type structures, among others, can occur in microemulsions. The droplet-type structure is conceptually more simple and is an extension of the emulsion structure that occurs at relatively high values of IFT. In this case, very small thermodynamically stable droplets occur, typically smaller than 10 nm (7). Each droplet is separated from the continuous phase by a monolayer of surfactant. Bicontinuous microemulsions are those in which oil and water layers in the microemulsion may be only a few molecules thick, separated by a monolayer of surfactant. Each layer may extend over a macroscopic distance, with many layers making up the microemulsion.

Compositions of injected micellar fluids can vary greatly. They include aqueous or oleic solutions of surfactant as well as complex mixtures containing components such as cosurfactants, cosolvents, or stabilizers, in addition to surfactant, oil, and brine. Regardless of the composition of the injected fluid, once in the reservoir the fluid system consists primarily of oil, water, and surfactant. The phase behavior of the fluid system can be quite complex but may be approximately described by means of pseudoternary diagrams in which the pseudocomponents are surfactant, brine, and oil (Figure 4). Depending on the system being studied, the pseudocomponents can range from pure substances to complex mixtures. For example, the oil may be a pure hydrocarbon or a crude-oil mixture. The surfactant can include cosurfactants and cosolvents, and the brine may include a variety of ionic constituents. The pseudoternary diagram is separated by a multiphase boundary into a single phase region above and a multiphase region below the phase boundary.

Phase Behavior.   Nelson and co-workers (23–25) and Healy et al. (26) have written extensively on phase behavior in micellar flooding. In Nelson's methodology, three different phase-behavior environments occur

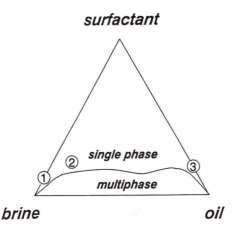

Figure 4. Pseudoternary diagram of oil–water–surfactant system with three compositions of interest. (Reproduced with permission from reference 40. Copyright 1977 Academic Press.)

(Figure 5). These are the type II(−), the type II(+), and the type III phase environments.

Both the type II(−) and the type II(+) phase environments have a maximum of two phases. In both cases, one phase only may be present at high surfactant concentrations. Types II(−) and II(+) phase environments are distinguished when the phases are plotted on a pseudoternary diagram. The tie lines in the two-phase region give either a negative slope for type II(−) behavior or a positive slope for type II(+) behavior. Winsor (27) assigned microemulsions occurring in the two-phase region of a type II(−) diagram as type I, and defined them as a microemulsion in equilibrium with excess oil. The microemulsion contains mostly brine and surfactant, and any oil present is solubilized in micelles. As such it is an oil-in-water (water-external) emulsion. The plait point of the pseudoternary diagram tends to be close to the oil apex. Microemulsions in the two-phase region of a type II(+) phase environment correspond to a Winsor type II emulsion. The microemulsion is in equilibrium with excess brine and contains mostly oil and surfactant. The brine is solubilized in micelles; thus this is a water-in-oil (oil-external) emulsion. In this case, the plait point of the pseudoternary diagram tends to be close to the brine apex.

The type III phase environment may contain a maximum of three phases. When this is the case, the emulsion present corresponds to Winsor type III, in which a microemulsion is in equilibrium with pure oil and pure brine phases. However, type II(−) behavior and type II(+) behavior may also be observed under certain conditions. In practice, type II(−) or II(+) behavior occurs when all of the brine or oil can be incorporated into the microemulsion or when insufficient surfactant is present to produce a measurable microemulsion.

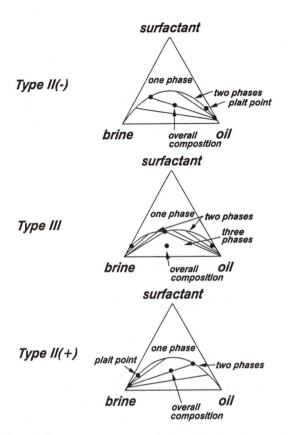

Figure 5. Phase-behavior environments. (Reproduced with permission from reference 24. Copyright 1978 Society of Petroleum Engineers.)

A number of factors can affect the phase type that is observed. These factors generally act by changing the partitioning of the surfactant between the brine and oil phases. In general, any change in the surfactant–oil–brine system that increases the solubility of surfactant in oil relative to brine will cause the phase environment type to shift from II(−) to III to II(+) as indicated in the following scheme:

increasing surfactant solubility in oil ⟶		
Nelson type II(−)	Nelson type III	Nelson type II(+)
Winsor type I	Winsor type III	Winsor type II
lower phase	middle phase	upper phase microemulsion

Increasing salinity generally decreases the solubility of the surfactant in the brine phase. This decreased solubility tends to shift the phase behavior from II(−) to III to II(+). Changing the oil type so that the surfactant is more soluble in the oil will also shift the phase behavior from left to right. By increasing the molecular weight of the hydrophobic part of the surfactant, solubility in the aqueous phase will generally decrease, and phase behavior will shift from left to right. Decreasing the amount of branching will have the same effect. Similarly, decreasing the polarity of the hydrophilic group will also decrease solubility in the aqueous phase. The effects of changes in surfactant structure on phase behavior have been discussed in detail (28).

The following changes will also shift phase behavior from II(−) to III to II(+):

- decrease in temperature for anionic surfactants
- increase in temperature for nonionic surfactants
- increase in alcohol concentration (alcohols of fewer than four carbons)
- decrease in alcohol concentration (alcohols of more than four carbons)
- increase in the divalent ion concentration of the brine

Solubilization Parameter.   The solubilization parameter (S) is defined as

$$S = \frac{V_o}{V_s} \quad \text{or} \quad \frac{V_w}{V_s} \qquad (4)$$

where V_o is the oil volume, V_s is the volume of surfactant, and V_w is the volume of water, all three measured in the microemulsion phase. Interfacial tension σ can be measured between the microemulsion and oil phases (σ_{mo}) or between the microemulsion and water phases (σ_{mw}). As either measure of interfacial tension decreases, the solubilization parameter increases. Healy and Reed (29) first showed that low interfacial tension correlates with high solubilization parameter, and Huh (30) showed it to be theoretically valid. Glinsmann (31) and Graciaa et al. (28) have validated the concept experimentally. This correlation is very useful because it enables the results of phase-behavior experiments to partially replace the experimentally more difficult measurement of interfacial tension.

Salinity Requirement Diagrams.   Maximum solubilization parameter occurs very close to the salinity at which maximum core-flood oil recovery is obtained. The salinity requirement of a chemical flooding system is defined by Nelson (*32*) as the salinity at which a type III microemulsion is at midpoint salinity. This condition occurs when the oil and brine concentrations in the microemulsion middle phase are equal. Many authors (*24, 25, 32–34*) have reported that oil recovery is at a maximum in a micellar flood when the system is near midpoint salinity. A salinity requirement diagram aids in the design and understanding of a micellar–polymer system.

To construct a salinity requirement diagram, 5 to 10 different brine salinities are prepared for at least three surfactant concentrations in screw-cap test tubes. Typically, surfactant concentration will range from 0 to 10 wt%, and salinity will vary according to the reservoir of interest. Sample tubes all contain an identical amount of brine, usually between 50 and 80% by volume. Sample tubes are mixed regularly for several days, then allowed to equilibrate. The equilibration process can take anywhere from several days to several months, depending on emulsion stability.

Figure 6 shows an idealized salinity requirement diagram. Within the type III phase environment, three phases occur in the area indicated, but two phases occur in the rest of the type III region. Type II(+) phase behavior occurs above the type III region, and the type II(−) behavior occurs below. Midpoint salinity is shown near the middle of the type III region.

Midpoint salinity generally decreases with decreasing surfactant concentration. As surfactant concentration decreases, the type III phase envi-

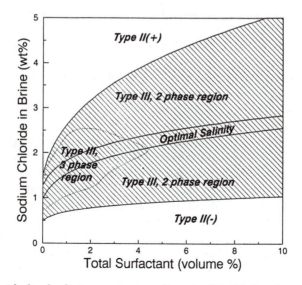

Figure 6. Idealized salinity requirement diagram. (Modified and reproduced with permission from reference 32. Copyright 1982 Society of Petroleum Engineers.)

ronment becomes narrower. This narrowing has important implications when carrying out a micellar flood in reservoir core. If a constant salinity flood is carried out in which the salinity is the same in the water-flood brine, the chemical slug, and the polymer drive, recovery will generally be inferior to a salinity gradient flood, even though both floods use the same optimum chemical composition. For instance, with a constant salinity flood designed to be at midpoint salinity, surfactant concentration will decrease in the front and rear mixing zones. The surfactant slug will then change from type III behavior to type II(+) behavior. If a salinity gradient is used, the following may occur. A high salinity water-flood, with salinity higher than midpoint salinity, is followed by a chemical slug with salinity at or just below midpoint salinity. The chemical slug is followed by a polymer drive that is much below midpoint salinity. In the front mixing zone, type III phase behavior quickly develops from initial type II(+) behavior. At the rear mixing zone, type III behavior is maintained before type II(−) behavior occurs. In the lower salinity polymer drive, surfactant that had partitioned into any remaining oil partitions back into the lower salinity aqueous phase and thereby keeps the system in type III behavior for a longer period. Typically, less surfactant adsorption occurs with a salinity gradient flood than with a constant salinity flood.

Process Design. The various applications of micellar flooding can be represented on a pseudoternary diagram (Figure 4). The injected slug used in aqueous surfactant flooding, indicated by point 1 on the diagram, has no added oil in the slug material. Some oil may be introduced from either surfactant or polymer manufacturing processes, but in very small amounts. An oil-in-water microemulsion injectant is used when the composition is represented by point 2. This type of system is used in a commercial (Maraflood) process. Compositions indicated by point 3 are termed soluble oil (35, 36) and can spontaneously emulsify water with oil remaining as the external phase. This type of composition forms the basis of another commercial (Uniflood) process.

Ideally, the injected micellar solutions will be miscible with the fluids that they are in contact with in the reservoir and can thus miscibly displace those fluids. In turn, the micellar solutions may be miscibly displaced by water. Highest oil recovery will result if the injected micellar solution is miscible with the reservoir oil. If there are no interfaces, interfacial forces that trap oil will be absent. Injection of compositions lying above the multiphase boundary initially solubilizes both water and oil and displaces them in a misciblelike manner. However as injection of the micellar solution progresses, mixing occurs with the oil and brine at the flood front, and surfactant losses occur because of adsorption on the reservoir rock. These compositional changes move the system into the multiphase region. The ability of the micellar fluid to recover oil is altered depending on the phase environment.

In the two-phase region, the type II(+) system has an oil-rich micellar phase in equilibrium with an excess brine phase. Surfactant is found almost exclusively in the oil-rich phase, and the concentration of surfactant in that phase can greatly exceed the concentration of surfactant in the injected chemical slug. In the type II(+) environment, the micellar phase remains miscible with the oil but is immiscible with the brine. Oil continues to be recovered by a misciblelike process. The opposite occurs if the phase environment is type II(−). The brine-rich micellar phase is immiscible with the oil phase, and oil recovery is by low IFT immiscible displacement.

Nelson and Pope (*24*) described these phase relationships and demonstrated that phases observed in laboratory test tube experiments also form and are transported in a porous medium subjected to a chemical flood at reservoir flow rates. Chemical floods continuously maintained in a type II(+) phase environment recovered substantially more residual oil than those maintained in a type II(−) environment.

Nelson and Pope concluded that chemical flood design should be such as to maintain as much surfactant as possible in the type III phase environment. This condition can be accomplished by designing the micellar fluid such that the initial phase environment of the immiscible displacement is type II(+). A negative salinity gradient is imposed, and it moves the phase environment to type III and, eventually, to II(−).

In practice, because of economic constraints, a finite chemical slug must be injected and effectively displaced through the reservoir. This step is accomplished by using a polymer drive fluid. The polymer drive fluid, being aqueous, is immiscible with type II(+) micellar fluids but miscible with type II(−). Thus, if a type II(+) system were to be displaced by a polymer drive fluid, phase trapping of the micellar solution could occur. The use of salinity gradient to produce a type II(−) environment at the trailing edge of the chemical slug allows miscible displacement by the polymer drive fluid.

Figure 7 shows the general sequence of a micellar–polymer flood. An initial preflush is sometimes used to lower salinity and divalent ion concentration in the reservoir. This preflush is followed by the surfactant slug, containing a surfactant–polymer mixture designed to produce a microemulsion with the crude oil. A polymer drive follows, and it prevents "fingering" of the brine into the surfactant slug. The polymer drive is often injected with a concentration gradient; polymer concentration decreases as the injection progresses.

Oil Bank Formation.   If a surfactant or surfactant-forming material is injected into a reservoir and mobilizes residual oil, then oil recovery is more efficient if the mobilized oil droplets can coalesce to form an oil bank.

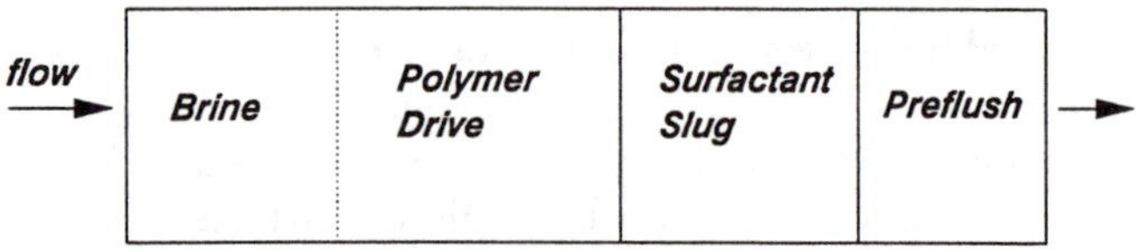

flow	Brine	Polymer Drive	Surfactant Slug	Preflush

Polymer Drive: 0 - 100% pore volume
                  200 - 2500 ppm polymer
                  biocides

Surfactant Slug: 5 - 20% pore volume
                  200 - 5000 ppm polymer
                  1 - 20% surfactant
                  0 - 5% cosurfactant
                  biocides

Preflush: 0 - 100% pore volume

Figure 7. Typical microemulsion injection scheme.

The oil bank that forms will exist at an oil saturation that is greater than the residual oil saturation. At the front of the bank, residual oil is taken up, while at the back, the capillary number must remain high to minimize oil entrapment. In this way, the oil bank grows larger and forms slightly ahead of the injected chemicals.

The formation and displacement of the oil bank depends upon the nature of the phases formed in the porous medium and their relative permeabilities, which may also change as a result of changes in wettability. Detailed discussion of these factors is beyond the scope of this chapter; Chapter 6 and references 37 and 38 address this topic.

Nonideal Behavior. The discussion of phase behavior up to this point represents the ideal case. A number of factors cause deviation from ideality. The phases present may include liquid crystals, gels, or solid precipitates in addition to the oil, brine, and microemulsion phases (*39, 40*). The high viscosities of these phases are detrimental to oil recovery. To control the formation of these phases, the practice has been to add low-molecular-weight alcohols to the micellar solution; these alcohols act as cosolvents or in some cases as cosurfactants.

The alcohol cosolvents or cosurfactants may partition between aqueous and oil phases in different proportions than the primary surfactant, and therefore, grouping these components in the surfactant pseudocomponent is inappropriate. Chromatographic separation of the components may occur during flow in the reservoir, and unwanted phases may form.

Ion exchange between injected brine and reservoir rock may result in locally high concentrations of divalent ions that can precipitate anionic surfactants. Together with diffusion and dispersion phenomena in the porous medium, precipitated surfactants can result in local immiscibilities and phase trapping.

Incompatibility of surfactant and polymer that is used in the micellar slug or chase fluid can occur because polymer may not be incorporated readily into micelles. Use of alcohols mitigates this difficulty.

Field Application. The micellar–polymer process for enhanced oil recovery has been used in many field trials. Petroleum sulfonates are the most commonly used surfactant (*41, 42*). Other surfactants have been used, such as ethoxylated alcohol sulfates (*43*) and nonionic surfactants mixed with petroleum sulfonates (*44*).

The Loudon field in Illinois, operated by Exxon, is an interesting example of micellar–polymer flooding design (*45, 46*). The reservoir is a moderate permeability sandstone with excellent properties for micellar–polymer flooding, except one: The salinity is very high, approximately 10.5 wt% total dissolved solids and 4000 ppm of divalent ions. Exxon has been studying the micellar–polymer process in this field for more than 10 years, and to date (mid-1991) has completed two pilot projects and has two others in progress. The sequence of the injected microemulsion and polymer drive is outlined in Figure 8. A microemulsion of 0.3 pore volumes, containing a relatively low surfactant concentration, 2.3 wt%, was used. The surfactant was a sulfate of a propoxylated ethoxylated tridecyl alcohol, of the following structure:

$$i\text{-}C_{13}H_{27}O(C_3H_6O)_m(C_2H_4O)_nSO_3Na$$

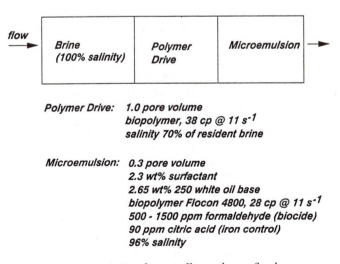

Figure 8. Loudon micellar–polymer flood.

where m and n are either 4 and 2 or 3 and 4, respectively. In practice, a mixture of these two surfactants was used, so that surfactant composition could easily be varied in the field to correct for any changes in injection brine composition. This type of surfactant is particularly suited to high salinities. If the number of ethoxyl groups is increased, optimal salinity increases, but if the number of propoxyl groups increases, the optimal salinity decreases. The mixture used had optimal salinity at the high salinity and hardness levels associated with this reservoir. The synthesis of this surfactant and its use in micellar–polymer flooding has been patented by Exxon (47).

In addition to the surfactant, a white oil was a component of the microemulsion. This oil was added in the minimum amount required to solubilize enough xanthan polymer to produce the target viscosity. In the absence of the white oil, the polymer could not be solubilized. The xanthan polymer itself was required for mobility control. To prevent biodegradation of the polymer, formaldehyde was added. Citric acid was also a component of the microemulsion, added to prevent the oxidation of ferrous ion present in the brine to ferric ion. The presence of ferric ion would lead to precipitation of iron compounds as well as cross-linking of the biopolymer.

A particularly interesting part of the pilot involved the treating of produced emulsions. Over the life of the pilot, 93% of the injected surfactant was produced at the production wells, and this situation led to serious emulsion problems. Heating the emulsion to a specific, but unreported, temperature caused the surfactant to partition completely into the aqueous phase and leave the crude oil with very low levels of surfactant and brine. The resulting oil was suitable for pipeline transportation. The critical separation temperature had to be controlled to within 1 °C. At higher temperatures, surfactant partitioned into the oil, and at lower temperatures, significant quantities of oil remained solubilized in the brine. Recovered surfactant was equivalent to the injected surfactant in terms of phase behavior, and had the potential for reuse.

The pilot area used for this test was relatively small, 0.71 acres. However, the test was a technical success, recovering 68% of the water-flood residual oil. The pilot began in 1982 and ended in November 1983. Since that time, Exxon has initiated two other micellar–polymer floods in the Loudon field, one a 40-acre pilot and the other an 80-acre pilot.

Principles of Alkaline Flooding

Alkaline flooding is an old concept, first patented by Atkinson (48) in 1927. Hydroxide ion in an alkaline solution reacts with acidic components present in some crude oils to produce petroleum soaps, which are generally sodium salts of carboxylic acids. These petroleum soaps are capable of adsorbing at the oil–water interface and lowering interfacial tension. Crude oils suitable for alkaline flooding generally have a total acid number (TAN) of 0.1 to 2 mg of KOH per gram of oil. The injection of alkaline solutions into a reservoir can improve oil recovery by several mechanisms. Emulsification and entrainment, emulsification and entrapment, and wettability reversal have been proposed (*49*, *50*).

Classical alkaline flooding uses only alkaline solutions and has several disadvantages. One of the most serious is that the petroleum soaps are very sensitive to increases in the ionic strength of the aqueous phase. In 1975, Burdyn et al. (*51*) began to address this problem with the addition of synthetic surfactants to the alkaline flooding process. In this case, a synthetic alkyl aryl sulfonate was added to sodium hydroxide solutions to provide low IFT behavior over a broader range of salinity than could be obtained with alkali alone. Burdyn also patented the use of polymer with alkali to increase the aqueous-phase viscosity. This increased viscosity had the effect of improving the sweep of the aqueous phase through the reservoir. Excessive alkali consumption also plagued the classical alkaline flooding process. Sodium hydroxide and sodium silicate solutions react with reservoir rock or precipitate in the presence of divalent cations. Buffered alkaline solutions such as sodium carbonate have increased in importance because of lower alkali–rock interactions. Currently, the alkali–surfactant–polymer process represents the state of the art in alkaline flooding.

Many of the basic concepts of micellar–polymer flooding apply to alkaline flooding. However, alkaline flooding is fundamentally different because a surfactant is created in the reservoir from the reaction of hydroxide with acidic components in crude oil. This reaction means that the amount of petroleum soap will vary locally as the water-to-oil ratio varies. The amount of petroleum soap has a large effect on phase behavior in crude-oil–alkali–surfactant systems.

Surfactant Mixing Rules. The petroleum soaps produced in alkaline flooding have an extremely low optimal salinity. For instance, most acidic crude oils will have optimal phase behavior at a sodium hydroxide concentration of approximately 0.05 wt% in distilled water. At that concentration (about pH 12) essentially all of the acidic components in the oil have reacted, and type III phase behavior occurs. An increase in sodium hydroxide concentration increases the ionic strength and is equivalent to an increase in salinity because more petroleum soap is not produced. As salinity increases, the petroleum soaps become much less soluble in the aqueous phase than in the oil phase, and a shift to over-optimum or type II(+) behavior occurs. The water in most oil reservoirs contains significant quantities of dissolved solids, resulting in increased IFT. Interfacial tension is also increased because high concentrations of alkali are required to counter the effect of losses due to alkali–rock interactions.

A solution to the problem has been to add a synthetic surfactant to modify the properties of the petroleum soap that is produced from reaction with hydroxide (51, 52). This process has been termed surfactant-enhanced alkaline flooding. The added synthetic surfactant is chosen to have a very high optimal salinity, and the resulting petroleum-soap–synthetic-surfactant mixture produces optimal phase behavior at intermediate salinities.

The mixing of a synthetic surfactant and a petroleum soap can be explained in terms of surfactant mixing rules proposed by Wade et al. in 1977 (53). These rules are based on previous studies (54) of the equivalent alkane carbon number (EACN) concept, which show that hydrocarbon behavior toward surfactants is additive and weighted by mole fraction according to the formula:

$$EACN_{avg} = \sum_i EACN_i X_i \qquad (5)$$

where $EACN_{avg}$ is the EACN of the mixture, $EACN_i$ is the EACN of component i, and X_i is the mole fraction of component i.

Cayias et al. (55) and others (31, 56) found that this relationship could be applied to crude oils. They found that crude oils in general behaved equivalently to n-alkanes in the range of pentane to decane, even with greatly differing oil types (55). Thus, the behavior of a crude oil toward a given surfactant can be described as the sum of the behavior of each of its hydrocarbon components toward that surfactant.

The behavior of mixed surfactant systems has been described in similar terms (53):

$$(N_{min})_{avg} = \sum_i (N_{min})_i X_i \qquad (6)$$

where $(N_{min})_i$ is the alkane carbon number of the interfacial tension minimum of surfactant i, $(N_{min})_{avg}$ is the alkane carbon number minimum of the IFT minimum of the surfactant mixture, and X_i is the mole fraction of surfactant i.

More simply stated, the behavior of a surfactant mixture toward a given oil can be described as the sum of the behavior of each of its components toward that oil. This hypothesis shows that the natural surfactant that is produced in alkaline flooding can be modified by an added synthetic surfactant in a predictable way.

Phase Behavior.   The use of phase-behavior diagrams in surfactant-enhanced alkaline flooding is more complicated than in micellar–polymer flooding for several reasons. One reason is that phase behavior is very sensitive to the water-to-oil ratio employed. From surfactant mixing rules, varying the amount of oil present will vary the amount of petroleum soap present, and the nature of the mixed surfactant will change. Another complication is that stable emulsions with high values of interfacial tension are much more likely to occur with the heavier oils used in the process, and these stabilized emulsions can lead to improper construction of phase-behavior diagrams. The third problem is that total surfactant concentration is much lower than is seen in micellar–polymer flooding. In some cases, IFT may be very low, and phase behavior can be in the type III environment, but a middle phase may not be readily apparent because of the low surfactant concentration.

Figures 9 and 10 show phase-behavior diagrams for David Lloydminster crude oil and the surfactant Neodol 25-3S in the presence of 1 wt% sodium carbonate. Phase-behavior measurements were carried out according to the method of Nelson et al. (52). The David Lloydminster oil field is near the Alberta–Saskatchewan border directly east of Edmonton. The oil has a density of 0.922 g/mL and a viscosity of 144 MPa·s at 23 °C. The region of optimal phase behavior is shown at a surfactant concentration of 0.1 wt% in Figure 9. The region of optimal phase behavior is shaded. Above this region, type II(+) behavior occurs, and type II(−) behavior occurs below the region of optimal phase behavior. Volume percent oil refers to the amount of oil present in the phase-behavior tube used. For a given oil-to-water ratio, a transition from type II(−) to type III to type II(+) occurs as salinity increases. As the amount of oil increases relative to the amount of aqueous phase, the same trend in phase behavior is seen.

Figure 10 shows the same system, but with a lower synthetic surfactant

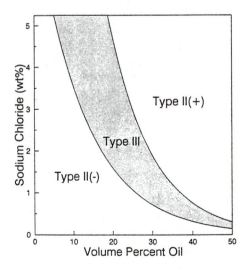

Figure 9. Activity map with David Lloydminster crude oil, 0.1 wt% Neodol 25-3S, and 1 wt% sodium carbonate.

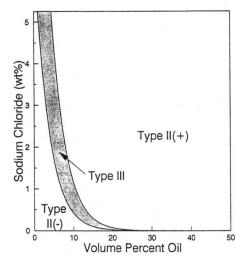

Figure 10. Activity map with David Lloydminster crude oil, 0.02 wt% Neodol 25-3S, and 1 wt% sodium carbonate.

concentration (0.02 wt%). The type III phase-behavior region is shifted to lower salinities and lower oil-to-water ratios. This shift is a direct result of changes in the petroleum-soap–synthetic-surfactant ratio as water-to-oil ratio varies. Nelson et al. (52) stated that these phase-behavior diagrams can be used in an equivalent fashion to the salinity requirement diagrams used for micellar–polymer flooding (32). He claimed that a surfactant-enhanced alkaline flood should be designed so that the flood begins at the optimum–over-optimum phase boundary. The residual oil saturation is used to determine the water-to-oil ratio in the diagram. This determination assumes that equilibration is rapid and does not address the possibility of petroleum soaps being extracted and concentrated in the flood front. A salinity gradient is applied when the alkaline agent is removed from the drive fluid. Nelson's results have been very promising in published laboratory core-flood experiments.

Dynamic Interfacial Tension.   Crude-oil–alkali systems are unusual in that they exhibit dynamic interfacial tension (Figure 11). A solution of 0.05 wt% sodium hydroxide in contact with David Lloydminster crude oil initially produces ultralow values of IFT. A minimum value is reached, after which IFT increases with time by nearly 3 orders of magnitude, measured in the spinning drop tensiometer. Taylor et al. (57) showed that dynamic interfacial tension can also occur in crude-oil–alkali–surfactant systems. Figure 11 shows interfacial tension versus time for a solution containing 1 wt% sodium carbonate, and the same solution containing 0.02 wt% of Neodol 25-

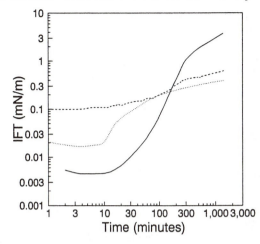

Figure 11. Interfacial tension versus time, David Lloydminster crude oil. Key: —, 0.05 wt% NaOH; . . ., 0.02 wt% Neodol 25-3S and 1 wt% Na$_2$CO$_3$; and - - -, 1 wt% Na$_2$CO$_3$.

3S, an ethoxylated alcohol sulfate. The addition of the synthetic surfactant greatly reduces the minimum IFT value obtained.

Dynamic IFT arises from the reaction of acidic components in the crude oil to form petroleum soaps. Reaction of acidic surface-active materials in the crude oil with sodium hydroxide in the aqueous phase is assumed to occur rapidly at the interface, but desorption of these species is taken to be slower. This slower desorption leads to a maximum in the concentration of surface-active species at the interface at some point in time and hence an interfacial tension minimum. Subsequently, IFT increases as equilibrium is approached (58).

In the spinning drop apparatus, the water-to-oil ratio is approximately 200 to 1. In a reservoir, the water-to-oil ratio would be much lower. Changes in this water-to-oil ratio are expected to affect the relative rates of desorption of surfactants from the oil–water interface. The significance of the dynamic IFT minimum for reservoir situations has been discussed by Rubin and Radke (58) and deZabala and Radke (59). They suggested that the IFT minimum for acidic crude oils is indicative of the lowest achievable reservoir equilibrium value. Taylor et al. (57) showed a correlation between minimum IFT and core-flood recovery efficiency in surfactant-enhanced alkaline flooding. Figure 12 shows a modification of the capillary number curve introduced in Figure 3. Capillary numbers from core-flood data are plotted by using both IFT minima and equilibrium values. Core floods were carried out in linear Berea sandstone cores. Agreement between published capillary number correlations is very poor when using equilibrium IFT values, as

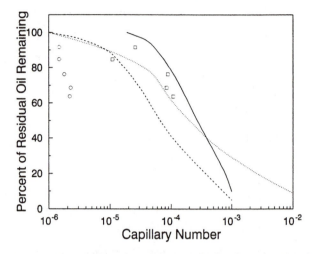

Figure 12. Capillary number correlation with dynamic interfacial tension. Key: . . ., Gupta and Trushenski (34); - - -, Taber (5); —, Chatzis and Morrow (6); □, minimum IFT; and ○, equilibrium IFT.

measured with the spinning drop instrument, but it is very good when using minimum IFT values.

Field Application. Field trials of classical alkaline flooding have been disappointing. Mayer et al. (60) indicated that only 2 of 12 projects had significant incremental oil recovery: North Ward Estes and Whittier with 6–8 and 5–7% pore volume, respectively. Estimated recovery from the Wilmington field was 14% with a classical alkaline flooding method (61). However, post-project evaluation of that field indicated no improvement over water-flooding (62).

A number of laboratory studies of the application of the alkali–surfactant–polymer flooding to various reservoir systems have been reported (63–67), but field application of this technology has been limited. Several field pilots are in progress or have been completed, but only one has been evaluated to date in the technical literature (68). This project is in the West Kiehl field in Wyoming operated by Terra Resources Inc.

The West Kiehl field is a medium permeability (350-mD; mD is millidarcies) sandstone. The reservoir brine contains 45,500 ppm of total dissolved solids, with about 450 ppm of divalent ions. The 24° API crude oil has viscosity of 19 mPa·s at a reservoir temperature of 49 °C. (API gravity is defined in the Glossary.)

The chemical slug used in the project was prepared in a fresh, relatively soft brine (800 ppm of total dissolved solids, 18 ppm hardness). On the basis of interfacial tension measurements (phase-behavior tests were not reported), a solution of 0.8 wt% Na_2CO_3 and 0.1 wt% Petrostep B100, a petroleum sulfonate, was selected. To this was added 0.1 wt% of a polyacrylamide polymer to provide the necessary mobility control. The project design called for injection of 0.25 pore volume of the alkali–surfactant–polymer solution followed by a similar volume of polymer solution acting as a mobility control buffer.

Early results in the West Kiehl field are very encouraging. It is anticipated that the use of the ASP process will increase oil recovery by an additional 15% over water-flood.

Other Applications

Emulsion Injection for Recovery of Heavy Oil. Oil-in-water emulsions may be useful as sweep improvement agents in heavy-oil reservoirs. To improve the mobility ratio occurring with high-viscosity oils, McAuliffe (69) and Schmidt et al. (70) proposed the use of stable oil-in-water emulsions. These authors conducted laboratory experiments with emulsions prepared by reaction of sodium hydroxide with a synthetic acidic oil. The theoretical background for emulsion blocking has been discussed in Chapter 6, and it forms the basis for one of several mechanisms of caustic flooding (71). These emulsions may form spontaneously during oil recovery processes (72), but can just as easily be prepared and injected as enhanced oil recovery fluids.

Fiori and Farouq Ali (73) proposed the emulsion flooding of heavy-oil reservoirs as a secondary recovery technique. This process is of interest for Saskatchewan heavy-oil reservoirs, where primary recovery is typically 2–8%. Water-flooding in these fields produces only an additional 2–5% of the original oil in place because of the highly viscous nature of the oil. In laboratory experiments, a water-in-oil emulsion of the produced oil is created by using a sodium hydroxide solution. The viscous emulsion formed is injected into the reservoir. Its high viscosity provides a more favorable mobility ratio and results in improved sweep of the reservoir. Important parameters include emulsion stability and control of emulsion viscosity.

Decker and Flock (74) investigated the application of emulsion injection for steam-flooding processes. In laboratory models, emulsions containing 5 vol% crude oil were effective in blocking channels created by steam injection during subsequent steam-injection cycles. Oil droplets in the emulsion were predominantly in the 1–2-μm range, but droplets as large as 10 μm were observed.

Matrix Acidization. Emulsion technology has been applied to the acid treatment of reservoir material in the region near the well bore. The pore structure of the region of the reservoir near the well bore may sometimes become plugged either by particulates from drilling fluids or by pre-

cipitation deposits caused by pressure or temperature changes during production. These pore-plugging materials reduce permeability in the region
near the well bore and hence reduce well productivity. Acid stimulation is
used routinely to increase well productivity by removing these unwanted
deposits. Wells in formations with naturally occurring low permeability can
also be stimulated by using the same process, but applied to the original rock
matrix. This process is referred to as matrix stimulation.

This matrix acidization process consists of injecting hydrochloric acid
(for limestones) or a hydrochloric acid–hydrofluoric acid mixture (for sandstones) into the formation pore space. The acid reacts with and dissolves
portions of the original rock matrix and thus increases permeability. The
depth that the acid penetrates into the formation is one of the factors that
determines the effectiveness of the treatment.

For carbonate reservoirs or carbonate cements in particular, acid consumption occurs very rapidly at elevated formation temperatures according
to the equation:

$$2H^+ + MCO_3 \rightleftarrows M^{2+} + CO_2 + H_2O \qquad (7)$$

where M is calcium or magnesium. The rate of dissolution is limited by mass
transfer, such that it depends on the rate at which acid diffuses to the surface
of the material being dissolved. The rate of mass transfer accompanying flow
through the rock matrix is high, so acid is consumed very quickly. Thus deep
penetration of the acid is difficult to achieve, and significant productivity
improvement cannot always be attained.

Dissolution of the rock matrix does not occur in a uniform radial manner. Because of permeability contrasts in the reservoir, dominant flow channels called worm holes can develop and extend into the formation in a
random fashion. The productivity increase of any well is then governed by
worm-hole direction and distance. The longer the worm hole, the better will
be the result.

Leak-off or loss of acid through the walls of worm holes often results in
worm holes being too short to provide significant productivity increase.
Therefore, effective stimulation often requires retardation of the mineral
dissolution rate. The use of microemulsions is one method to accomplish this
retardation. The hydrochloric acid is injected as an water-in-oil
microemulsion. The diffusion rate of the dispersed aqueous acid to the rock
surface is slower than molecular diffusion of acid from a totally aqueous
system. Thus the rate of limestone dissolution is retarded with the
microemulsion system.

The degree of retardation is dependent on the micellar structure of the
system. Hoefner and Fogler (75) described one such microemulsion system
containing cetylpyridinium chloride and butanol as the surfactant–
cosurfactant in a 35:65 weight ratio. In this system, dodecane was used as the

oil phase, and hydrochloric acid as the aqueous phase. To maintain hydrochloric acid as the dispersed phase, it was necessary to use a composition that sits near the oil apex of the pseudoternary diagram. The diffusivity of the acid from the microemulsion system was 2 orders of magnitude less than from the aqueous acid alone.

Summary

Crude oil becomes trapped in porous media as a result of capillary forces. The reduction of these forces is required for the recovery of residual oil, and this is the basis of enhanced oil recovery. In practice capillary forces are reduced primarily by lowering interfacial tension between oil and water phases, although increasing the viscosity of the water is also important. Lowering interfacial tension leads to the formation of emulsions and microemulsions, which are of great importance in enhanced oil recovery techniques.

Micellar–polymer flooding and alkali–surfactant–polymer flooding both rely on the injection into a crude-oil reservoir of surfactants or surfactant-forming materials. Emulsions may be injected into the reservoir, or they may be formed in the reservoir, but their properties will change as they travel through the reservoir to eventually flow from a producing well after weeks or months.

Micellar–polymer flooding is a technically well-developed process. Phase compositional aspects of microemulsion design are relatively well understood, and several technically successful field trials have been carried out. Micellar–polymer floods can be designed and carried out with a good chance of success. However, the process is too expensive. This high cost is due primarily to the high concentrations of synthetic surfactants required. The problem is further compounded because these synthetic surfactants are made from petrochemicals, a fact that ties their price to the price of crude oil.

Alkali–surfactant–polymer (ASP) flooding shows promise to become economically more attractive than micellar–polymer flooding. The process is inherently less expensive because the concentration of synthetic surfactant is significantly lower. However, the ASP process is more complex and technically less developed than micellar–polymer flooding. In addition, the disappointing history of field trials of classical alkaline flooding has left many researchers skeptical of the process in general. But with the use of high optimal salinity surfactants to lower interfacial tension at realistic reservoir salinities, the use of buffered alkali to reduce alkali–rock interactions, and the addition of polymers to the system to increase displacing phase viscosity, many of the problems associated with classical alkaline flooding have been addressed. Areas that require further investigation include the effect of dynamic interfacial tension in the reservoir, conciliation of interfacial tension and phase-behavior measurements, and computer simulation of the ASP process.

List of Symbols and Abbreviations

ASP	alkali–surfactant–polymer
EACN	equivalent alkane carbon number
$EACN_{avg}$	EACN of a mixture of components
$EACN_i$	EACN of component i
IFT	interfacial tension
N_c	capillary number
$(N_{min})_{avg}$	alkane carbon number of the IFT minimum of a surfactant mixture
$(N_{min})_i$	alkane carbon number of the IFT minimum of surfactant i
%OOIP	percent of original oil in place
P_c	capillary pressure
P_o	pressure in oil phase
P_w	pressure in water phase
r	radius of curvature of the interface
R	half of drop width
S	solubilization parameter
TAN	total acid number of a crude oil (mg KOH per gram of oil)
v	Darcy flow velocity
V_o	oil volume in microemulsion phase
V_s	surfactant volume in microemulsion phase
V_w	water volume in microemulsion phase
X_i	mole fraction of component i

Greek

θ	contact angle
μ	viscosity
ρ	density
σ	interfacial tension
σ_{mo}	interfacial tension between the microemulsion and oil phases
σ_{mw}	interfacial tension between the microemulsion and water phases
ω	angular velocity

References

1. Moore, T. F.; Slobod, R. L. *Prod. Mon.* **1956**, *20*, 20–30.
2. Chatzis, I.; Dullien, F. A. L. *J. Colloid Interface Sci.* **1983**, *91(1)*, 199–222.

3. Lake, L. W. In *Enhanced Oil Recovery;* Prentice Hall: Englewood Cliffs, NJ, 1989; p 65.
4. Chatzis, I.; Morrow, N. R.; Lim, H. T. *Soc. Pet. Eng. J.* **1983,** *23(2),* 311–326.
5. Taber, J. J. In *Surface Phenomena in Enhanced Oil Recovery;* Shah, D. O., Ed.; Plenum: New York, 1981; pp 13–52.
6. Chatzis, I.; Morrow, N. R. *Soc. Pet. Eng. J.* **1984,** *24(5),* 555–562.
7. Aveyard, R.; Vincent, B. *Prog. Surf. Sci.* **1977,** *8(2-A),* 59–102.
8. Cayias, J. L.; Schechter, R. S.; Wade, W. H. In *Adsorption at Interfaces;* Mittal, K. L., Ed.; ACS Symposium Series 8; American Chemical Society: Washington, DC, 1975; pp 234–247.
9. Dorshow, R. B.; Swofford, R. L. *Colloids Surf.* **1990,** *43,* 133–149.
10. Vonnegut, B. *Rev. Sci. Instrum.* **1942,** *13,* 6–9.
11. Shah, D. O.; Chan, K. S.; Bansal, V. K. *Proceedings of AIChE 83rd National Meeting;* American Institute of Chemical Engineers: Houston, TX, 1977; p 98.
12. Wasan, D. T.; Gupta, L.; Vora, M. K. *AIChE J.* **1971,** *17(6),* 1287–95.
13. Flumerfelt, R. W.; Oppenheim, J. P.; Son, J. R. *Interfacial Phenomena in Enhanced Oil Recovery;* Wasan, D.; Payatakes, A., Eds.; AIChE Symposium Series 212; American Institute of Chemical Engineers: New York, 1982; Vol. 78, pp 113–126.
14. Goodrich, F. C.; Allen, L. H.; Poskanzer, A. *J. Colloid Interface Sci.* **1975,** *52(2),* 201–212.
15. Wasan, D. T.; Shah, S. M.; Aderangi, N.; Chan, M. S.; McNamara, J. J. *Soc. Pet. Eng. J.* **1978,** *18(6),* 409–17.
16. Blair, C. M. *Chem. Ind.* **1960,** *(5),* 538–544.
17. Reisberg, J.; Doscher, T. M. *Prod. Mon.* **1956,** *11,* 43–50.
18. Chiang, M. Y.; Chan, K. S.; Shah, D. O. *J. Can. Pet. Technol.* **1978,** *17(4),* 61–68.
19. Bansal, V. K.; Chan, K. S.; McCallough, R.; Shah, D. O. *J. Can. Pet. Tech.* **1978,** *17(1),* 69–72.
20. Morrow, N. R; Lim, H. T.; Ward, J. S. *SPE Form. Eval.* **1986(2),** 89–103.
21. Morrow, N. R. *J. Pet. Technol.* **1990,** *42(12),* 1476–84.
22. Shinoda, K.; Lindman, B. *Langmuir* **1987,** *3(2),* 134–149.
23. Nelson, R. C. *Chem. Eng. Prog.* **1989,** *3,* 50–57.
24. Nelson, R. C.; Pope, G. A. *Soc. Pet. Eng. J.* **1978,** *18(5),* 325–38.
25. Nelson, R. C. In *Surface Phenomena in Enhanced Oil Recovery;* Shah, D. O., Ed.; Plenum: New York, 1981; pp 73–104.
26. Healy, R. N.; Reed, R. L.; Stenmark, D. G. *Soc. Pet. Eng. J.* **1976,** *16(3),* 147–160.
27. Winsor, P. A. *Solvent Properties of Amphiphilic Compounds;* Butterworth: London, 1954.
28. Graciaa, A.; Fortney, L. N.; Schechter, R. S.; Wade, W. H.; Yiv, S. Presented at the Second Joint SPE/DOE Symposium on Enhanced Oil Recovery, Tulsa, OK, 1981; paper SPE 9815.
29. Healy, R. N.; Reed, R. L. *Soc. Pet. Eng. J.* **1974,** *14(5),* 491–501.
30. Huh, C. *J. Colloid Interface Sci.* **1979,** *71(2),* 408–426.
31. Glinsmann, G. R. Presented at the 54th Annual Fall Technical Conference and Exhibition of the Society of Petroleum Engineers and the American Institute of Mining, Metallurgical, and Petroleum Engineers, Las Vegas, NV, September 23–26, 1979; paper SPE 8326.
32. Nelson, R. C. *Soc. Pet. Eng. J.* **1982,** *22(2),* 259–270.
33. Healy, R. N.; Reed, R. L. *Soc. Pet. Eng. J.* **1977,** *17(2),* 129–139.
34. Gupta, S. P.; Trushenski, S. P. *Soc. Pet. Eng. J.* **1979,** *19(2),* 116–128.
35. Holm, L. W. In *Improved Oil Recovery by Surfactant and Polymer Flooding;*

Shah, D. O.; Schechter, R. S., Eds.; Academic Press: New York, 1977; pp 453–485.

36. Holm, L. W. *J. Pet. Technol.* **1971,** *23(12),* 1475–483.
37. Larson, R. G.; Davis, H. T.; Scriven, L. E. *J. Pet. Technol.* **1982(2),** 243–258.
38. Pope, G. A. *Soc. Pet. Eng. J.* **1980,** *(6),* 191.
39. Miller, C. A.; Mukherjee, S.; Benton, W. J.; Natoli, J.; Qutubuddin, S.; Fort, T., Jr. In *Interfacial Phenomena in Enhanced Oil Recovery;* Wasan, D.; Payatakes, A., Eds.; AIChE Symposium Series 212; American Institute of Chemical Engineers: New York, 1982; Vol. 78, pp 28–41.
40. Reed, R.; Healy, R. In *Improved Oil Recovery by Surfactant and Polymer Flooding;* Shah, D. O.; Schechter, R. S., Eds.; Academic Press: New York, 1977; pp 383–437.
41. Lorenz, P. B.; Trantham, J. C.; Zornes, D. R.; Dodd, C. G. Presented at the SPE/DOE Fourth Symposium on Enhanced Oil Recovery, Tulsa, OK, April 15–18, 1984; paper SPE/DOE 12695.
42. Whiteley, R. C.; Ware, J. W. *J. Pet. Technol.* **1977(8),** 925–932.
43. Fanchi, J. R. Carroll, H. B. *SPE Reservoir Eng.* **1988,** *3(2),* 609–616.
44. Raterman, K. T. *SPE Reservoir Eng.* **1990,** *5(4),* 459–466.
45. Maerker, J. M.; Gale, W. W. Presented at the Joint SPE/DOE 7th Symposium on Enhanced Oil Recovery, Tulsa, OK, April 22–25, 1990; paper SPE/DOE 20218.
46. Reppert, T. R.; Bragg, J. R.; Wilkinson, J. R.; Snow, T. M.; Maer, N. K., Jr.; Gale, W. W. Presented at the Joint SPE/DOE 7th Symposium on Enhanced Oil Recovery, Tulsa, OK, April 22–25, 1990; paper SPE/DOE 20219.
47. Gale, W. W.; Puerto, M. C.; Ashcraft, T. L.; Saunders, R. K.; Reed, R. L. U.S. Patent 4,293,428, 1981.
48. Atkinson, H. U.S. Patent 1,651,311, 1927.
49. Johnson, C. E., Jr. *J. Pet. Technol.* **1976(1),** 85–92.
50. Castor, T. P.; Somerton, R. S.; Kelly, J. F. In *Surface Phenomena in Enhanced Oil Recovery;* Shah, D. O., Ed.; Plenum: New York, 1981.
51. Burdyn, R. F.; Chang, H. L.; Cook, E. L. U.S. Patent 4,004,638, 1977.
52. Nelson, R. C.; Lawson, J. B.; Thigpen, D. R.; Stegemeier, G. L. Presented at the Joint SPE/DOE 4th Symposium on Enhanced Oil Recovery, Tulsa, OK, April 15–18, 1984; paper SPE 12672.
53. Wade, W. H.; Morgan, J. C.; Jacobson, J. K.; Schechter, R. S. *Soc. Pet. Eng. J.* **1977,** *17(2),* 122–128.
54. Cash, L.; Cayias, J. L.; Fournier, G.; MacAllister, D.; Schares, T.; Schechter, R. S.; Wade, W. H. *J. Colloid Interface Sci.* **1977,** *59(1),* 39–44.
55. Cayias, J. L.; Schechter, R. S.; Wade, W. H. *Soc. Pet. Eng. J.* **1976,** *16(6),* 351–357.
56. Pushpala, S. M.; Michnik, M. J. Presented at the International Symposium on Oilfield and Geothermal Chemistry, Denver, CO, June 1–3, 1983; paper SPE 11774.
57. Taylor, K. C.; Hawkins, B. F.; Islam, M. R. *J. Can. Pet. Technol.* **1990,** *29(1),* 50–55.
58. Rubin, E.; Radke, C. J. *Chem. Eng. Sci.* **1980,** *35,* 1129–38.
59. deZabala, E. F.; Radke, C. J. *SPE Reservoir Eng.* **1986,** *1(1),* 29–43.
60. Mayer, E. H.; Berg, R. L.; Carmichael, J. D.; Weinbrandt, R. M. *J. Pet. Technol.* **1983,** *35(2),* 209–221.
61. Kuuskraa, V. A. Presented at the SPE/DOE 5th Symposium on Enhanced Oil Recovery, Tulsa, OK, April 20–23, 1986; paper SPE/DOE 14951.
62. Dauben, D. L.; Easterly, R. A.; Western, M. M. "An Evaluation of the Alkaline Waterflooding Demonstration Project, Ranger Zone, Wilmington Field, Cali-

fornia"; U.S. Department of Energy; National Technical Information Service: Springfield, VA, 1987; DOE/BC-10830–5.

63. Krumrine, P. H.; Falcone, J. S., Jr. Presented at the International Symposium on Oilfield and Geothermal Chemistry, Denver, CO, June 1–3, 1983; paper SPE 11778.

64. Martin, F. D.; Oxley, J. C.; Lim, H. Presented at the 60th Annual Technology Conference and Exhibition of the Society of Petroleum Engineers, Las Vegas, NV, September 22–25, 1985; paper SPE 14293.

65. Saleem, S. M.; Faber, M. J. *Rev. Tec. INTEVEP* **1986,** *6,* 133–142.

66. Saleem, S. M.; Hernandez, A. J. *Surf. Sci. Technol.* **1987,** *3,* 1–10.

67. Hawkins, B.; Taylor, K.; Nasr-El-Din, H. Presented at the CIM/AOSTRA Technical conference, Banff, Alberta, Canada, April 21–24, 1991; paper CIM/ AOSTRA 91-28.

68. Clark, S. R.; Pitts, M. J.; Smith, S. M. Presented at the Society of Petroleum Engineers Rocky Mountain Regional Meeting, Casper, WY, May 11–13, 1988; paper SPE 17538.

69. McAuliffe, C. D. *J. Pet. Technol.* **1973,** *25,* 727–733.

70. Schmidt, D. P.; Soo, H.; Radke, C. J. *Soc. Pet. Eng. J.* **1984,** *24,* 351–360.

71. Castor, T. P.; Somerton, W. H.; Kelly, J. F. In *Surface Phenomena in Enhanced Oil Recovery;* Shah, D. O., Ed.; Plenum: New York, 1981; pp 249–291.

72. Cash, R. L., Jr.; Cayias, J. L.; Hayes, M.; MacAllister, D.J.; Schares, T.; Schechter, R. S.; Wade, W. H. Presented at the SPE 50th Annual Technical Conference and Exhibition, Dallas, TX, Sept. 28–Oct. 1, 1975; paper SPE 5562.

73. Fiori, M; Farouq Ali, S. M. Presented at the 40th Annual Technical Meeting of the Petroleum Society of CIM, Banff, Alberta, Canada, May 28–31, 1989; paper 89-40-43.

74. Decker, R. M.; Flock, D. L. *J. Can. Pet. Technol.* **1988,** *27(4),* 69–78.

75. Hoefner, M. L.; Fogler, H. S. *Chem. Eng. Prog.* **1985,** *81(5),* 40–44.

RECEIVED for review December 18, 1990. ACCEPTED revised manuscript May 6, 1991.

8

Pipeline Emulsion Transportation for Heavy Oils

D. P. Rimmer*, A. A. Gregoli, J. A. Hamshar, and E. Yildirim

Canadian Occidental Petroleum, Ltd., 1500, 635 8th Avenue, S.W., Calgary, Alberta, Canada T2P 3Z1

Oil-in-water emulsions provide a cost-effective alternative to heated pipelines or diluents for transportation of heavy crude oil or bitumen. A typical "transport emulsion" is composed of 70% crude oil, 30% aqueous phase, and 500–2000 ppm of a stabilizing surfactant formulation. The resulting emulsion has a viscosity in the 50–200-cP range at pipeline operating conditions. Nonionic surfactants have the advantage of relative insensitivity to the salt content of the aqueous phase. The ethoxylated alkylphenol family of surfactants has been used successfully for the formation of stable emulsions that resist inversion. Correlations have been developed for prediction of emulsion viscosity as a function of emulsion life and process conditions. The cost of stabilizing surfactants is estimated at $0.50 to $1.00 per barrel of crude oil for a transportation distance of 200 to 400 miles.

EONS OR DISPERSIONS OF HEAVY CRUDE OIL in water or brine have been used in several parts of the world for pipeline transportation of both waxy and heavy asphaltic-type crude oils. The hydrodynamically stabilized dispersion transportation concept is described by the Shell Oil Corporation core flow technology (1). The use of surfactants and water to form oil-in-water emulsions with crude oils is the subject of a long series of patents and was proposed for use in transporting Prudhoe Bay crude oil (2). Furthermore, surfactants may be injected into a well bore to effect emulsification in the pump or tubing for the production of heavy crude oils as oil-in-water emulsions (3, 4).

*Corresponding author. Current address: Oxy USA, Inc., Box 3908, Tulsa OK 74102

0065–2393/92/0231–0295 $06.00/0
© 1992 American Chemical Society

The use of oil-in-water emulsions to reduce the viscosity of heavy crude oils and bitumens and thus permit their transportation by conventional pipeline has been under development by Canadian Occidental since the fall of 1984. The benefits of these emulsions may be applied to pipeline transportation, to the combustion of heavy fuels, to increase the production rates of heavy-crude-oil wells, and to improve secondary recovery of heavy crude oil and bitumen. In this chapter, the emphasis is on discussion of the general characteristics of oil-in-water emulsions as related to their application for pipeline transportation. The incentive for developing this technology is to provide an alternative to the use of diluents or the application of heat for viscosity reduction in pipelines for heavy crude oil. The viscosity range for oil-in-water emulsions as compared to undiluted heavy crude oils and bitumens is illustrated in Figure 1. Also indicated in the figure is the viscosity specification for typical pipelines for heavy crude oil. As noted, the emulsion viscosity is well below the required level and provides operating benefits compared to normal operations in which viscosity reduction is achieved by use of diluents.

The use of oil-in-water emulsions in major pipeline systems represents a

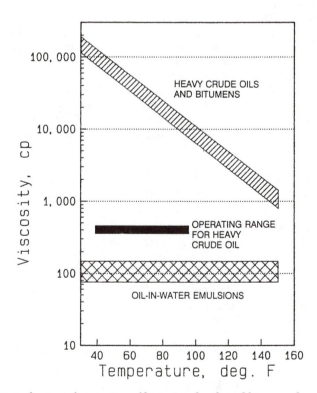

Figure 1. Reduction of viscosities of heavy crude oils and bitumens by conversion to oil-in-water emulsions.

radical departure from conventional practice. As a result, a number of possibilities are causes for concern, including the possibility of freezing, corrosion, emulsion separation or inversion, custody transfer, water separation, and treatment. Such issues may be satisfactorily handled and will be discussed.

Oil-in-water emulsions for pipeline transportation of heavy crude oils may be considered a developing technology that is not yet in wide commercial use. Several companies have ongoing programs in this area and are competing in marketing of the processes and the surfactant formulations involved. Therefore, much of the information relating to this technology is confidential. In this chapter, the topic is discussed on the basis of our experience in development and testing of the emulsion transportation technology.

Canadian Occidental's interest in oil-in-water emulsions is related to marketing and transportation of Athabasca bitumen and heavy Alberta crude oils. A laboratory and pilot-plant development program was initiated in late 1984 at the Occidental Center (formerly the Cities Service Technology Center) in Tulsa, Oklahoma. The program has included the following features:

- development of surfactant systems for preparation of stable oil-in-water emulsions
- evaluation of emulsion preparation systems and selection of optimal conditions for continuous-emulsion preparation
- development of laboratory tests for evaluating the stability and pipelining life of oil-in-water emulsions
- construction of an emulsion pilot plant and testing of the rheological properties and pipeline stability of oil-in-water emulsions
- completion of two field tests to demonstrate the technology

Process Design and Operation

Emulsions designed for pipeline transportation are composed of a continuous phase consisting of water or brine, droplets of the heavy crude oil to be transported, and additives generally consisting of chemical surfactants. The purpose of the surfactants is to provide sufficient stability to the hydrocarbon droplets so that they do not coalesce or absorb water or brine during the pipelining operation. The aqueous phase typically comprises approximately 25–35 wt% of the total emulsion, and the actual concentration is selected so that the minimum quantity of water is used while meeting desired viscosity specifications. The principal components of an emulsion pipeline system are illustrated in Figure 2. As the figure indicates, the system is relatively simple

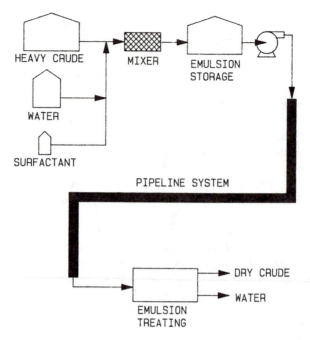

Figure 2. Facilities required for a heavy-crude-oil emulsion transportation system.

and does not require extensive modifications to the pipeline system itself. The principal steps included in operation of a transport emulsion system include preparation of the oil-in-water emulsion, storage and pumping of the emulsion, and finally breaking of the emulsion for recovery of the dry heavy crude oil. The details involved in each phase of the operation are discussed in the following sections.

Emulsion Preparation.   Preparing a transport emulsion is a fundamentally simple operation that includes the steps of forming a water–brine solution of the emulsion-stabilizing composition followed by a shearing process in which the crude oil and aqueous phases are metered to a specific mixing device.

Each developer of transport emulsion technology selects specific surfactant formulations for particular applications. The primary functions of the surfactant are to reduce the interfacial tension between the crude oil and aqueous phases, to provide stability to the individual oil droplets formed during the shearing process, and to prevent subsequent coalescence of the droplets. The surfactant molecules collect at the phase boundaries and provide resistance to coalescence of the oil droplets by establishing mechanical, steric, and electrical barriers (5).

A wide range of surfactant types may be used to form and stabilize transport emulsions. Nonionic surfactants have the advantage of relative insensitivity to the salt content of the aqueous phase being employed (6). The group of surfactants known as ethoxylated alkylphenols, represented by the formula,

$$R-C_6H_4-(CH_2CH_2-O)_x-H$$

where R can be any hydrocarbon and x is the number of ethylene oxide units, is particularly useful in the formation and transportation of heavy-crude-oil emulsions.

Whatever the specific formulation used, the final concentration of the surfactant is selected on the basis of the characteristics of the heavy-crude-oil–brine system and the conditions to which the emulsion will be subjected. The principal factor influencing the quantity of surfactant required is the length of the pipeline system in which the emulsion will be pumped. The concentration of surfactant based on the total emulsion may range from 200 to 5000 ppm, depending on specific system characteristics.

Major equipment required for preparation of transport emulsions includes heated tankage for crude oil and brine; injection pumps for crude oil, brine, and surfactant; premixing and mixing devices; and emulsion storage tanks. Minimal instrumentation is also required to monitor flow rates and temperatures. The basic method for emulsion preparation is to heat the crude oil and brine solutions to the desired operating temperature, dissolve the surfactant into the brine, and simultaneously pump the crude oil and brine through a mixing device in the desired proportions. Typical emulsion formation temperatures are in the 50–90 °C range.

Crude oil and brine pumps may be centrifugal or positive displacement, but must be capable of providing steady flow to the mixing device because emulsion properties are highly dependent on the resulting crude-oil–brine ratio. Surfactant may be dissolved in the brine phase on a batch or continuous basis. Static mixers provide a simple method for the preparation step because they require no moving parts, are easy to scale up, and provide an mixing intensity that is suited to preparation of transport emulsions.

The techniques used in the preparation of a stable oil-in-water emulsion for pipeline transportation are illustrated by the results of a field test in which an Athabasca bitumen was emulsified and pumped through a 3-in. × 4000-ft. pipe-loop system for a total distance of approximately 500 miles. The emulsion in this case comprised 75% by weight of the 8.3° API bitumen and 25% of a synthetic brine containing 1.7% NaCl. (API gravity is defined in the Glossary.) The surfactant used was a mixture of two ethoxylated nonylphenol surfactants; the first component contained an average of 40 ethylene oxide units per molecule, and the second component contained 100 units. Approximately 1500 ppm of the surfactant mixture, based on the total emulsion, was used in preparation of the emulsion. The emulsion was formed by heating both the brine and bitumen to 180 °F and pumping the combined streams through a 2-in. static mixer at a rate of about 10 ft/s. This operation produced an emulsion with an average droplet diameter of 27 μm and a viscosity near 120 cP at ambient conditions. The emulsion was introduced directly into the pipe-loop system for rheology and stability testing and was stable throughout the operating test period of approximately 1 week.

Emulsion Pipeline Operations.   Prediction of pipeline pressure gradients is required for operation of any pipeline system. Pressure gradients for a transport emulsion flowing in commercial-size pipelines may be estimated via standard techniques because chemically stabilized emulsions exhibit rheological behavior that is nearly Newtonian. The emulsion viscosity must be known to implement these methods. The best way to determine emulsion viscosity for an application is to prepare an emulsion batch conforming to planned specifications and directly measure the pipe viscosity in a pipe loop of at least 1-in. inside diameter. Care must be taken to use the same brine composition, surfactant concentration, droplet size distribution, brine–crude-oil ratio, and temperature as are expected in the field application. In practice, a pilot-plant run may not be feasible, or there may be some disparity between pipe-loop test conditions and anticipated commercial pipeline conditions. In these cases, adjustments may be applied to the best available viscosity data using adjustment factors described later to compensate for disparities in operating parameters between the measurement conditions and the pipeline conditions.

After the emulsion viscosity is estimated, friction factor charts may be used directly to determine the flow regime (laminar or turbulent) and the pressure gradient. Emulsion viscosity may be used as an input to a standard pipeline model. Nevertheless, it is strongly recommended that pilot-plant testing be completed on new crude oils before commercial application.

Direct measurement of emulsion viscosity at pipeline conditions is recommended, especially if laminar flow operation is expected. Viscosity is of lesser significance in turbulent flow.

For practical purposes, emulsion viscosities may be adjusted for variations in temperature, water content, and droplet size distribution according to a sensitivity formula of the following type:

$$\mu = \mu_1(\text{TAF})\left[\frac{\text{WAF}_2}{\text{WAF}_1}\right]\left[\frac{\text{PSAF}_2}{\text{PSAF}_1}\right] \qquad (1)$$

where μ is emulsion viscosity (in cP; 1 cP = 0.001 Pa·s), TAF is the adjusting factor for temperature difference, WAF is the adjusting factor for water content, PSAF is the adjusting factor for droplet size, subscript 1 refers to conditions at which viscosity is known, and subscript 2 refers to conditions of application. This formula may be used for field applications in which the viscosity is known from experiments, but it must be adjusted to actual conditions.

The adjustment factor for temperature is based on a temperature difference, $(T_2 - T_1)$. If the temperature difference is negative, then TAF > 1, and the inverse of the TAF from the correlations must be used.

The temperature sensitivity of emulsion viscosity may be described as a percentage change in viscosity per unit temperature change. The temperature-adjusting factor varies for different emulsions, depending on the base crude-oil content, brine content, surfactant, and other variables and generally ranges from about 1.8 to 3.6 cP/°C. (The variation in the viscosity of water with temperature is approximately 2.2 cP/°C.)

To apply this factor, the percentage viscosity change must be compounded, as with interest rates. For example, a correction of 20 °C based on a factor of 2.5% per °C would be calculated as follows:

$$\text{TAF} = \frac{1}{1.025^{20}} = 0.61 \tag{2}$$

Aged emulsions containing a substantial portion of large (>200 μm) droplets exhibit a lower temperature–viscosity sensitivity, and this effect must be considered in calculating pressure gradients. Adjustment factors shown are for temperature increases (lower viscosity). The inverse of the factor applies to temperature decreases (higher viscosity).

The viscosity of an oil-in-water emulsion is sharply dependent on water content. Viscosity adjustment factors for water content may be obtained from a correlation such as that shown in Figure 3. In this figure, the adjustment factor is defined as 1.0 at the base level of 30% water. The actual correlation to be used is dependent on the base crude-oil content and other factors.

A portion of the water in an emulsion can be dispersed within the oil droplets. This portion of the total water should be treated as oil when estimating emulsion viscosity. Generally, added water is present in the continuous phase. If the crude oil contains water prior to emulsion formation, this water may be present in either the continuous (water) phase or the dispersed (oil) phase after emulsion formation, depending primarily on the water droplet size in the crude oil. In order to predict how much of the water in the crude oil will be freed into the continuous phase, emulsion preparation experiments with the actual crude oil to be used are necessary.

Viscosity adjustment factors for droplet size distribution may be determined by a correlation such as that shown in Figure 4. Mean droplet size is defined on a volume basis. Dispersity is an index of wideness of the droplet size distribution. It is defined for this purpose as the ratio of volume-mean droplet size to population-mean droplet size.

As an emulsion ages, droplet coalescence occurs and leads to increased

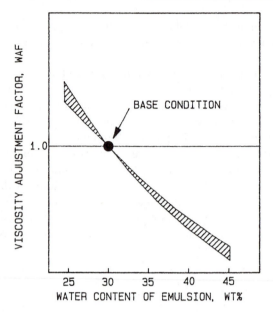

Figure 3. Viscosity adjustment factors for water content variations based on emulsions containing 30% water.

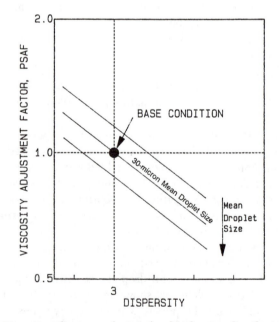

Figure 4. Viscosity adjustment factors for droplet size distribution based on viscosity at 30-μm mean droplet size and dispersity of 3.0.

droplet size and dispersity. These increases, in turn, cause reduced emulsion viscosity. Emulsion viscosity may be related directly to aging by an equation of the form:

$$\mu = \mu_0 \exp{(-k\theta)} \tag{3}$$

where μ is the viscosity (subscript 0 indicates initial viscosity), θ is the time of emulsion transport, and k is a constant characterizing the emulsion degradation rate. This equation can also be written in terms of distance traveled rather than time of transport.

The value of k may be measured experimentally for a given emulsion. The emulsion degradation rate decreases as pipe diameter increases. A conservative assumption for calculation of viscosities for pipeline design is that the degradation rate is proportional to the surface-to-volume ratio $(1/d)$, where d is the pipe diameter.

Emulsion aging rates increase with temperature. Aging rates in turbulent flow appear to become arrested after a certain point, generally being less than the rates observed in laminar flow. Aging rates are suppressed by increased surfactant concentration as a result of the anticoalescence action of the surfactant.

The viscosity of an oil-in-water emulsion generally varies in proportion to the continuous-phase viscosity. If concentrated brines or brines containing additives are to be used, then the continuous-phase viscosity may be substantially greater than that of water, and a correction should be applied. Specific adjustment factors for this effect may be estimated as the ratio of viscosities of the brines in the known and unknown emulsions.

The surfactant concentration is normally not high enough to substantially affect the continuous-phase viscosity. However, changes in surfactant concentration for a pipeline application generally cause an indirect effect on rheology by way of their effects on emulsion preparation and on the emulsion aging rate. Generally, an increase in surfactant concentration results in a smaller initial droplet size and slower emulsion aging. Both of these conditions tend to increase viscosity.

Monitoring Emulsion Aging. The surfactants used in transport emulsions may gradually lose their ability to stabilize the oil droplets. As the oil droplets coalesce, a two-phase mixture is formed, and it remains pumpable with no significant change in effective viscosity. This process is referred to as emulsion failure. An alternative to this process is inversion of the emulsion, in which a water-in-oil emulsion is formed with a potentially very high viscosity. Proper selection of the surfactant formulation can prevent the occurrence of emulsion inversion.

Indicators of emulsion aging that may be monitored include droplet size growth, viscosity decline, surfactant loss, and reduction of shear stability.

Droplet growth rates and viscosity decline rates both are exponential processes, following a straight line on a semi-log plot (log μ or log d_p vs. time), where d_p is the mean droplet diameter. Emulsion failure is also associated with a certain minimum viscosity, depending on water content, crude-oil content, temperature, etc. Viscosity and mean droplet size may be projected to estimate the time remaining before emulsion failure. The ultimate droplet size and viscosity should be determined experimentally for the same formulation in a pilot-plant pipe loop.

The emulsion surfactant concentration generally declines gradually as the emulsion approaches failure, culminating in a sudden sharp drop at or after the time of emulsion failure. Changes of surfactant concentration tend to lag behind changes evident from droplet size and other indicators, and this situation makes surfactant concentration analysis ineffective as an indicator of approaching emulsion failure.

Emulsion life expectancy for a formulation may be conservatively scaled up from 2-in. pipe-loop tests at the same velocity by demonstrating that the emulsion will survive transport for the desired actual distance in the pilot plant. Pilot-plant transport is a more severe test of emulsion life than transport in larger lines. The conservative nature of this scale-up criterion tends to dictate specification of some excess surfactant for a large-scale application beyond the minimum quantity required.

Effects of Pumps and Valves. The flow of emulsions through pipeline pump and valves could potentially affect the emulsion properties.

Pumps. Impeller tip speed is a useful guide to relate centrifugal pumps in terms of the energy they may impart on an emulsion. Several pumps have been tested on emulsion service with tip speeds up to 200 ft/s, compared to typical pump-station applications of approximately 300 ft/s. The results of these tests show that emulsion shear stability is unchanged after several passes through a centrifugal pump, typical of multistage pump-station application. Some undersized material is formed at the expense of oversized. These results indicate that commercial pump applications should not be a problem. Testing has been limited, however, and thus prior to any commercial application, the specific pump characteristics should be compared against the pumps already tested. Pump tip speeds and the pump modeling law that relates pump geometries should be reviewed.

Passage of emulsion through a centrifugal pump at abnormally low rates and at a high back pressure can shorten the emulsion shear stability. Gear pumps are low-shear devices and do not adversely affect the emulsion.

Valves. Limited laboratory testing shows that emulsions can be let down across a pressure differential of 1000 ψ (6895 kPa), typical of commercial applications, with marginal reduction in shear stability. Velocities across the valve port reached 130 ft/s (39.6 m/s).

Effects of Pipeline Shutdown and Restart.   When a pipeline containing emulsion is stopped and allowed to remain shut-in for one or more days, there is generally a higher than normal pressure gradient on restarting flow. In pilot-plant experiments, the pressure gradient substantially returned to its normal value within 5 min. The increased pressure gradient is due to creaming or stratification of the emulsion. Additional energy is required on restart to redisperse the material uniformly.

Restarting of a phase-separated mixture (failed emulsion) is similar to restarting an emulsion, but the starting pressure surge is substantially greater as a result of total separation of phases.

Fluid Being Restarted	Pressure Surge Range, Percent of Steady-State Pressure
Emulsion (30% water)	100–250, 200 typical
Emulsion (38% water)	100–150, 130 typical
Phase-separated mixture (30%)	400–800, 700 typical

Corrosion Considerations.   Corrosion rates are dictated by the properties of the brine being used. Pilot-plant testing with an electrochemical corrosion rate probe indicated corrosion rates of less than 5 mils/year for emulsions flowing in pipes. The corrosion rate declined over time, presumably because of formation of an oil layer on the metal. In all pilot and field tests that we conducted, pipe walls have always shown a thin (approximately 0.001-in.) layer of crude oil on the wall after emulsion runs.

Corrosion rates for live brines containing CO_2 or H_2S are expected to be higher than those measured on dead brines in pilot-plant testing. On-line corrosion monitoring in the field is indicated on a case-by-case basis. Corrosion rates for emulsions are not expected to be any worse than those for crude oils containing brine.

Demulsification.   The final part of the emulsion transportation system is demulsification or breaking of the oil-in-water emulsion to recover dry crude oil. The equipment and process conditions required for this operation are the same or similar to those used for a conventional crude-oil dewatering process.

The techniques used for demulsification of a transport emulsion may include raising the temperature of the emulsion, addition of emulsion-breaking additives, addition of diluents to reduce the viscosity of the heavy crude oil, and the use of equipment designed to promote coalescence of the crude-oil droplets. Raising the temperature of the emulsion increases the difference in density between the hydrocarbon and aqueous phases and encourages creaming of the emulsion. The reduced viscosity of the hydrocarbon phase at high temperatures also improves the operability of the demulsification process. The viscosity of the crude oil may also be reduced by addition of diluents if this operation is appropriate for downstream processing. The use of demulsifying additives is designed to counteract the effects of the emulsifying surfactants. The surfactants used in the stabilization of emulsions of heavy crude oils generally have a high HLB (hydrophilic–lipophilic balance). For demulsification operations, the effectiveness of these surfactants may be counteracted by addition of surfactants with a low HLB. Many commercially available products with proprietary compositions are available for this purpose. Ethoxylated alkylphenols, which are used in the emulsification process, may also be used for demulsification if components are selected with a low number of ethylene oxide groups.

The basic procedure for demulsification of a heavy-crude-oil transport emulsion then consists of the following steps:

1. Raise the emulsion temperature to 190 to 250 °F.
2. Add demulsification surfactants.
3. Possibly add diluents for viscosity reduction.
4. Provide residence time sufficient for separation of the oil and water phases.

The time required for separation in step 4 depends on the density difference between the oil and water phases, the treating equipment, and the treating temperature.

In some cases it may be desirable to perform the demulsification process in two stages. In the first stage the bulk of the water may be removed at a minimum process severity, as already described. A second process stage at a higher temperature and possibly at elevated pressure may then be used for final dry crude-oil recovery.

Changes in process control procedures for the demulsification operation may be required for oil-in-water emulsions. Interface detection instruments must be able to detect the difference in water and an oil-in-water emulsion. Adjustment of control levels in separation vessels may be required for proper operation.

In some cases, minimal effort is required for the demulsification process. For example, in field tests, adequate separation of a bitumen emulsion could be achieved without the use of demulsifiers by raising the temperature of the emulsion to 190 °F and providing 24 to 48 h of residence time in quiescent storage tanks. However, proper selection of demulsification chemicals is essential when treating the emulsions in conventional equipment on a continuous-flow basis.

The water separated in the demulsification step is similar in character to water produced in a conventional refinery desalting process. Although no detrimental effects are anticipated, the impacts of surfactants in the water or downstream processing units must be evaluated for each specific case. The water phase may contain surfactant fragments that could require treatment or removal prior to disposal. Refineries, however, use a variety of chemicals, catalysts, additives, etc., many of which end up in waste streams and require treatment.

Storage, Maintenance, and Special Requirements. If oil-in-water emulsions must be stored in tanks either before or after pipelining, agitation is required to prevent creaming in the tank. Creaming refers to the concentration of oil droplets on the surface of the fluid that can result in a thick skin or crust that may not readily be dispersed into the bulk of the emulsion. Slow agitation just sufficient to continually roll the tank contents will prevent creaming. Excessive agitation should be avoided to prevent shear degradation of the emulsion.

Maintenance requirements should be the same for an emulsion pipeline as for a conventional petroleum pipeline. Similarly, no unusual maintenance is expected for the emulsion preparation or demulsification parts of the system.

If the pipeline used for emulsion transportation is a common carrier, special procedures may be necessary for metering and custody transfer. On-line instruments for measurement of emulsion water content may be required in such an application.

Economics

The economic analysis of an emulsion pipeline transportation system is highly site specific and depends on several factors that cannot be specified for a general case. However, example cases are presented to illustrate typical costs associated with use of the technology.

Surfactant Cost. A major cost associated with using oil-in-water emulsions is the cost of the surfactants used to stabilize the oil droplets within the emulsion. This cost will depend upon the surfactant formulation chosen for the specific application, the transportation distance involved, and in some cases the type of crude oil being emulsified. On the basis of the formulation that we typically use and current market prices, the estimated surfactant cost to transport heavy crude oil as an emulsion for a distance of 200 to 400 miles (322 to 644 km) is approximately $0.50 to $1.00 per barrel of crude oil shipped. For greater pipeline lengths, up to 1500 to 2000 miles, the surfactant cost may increase by 50–100% relative to the shorter distances. These costs are based on conservative estimates of the quantities of surfactants required and may likely be reduced after optimization for a particular application.

Emulsion Transportation versus Diluent Recycle. The prevailing method in use for transportation of heavy crude oil and bitumen in Alberta is to dilute the crude oil with approximately one part of a light hydrocarbon diluent, typically natural gas condensate, to two parts of crude oil. This quantity of diluent is generally sufficient to reduce the crude-oil viscosity enough so that minimum pipeline specifications may be met. However, a potential shortage of condensate diluent may limit the use of this method for heavy-crude-oil transportation. An alternative to the once-through use of diluent is to recover the diluent at the pipeline end by fractionation and recycle it to the start of the pipeline. This method is compared in the following section with the use of oil-in-water emulsions for the same hypothetical application. The basis for this example is a 24-in., 200-mile pipeline designed to transport 300,000 barrels per day of blend or 200,000 barrels per day of undiluted heavy crude oil. A parallel line is assumed for return of separated diluent. Other bases and assumptions used in the evaluation are as follows:

Emulsion properties	
Gravity	10° API (sp. gr. = 1)
Viscosity	100 cP
Crude-oil concentration	70%
Flow rate	286,000 barrels per day (45,474 m³/day)
Surfactant cost	$0.75 per barrel of crude oil
Water-disposal cost	$0.20 per barrel of water
Tariffs	
For blend	$1.00 per barrel
For diluent	$0.50 per barrel
For emulsion	$0.50 per barrel
Fuel	$5.00 per million Btu
Electricity	$0.06 per kilowatt hour
Capital related costs	25% of total installed cost

The reduced tariffs for emulsions compared to blend are assumed because of the 75% reduction in viscosity for emulsions versus blend and the resultant decrease in pumping costs.

The calculated crude-oil transportation cost for the emulsion case was based on estimates of the required capital investments for emulsification and oil recovery facilities, operating costs including pipeline tariffs, surfactant costs, and water disposal. The costs for the recycle blend case included the normal pipeline operating costs plus the costs of separating and pumping the diluent back to the start of the pipeline. The costs assumed for the emulsion case were based on the assumption that the costs of crude-oil production facilities will not be influenced by the subsequent conversion of the produced crude oil to an oil-in-water emulsion. In some cases, these costs may be reduced by reducing the normal crude-oil dewatering requirements necessary to meet usual pipeline specifications (<0.5% H_2O). In any case however, crude-oil drying operations will still be required to avoid pumping excess water in the pipeline system.

An alternative to the case described is one in which the water separated from the emulsion is recycled to the start of the pipeline for reuse. Water recycle would eliminate the problems and expense of water disposal and would reduce the required quantity of surfactant, because a portion of the surfactant would remain in the separated water. For this case it is assumed that a 50% recovery of surfactant could be achieved after demulsification.

The disposal and recycle cases were compared to an alternative transportation method that includes fractionation of a light diluent fraction from a heavy-crude-oil–diluent blend and recycling of the diluent to the start of the pipeline. For this case, an 85% recovery of diluent from the blend was assumed together with a loss in value of $5 per barrel of unrecovered diluent. The estimated transportation costs (dollars per barrel of crude oil) for the three cases are summarized as follows:

- emulsion with water disposal, $2.04

- emulsion with water recycling, $1.97

- heavy-crude-oil–diluent blend, $2.74

A detailed breakdown of the capital and operating costs estimated for the three cases is shown in Table I. The analysis just presented is for illustration only, and the relative costs for any specific application require further evaluation.

Economic Effects of Emulsion Water Concentration. An important parameter in preparation of an oil-in-water emulsion used for heavy-crude-oil transportation is the concentration of water or brine used. The economic effects of this variable may be evaluated to determine the optimal value for minimization of the crude-oil transportation cost. Changes in the water content of a transported emulsion mainly affect the fluid viscosity, the total fluid flow rate, and the volume of water requiring disposal or recycling. Other system components are also affected, including raw water handling, emulsion formation, and demulsification. For this illustration, only the pipeline pumping costs and water-disposal costs are considered. In a commercial pipeline system, the selection of the emulsion water content may be based on tariffs rather than on pumping costs. This analysis is based only on the actual pumping costs incurred.

Table I. Transportation Costs for an Emulsion-Transport System Compared to Diluent Recycling

	Emulsion System		Diluent Recycle
Item	Water Disposal	Water Recycle	Diluent Recycle
Capital Costs, million dollars			
Emulsification system	3.5	3.5	
Emulsion separation	33.6	33.6	
Diluent recovery			72.0
Total	37.1	37.1	72.0
Operating Costs, dollars per barrel			
Surfactants	0.75	0.56	
Fuel	0.20	0.20	0.52
Tariffs	0.71	0.92	1.50
Water disposal	0.09		
Maintenance and labor	0.07	0.07	0.10
Diluent makeup			0.37
Miscellaneous	0.10	0.10	
Subtotal	1.92	1.85	2.49
Capital-related costs	0.12	0.12	0.25
Total	2.04	1.97	2.74

The emulsion water concentration has a double effect on pumping costs. For a fixed flow rate of crude oil, as the water content increases, the total volume flowing increases, which increases the cost of pumping. However, additional water in the emulsion also reduces its viscosity and thereby lowers pumping costs. These two factors tend to offset each other for emulsion crude-oil concentrations in the 50–70% range. At higher oil concentrations, the viscosity increases more rapidly, and this situation reduces the incentive for further water reductions. Correlations of emulsion viscosity as a function of water concentration, such as were described earlier, are required to perform this analysis. Calculations were performed for a hypothetical pipeline system with the following characteristics:

- length, 200 miles
- diameter, 12 in.
- crude-oil gravity, 10° API
- crude-oil flow rate, 50,000 barrels per day
- pump efficiency, 67%
- electricity cost, $0.05 per kilowatt hour
- water disposal, $0.20 per barrel

The results of this analysis are shown in Figure 5. The emulsion pumping cost is minimized for the hypothetical pipeline system at an emulsion crude-oil concentration in the 60–62% range, and the water-disposal costs naturally decrease continuously as emulsion water content is decreased. The total cost for this case reaches a minimum at about 76% crude oil in the emulsion.

This evaluation illustrates that emulsion water concentration should generally be reduced until the viscosity begins to be significantly increased. In selecting the actual optimum concentration for a specific case, other factors such as the effect of water concentration on emulsion stability, the effect of total flow rate on pump efficiency, and the cost and availability of water should also be considered.

Conclusions

Oil-in-water emulsion technology for transportation of heavy crude oils and bitumens provides a viable alternative for the use of diluents or heated pipelines. Surfactant formulations that have been developed provide stable

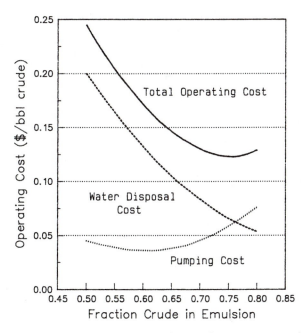

Figure 5. Total operating costs, minimized in emulsions containing high concentrations of crude oil.

operation and adequate emulsion life. Methods are available for the formation of emulsions with desired properties for efficient pipeline operation.

Further development of emulsion transport technology is dependent upon future economic factors such as increases in the price of heavy crude oil and potential shortages of diluent. Commercial operation of an emulsion transport system is required to determine the long-term technical and economic viability of this technology.

References

1. *Oil Gas J.* **1972,** *2(17),* 37.
2. Marsden, S. S.; Rose, S. C. *Oil Gas J.* **1971,** *10(11),* 24.
3. Simon, R.; Poynter, W. G. Presented at the 43rd Annual Meeting of the Society of Petroleum Engineers of American Institute of Mining, Metallurgical, and Petroleum Engineers, Santa Barbara, CA, 1968; paper SPE 2174.
4. Simon, R.; Poynter, W. G. *J. Pet. Technol.* **1968,** *20,* 1349.
5. Rosen, M. J. *Surfactants and Interfacial Phenomena;* Wiley: New York, 1978, pp 226–235.
6. *Nonionic Surfactants;* Schick, M. J., Ed.; Surfactant Science Series; Dekker: New York, 1966, Vol. 2.

RECEIVED for review December 18, 1990. ACCEPTED revised manuscript June 14, 1991.

9

Commercial Emulsion Breaking

Richard Grace

Nalco Canada Inc., 3464 78th Avenue, Edmonton, Alberta, Canada T6B 2X9

*This chapter's purpose is to share the qualitative perspective of emul-
sion breaking held by the petroleum industry at the production or
refining operations level. Incorporation of site-specific data is
avoided in favor of broader conceptual information to emphasize that
each commercial facility must be viewed as unique when developing
emulsion-breaking goals and methods. Theories of emulsions and
demulsification, variability of applied chemicals, and limitations of
present demulsifier selection techniques are presented. This chapter
reinforces the belief that a qualitative view of emulsion breaking is
essential at this time for the petroleum industry, unless the number of
variables is reduced, typically by studying each commercial facility
or crude oil as a unique case. If such study is economically feasible,
valid models for predicting emulsion-breaking performance may be
developed and applied on a wider scale.*

EMULSIONS OF OIL AND WATER are one of many problems directly associated with the petroleum industry, in both oil-field production and refinery
environments. Whether these emulsions are created inadvertently or are
unavoidable, as in the oil-field production area, or are deliberately induced,
as in refinery desalting operations, the economic necessity to eliminate
emulsions or maximize oil–water separation is present. Furthermore, the
economics of oil–water separation dictate the labor, resources, and monies
dedicated to this issue. Before we describe the methods and economics of
emulsion breaking at commercial facilities, we will restate several key concepts concerning emulsions and the petroleum industry.

Emulsions and Demulsification

Theory of Emulsions. An emulsion is a mixture of two immiscible
liquids, one of which is dispersed as droplets in the other. For the petroleum

0065–2393/92/0231–0313 $07.75/0

industry, the two liquids are usually crude oils or refined hydrocarbon products and water. The presence of water in crude-oil systems is a part of the production from oil wells. Most producing wells will produce water and oil simultaneously at some point in their life spans, either as a result of natural formation conditions or as an effect of secondary or tertiary production methods. Within the refining industry, water is either present as a result of water contamination present in the crude oil, induced into the crude oil to "wash" contaminants from it, or the result of steam injection to improve fractionation. Through a variety of mechanisms an emulsion may form from this oil–water mixture.

Emulsions formed in the petroleum industry are predominantly water-in-oil or regular emulsions, in which the oil is the continuous or external phase and the dispersed water droplets form the dispersed or internal phase. Reverse emulsions, or oil-in-water emulsions, are formed when water constitutes the continuous phase and oil constitutes the dispersed phase. It is not unusual to find both regular and reverse emulsions occurring together. More complex emulsions have also been noted when reverse emulsions exist within the internal phase of a regular emulsion. The complexity of these emulsions may allow many alternating internal–external phases, although these are very rare.

A stable emulsion is one that is unable to resolve itself in a defined time period without some form of mechanical or chemical treatment. Three basic conditions must exist before the formation of a stable emulsion occurs:

1. Two immiscible liquids must be present. This condition is met by the simultaneous presence of oil and water in many petroleum industry environments.

2. An emulsifying agent must be present to form stable oil-and-water emulsions (in much the same manner that the normally immiscible combination of oil and vinegar is emulsified by egg whites to form the stable emulsion, mayonnaise). The type of oil and water emulsion formed is dependent on the type of emulsifying agents present. Emulsifying agents that are more soluble, dispersible, or wettable in or by oil favor the development of oil as the external phase and hence a water-in-oil emulsion. Emulsifying agents that are more soluble, dispersible, or wettable in water favor the development of oil-in-water emulsions. Commonly occurring emulsifying agents found in petroleum emulsions are asphaltenes, resinous substances, oil-soluble organic acids (such as naphthenic acid), finely divided carbonate scales, silica, clays, metal sulfates, metal sulfides, or chemical additives. These substances usually stabilize droplet interfaces between external and internal phases of the emulsion.

3. Mixing energy or agitation must be supplied to the mixture of oil and water to disperse one liquid within the other. In general, the greater the agitation or energy applied, the more stable the emulsion. This stability is a result of the reduction in droplet size of the internal phase.

Each of these factors is variable. The composition and properties of oil, water, and associated contaminants vary widely from source to source. The type and amount of emulsifying agents present within different oils and waters also vary widely. The amount of agitation or energy that an oil–water mixture is subjected to is dependent on the fluid types, pressures, velocities, and mechanical parameters present at each commercial petroleum facility. Although the variations of each component are considerable, the sum total of their effects produces an almost infinite variety of emulsions and environments in which to "break" emulsions. As a result, determining the most cost-effective method of breaking an emulsion is generally a site-specific venture. Theories and methods of demulsification are used as a base from which to develop a tailor-made emulsion-breaking program that addresses the goals of either the oil producer or the refiner in the most cost-effective manner.

Theories of Demulsification. Within commercial emulsion breaking, a number of general rules help to form the basic philosophy of how emulsions behave:

1. Petroleum emulsions are composed primarily of immiscible liquids. Separation should be the natural tendency of these liquids, providing a density differential between the liquids exists.

2. The rate of gravitational settling or rising is dependent on the surface tension of the droplets that form the internal phase of the emulsion. Large droplets have less surface tension as a function of mass than small droplets; therefore, anything that can be done to increase droplet size, or coalescence, will increase the rate of separation.

3. An emulsion is stable within a given environment. Altering the environment may affect the stability of an emulsion and thus allow separation of the phases.

4. A stable emulsion exists only when emulsifying agents are present. Elimination, alteration, or neutralization of the emulsifying agents will allow immiscible liquids to separate.

From these four generalizations it becomes apparent that a number of options exist in emulsion breaking. Any single change in these areas may result in the resolution of an emulsion. How these various factors affect emulsion stability is presented in brief as follows.

Viscosity.   An oil with a high viscosity has the ability to hold up more and larger water droplets than an oil with a lower viscosity. The viscosity of an oil can be reduced by the application of heat, the addition of a diluent, or the addition of chemicals. Lowering the viscosity increases both the rate at which water droplets settle and the mobility of water droplets and thereby leads to collisions, coalescence, and a further increase in the rate of separation.

Density Differential.   The difference in densities of the two liquid phases may be increased. Heating the emulsion typically decreases the density of the oil at a greater rate than that of water and thus allows more rapid settling of the water. Heavier oil is typically more difficult to dehydrate than light oil, as its density is closer to that of water. The density of the water is also important; fresh water will tend to separate from oil at a slower rate than salt water.

Water Percentage.   The relative proportion of oil and water affects the stability of an emulsion. In a regular emulsion, the maximum stability of an emulsion will occur at a set ratio of water to oil. Typically this maximum is found at low water percentages as these droplets have a much smaller chance of colliding with other water droplets and coalescing. Increasing the water percentage may destroy the stability of an emulsion.

Age of Emulsion.   Stabilities of emulsions generally increase with age. Oxidation, photolysis, evaporation of light ends, or bacterial action may increase the ratio of emulsifying agents within an oil. (Light ends are low-molecular-weight, low-density hydrocarbons, such as pentane, hexane, and butane, that will vaporize xylene significantly over time.) Breaking emulsions as soon as possible after emulsion formation will eliminate or reduce the effects of aging.

Control of Emulsifying Agents.   Emulsifying agents are necessary to create emulsions. The elimination, alteration, or neutralization of these materials allows for resolution or prevention of emulsions. Elimination of emulsifying agents may include corrosion inhibition programs to reduce the amount of iron sulfide available, careful selection of corrosion inhibitors to avoid emulsification tendencies, or elimination of incompatible crude oils from crude-oil blends. An incompatible crude-oil blend is one that, when blended, results in the precipitation of asphaltenes. This precipitation most commonly occurs when an asphaltic crude oil is blended with a paraffinic crude oil. Alteration of emulsifying agents would include such measures as the addition of an asphaltene dispersant to "tie up" asphaltene polar sites, addition of paraffin crystal modifiers to prevent large paraffin crystals from stabilizing emulsions, or raising treating temperatures above the paraffin cloud point of a crude oil. Neutralization of emulsifying agents generally relates to the neutralization of polar charges associated with the film of emulsifying agents formed around the emulsified droplets. Neutralization is the function carried out by commercial emulsion breakers or coagulants that promote coalescence and thereby accelerate settling by gravity.

Agitation Control. Measures that reduce or eliminate agitation of an oil-and-water mixture will reduce emulsion stability or prevent emulsion formation.

Performance Parameters in Production and Refining Operations

Within the petroleum industry, emulsion of oil and water may be associated with every stage of production, transportation, or refining. The extent of emulsification and the economic impact of contaminants associated with emulsions in hydrocarbon-processing equipment will determine what treating methods, if any, are necessary to produce desired hydrocarbon specifications. An understanding of the impact of oil contaminants and incomplete demulsification on key hydrocarbon processing areas is required. The prime areas of concern are typically

1. hydrocarbon dehydration
2. inorganic solids and salt removal
3. effluent or produced-water quality
4. oil–water interface control
5. treating temperatures

In a broader sense, these areas of concern can be divided into product quality issues, operability issues, and energy conservation. Although these concerns are valid for both the oil producer and refiner, the product quality goals of each portion of the oil industry may not be synergistic.

Producers and refiners tend to view demulsification performance parameters from the following perspective.

Oil Dehydration. Oil production companies must be able to bring their product to market. For most large-scale production, this goal requires entry into a crude-oil pipeline system. Produced oil must meet or exceed pipeline specifications. In Canada, the oil may not contain more than 0.5%

basic sediment and water (BS&W) as determined by a standard BS&W test (ASTM D 96) or a variation of this test.

Pipeline specifications for water content are needed for a variety of reasons. Among these are the refiner's need for a predictable and high quality of raw materials, the desire of the pipeline company to reduce corrosion potential by eliminating the electrolyte (water) from the corrosion process, and the desire of the pipeline company to construct pipelines on the basis of their capacity to deliver marketable hydrocarbon products rather than waste materials. Failure by the oil producer to meet pipeline specifications for any extended period of time will result in the pipeline company refusing to accept produced oil. This outcome alone forces the oil producer to ensure that emulsions are resolved to reduce BS&W in oil to 0.5% or less. Producers pay pipeline tariffs according to their BS&W content, so that reductions in BS&W below 0.5% reduce the cost of transporting their product to market. If each reduction in BS&W below 0.5% is less than the costs necessary to achieve the new BS&W standard, the oil producer will pursue that standard.

Within the refining environment, the field of crude-oil dehydration is viewed somewhat differently. The first process a crude oil (or blend of crude oils) is subjected to is the desalting process. This process was developed with the expectation that a crude oil will have a known water content (less than 0.5%) and a soluble inorganic chloride salts content associated with this water (formation waters from oil-field production may have salt contents approaching 300,000 mg/L). Salts may also occur in crystalline form dispersed within the oil. As these salts have considerable negative effects in the downstream processes of the refinery, it is desirable to remove them.

This salt removal is accomplished by injecting a relatively fresh water (typically 3–8% of the crude-oil volume) into the crude-oil charge line to extract the salt or "wash" the oil. (The charge line is the line that transfers crude oil from storage or pipeline through preheat exchangers to the desalter vessel.) The more thorough the contact of the water with the crude oil, the higher the potential for extracting salts. Hence, agitation is usually applied with mix valves or in-line static mixers to promote thorough mixing. This agitation usually creates an emulsion. The separation of the oil and water phases then takes place in a desalting vessel or treater.

A problem may occur here in that the separation of oil and water may not be complete by the time the oil exits the desalter on its way to an atmospheric fractionation unit. Any water that remains with the crude oil will have to be heated to atmospheric fractionator inlet temperature, typically 290–370 °C.

This requirement provides a significant cost in fuel gas to heat the water. At a 20,000-m³/day refinery with a desalter outlet temperature of 100 °C and a crude-oil heater outlet temperature of 315 °C, each tenth of a percent of water carryover will result in 20,000 kg of water being heated 215 °C to form superheated steam. If very high amounts of water carryover are present, the expansion of the steam, once it enters the fractionator, may damage the vessel. Incomplete dehydration will also reduce desalting efficiency as the salts are carried into the system with the excess water. Refiners typically attempt to limit water carryover to 0.2% in light crude oils and 0.4% in heavy crude oils by attempting to fully resolve emulsions within the desalter.

Inorganic Solids and Salt Removal. For the oil producer, the driving force for the removal of inorganic solids and salts from produced oil is to achieve pipeline specifications. Any combination of basic sediments (primarily inorganic solids) and water greater than 0.5% will prevent shipment of product. The removal of a majority of salts is usually accomplished as a side-effect of removing excess water from the crude oil. This removal necessitates disposal of the produced water by the producer. Typically, disposal is accomplished through deep disposal wells or water-flood injection wells. Deep-well disposal is viable only in certain regions because of environmental concerns or regulations. The removal of solids such as sands, silts, clays, and corrosion products must also occur if these are present in significant quantities. Heavy-oil production typically contains far greater quantities of inorganic solids than does light-oil production.

Pipeline requirements for reducing inorganic solids content are needed for a variety of reasons. Again, the refiner wishes to receive a product of predictable and high quality. The pipeline company wishes to transport oil, not inorganic solids, and can reduce maintenance costs by controlling erosion of mechanical parts caused by solids. Controlling solids also helps to mitigate underdeposit types of corrosion within a pipeline.

The oil producer, on the other hand, must dispose of any solids removed from an emulsion-treatment system by land farming (placement of waste solids in an approved landfill area), shipment to an approved waste facility, re-injection through a disposal well or water-flood system, or shipment to pipeline. These options are governed by the amount of oil associated with the solids, which is directly related to emulsion-breaking capabilities and the amount of solids present. In many cases it may be desirable for an oil producer to blend a portion of oil-wet solids into shipments for pipeline if the specification of less than 0.5% BS&W is not exceeded. This method of solids disposal may be the most cost-effective available to the producer.

The refiner's position on inorganic solids and salt removal is that as much of these contaminants (as is cost-effective) should be removed from the incoming crude oil into the wash water by the desalting process. Excess chloride salts become catalyst poisons that promote excessive catalyst consumption or reduce conversion in the cracking and treating processes. Chloride salts also compromise the reliability of refinery overheads where, because of hydrolysis upon heating, they form highly corrosive hydrochloric acid in the overhead system. (The refinery overhead is the equipment, such as exchangers, accumulators, reflux equipment, and lines, that is used to condense hydrocarbons and water vaporized in a fractionation column and exiting from the top of the column.) Chloride salts also cause fouling, which restricts refinery run lengths or aggravates corrosion. The economics of salt removal are very complicated. A salt content of less than 2.85 g/m³ (1 lb/1000 barrels) is a common refinery target for desalted crude oil, although this target varies considerably from site to site. Prime modifiers to this target are desalting equipment available and type of crude-oil stock processed.

The removal of inorganic solids is also important to the refiner. By efficient emulsion breaking at the desalter, the refiner can expect to significantly reduce the amount of solids carried through the desalter into the downstream refining units. Inorganic solids removal of 80% or greater is a typical base standard, although this amount varies greatly with each refiner, the type of separation equipment available, and the crude oils involved.

Inorganic solids have numerous detrimental effects on the refiner. As with chloride salts, many solids (especially metal-containing solids) may poison catalysts and thereby increase catalyst consumption or reduce conversion of hydrocarbons. Solids may act as foulants or promote reactions that create fouling, which restricts heat transfer or run lengths in exchangers and furnaces. Excess solids may reduce the product quality of residual fuels or coke. Solids may also promote foaming and fouling in fractionators, which in turn may restrict throughput or induce use of antifoams. All of these effects usually occur simultaneously in an integrated refinery. Accurate calculations on the economic impact of inorganic solids are extremely complicated and difficult to perform, although such calculations can be the prime justification for improving desalter operation. Refiners may dispose of removed inorganic solids by land farming, shipment to approved waste-disposal facilities, or disposal to the water-handling system.

Effluent Water Quality.   When resolving emulsions in the petroleum industry, water is always produced. In some areas the goals of the refiner and oil producer are identical in effluent water quality. Both refiner and producer want to minimize the amount of oil in the water systems, as this represents either lost product, increased capital, or increased operating costs to recover the oil from the water. The costs of not achieving complete oil–water separation the first time in a desalter or oil-field treater are reflected by the need for equipment such as water-skim tanks, water-filtration equipment, various flotation units, clarifiers, settling ponds, biox units, and accumulated-oil treating facilities in water-disposal systems. Each of these units will have significant operating costs above and beyond the capital costs. (A biox unit is a basin, pond, or vessel where biological activity digests organic materials such as hydrocarbons carried with the water charged to the unit.)

Oil producers will typically set standards for oil-in-water content ranging from less than 10 ppm in very light crude oils to several hundred parts per million in very heavy crude oils. These specifications are usually site-specific and are dependent on equipment available and crude-oil type. Oil producers in Canada usually have the advantage of disposal wells or water-flood schemes in which produced water is disposed. Failure to meet self-imposed oil-in-water limits usually results in loss of hydrocarbon product back to the formation. For an oil production facility that disposes of 1000 m³ of water per day with an oil content of 1000 ppm, 365 m of oil is lost per year. At $25 (Canadian) per barrel, this amount of oil translates to a product loss worth approximately $57,000 per year, plus any maintenance costs and well stimulation costs to restore injectivity lost as a result of formation plugging from oil-wet solids. Oil-wet solids in water-flood systems may damage formation permeability and reduce recovery.

In the Canadian refining industry, only refineries in the western provinces have access to disposal wells for produced water. Their concerns are similar to those mentioned with the oil producers. In other areas of the country, effluent waters must be treated to a standard that will allow for discharge to the environment. The typical standard for allowable oil-in-water content is less than 10 ppm. Failure to consistently achieve this specification may result in fines, shutdown, and poor public perception of the offending refiner.

Oil–Water Interface Control.   In any petroleum processing unit in which emulsions are resolved, an interface between oil and water must occur. The quality of this interface is directly related to the efficiency of demulsification in either a refinery desalter or an oil-field free-water knock-out or treater. The sharper the transition between clean water and clean oil (or the tightness of the interface), the better the ability to control oil and water retention times and quality and operate the vessel.

Broad transitions between oil and water are usually the result of (1) incomplete resolution of the emulsion, (2) failure of an emulsion-treating program to resolve emulsions caused by one or more specific emulsifying agents, (3) excess or incompatible treating chemicals, (4) buildups of oil-wet solids, (5) buildups of insoluble organic materials such as paraffins or asphaltenes, or (6) any combination thereof. With a broad interface ("rag", "cuff", or "pad" layer) present, level-sensing equipment and water dumps may operate incorrectly or malfunction, and thereby divert water to clean-oil outlets and oil to water outlets. Electrostatic grids may short out if the pad (usually high in water content) makes contact with the grids. Emulsion pads also occupy space within a treating vessel and promote channeling, which affects the retention time of a vessel. Concentration of polar hydrocarbons, emulsifying agents, and solids usually occurs in interface "pads". These molecules have a high tendency to be dispersed in the water phase, become concentrated through oil-in-water recovery processes, and if recycled, upset the vessel to further aggravate emulsion treatment.

In general, emulsion pads limit the performance of emulsion-breaking equipment and chemicals. Both refiners and oil producers view large emulsion pads as symptoms of inefficient emulsion treatment. If the emulsion pad can be eliminated, performance in all aspects of emulsion breaking should improve.

Treating Temperature.   Treating temperatures may have significant impacts on both refinery and oil production emulsion-breaking processes. In general, the higher the temperature, the greater the ability to resolve emulsions; however, increases in treating temperature affect many other factors negatively. In the refinery desalter the range of treating temperatures available is limited by plant design. Heating of the crude oil up to desalting temperatures is the result of exchanger efficiencies, throughputs, and plant design.

Treating temperatures are almost always designed to surpass paraffin crystal melting points (50–65 °C) and generally approach the boiling point of water at the specific unit pressures. The treating temperatures are usually set to provide the greatest amount of dehydration possible. As temperatures increase, the ability to resolve emulsions increases, but so does the solubility of water in hydrocarbons. Temperatures that will provide the greatest measure of crude-oil dehydration must be selected. This temperature choice will vary with the composition of each crude oil and with treating vessel pressure.

Within the oil-field production facilities, much greater control of temperature may be exercised. The end goal of the producer is to select a treating temperature that, in combination with other factors, provides the most cost-effective method of meeting pipeline specifications and any site-specific standards with respect to oil-in-water and interface quality.

In general, the less heat is applied the greater the cost savings. As heat is applied at production facilities by fuel-gas-fired heaters, any increase in heat is reflected in fuel-gas consumption. The addition of heat also boils lighter hydrocarbon fractions from the crude oil; less product at a lower API gravity (defined in the Glossary) results. Addition of heat also accelerates rates of corrosion and increases the likelihood of scale formation on vessel internals, particularly the fire tubes.

Methods of Emulsion Breaking

To maximize the cost-effectiveness of petroleum production and refining processes and to achieve required quality parameters of oil, water, and crude-oil contaminants, it is often necessary to resolve emulsions to promote complete separation of the oil, water, and inorganic solids present. Breaking emulsions implies breaking emulsifying films around droplets of water or oil so that coalescing and gravitational settling may occur. Any or all of the following methods of aiding this process may be employed:

1. Providing low-turbulence and low-velocity environments (treating vessels) that allow gravitational separation and removal of oil, water, and solids. Gas may also be removed.
2. Increasing the temperature of the emulsion.
3. Applying chemicals designed to break emulsions.
4. Applying electrical fields that promote coalescence.
5. Changing the physical characteristics of an emulsion by the addition of diluents or water.

Because of the wide variety of potential emulsion and crude-oil types, mechanical configurations, treating chemicals, throughputs, and product specifications, each individual emulsion-breaking application is generally unique in its selection of specific emulsion-breaking methods and environments. Furthermore, all emulsions change in composition with time, and human beings control the process of emulsion breaking; these two factors re-emphasize this point.

The selection of mechanical equipment type is based primarily on previous experience. Within the refinery desalting process the selection of mechanical equipment usually falls within two broad areas, one-stage desalters (Figure 1) and multistage desalters (Figure 2). Each desalter vessel is sized

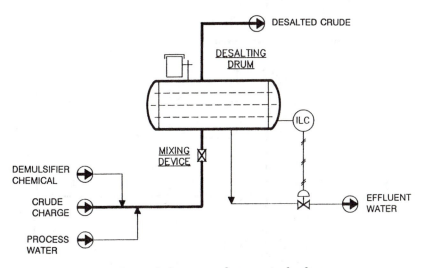

Figure 1. One-stage electrostatic desalter.

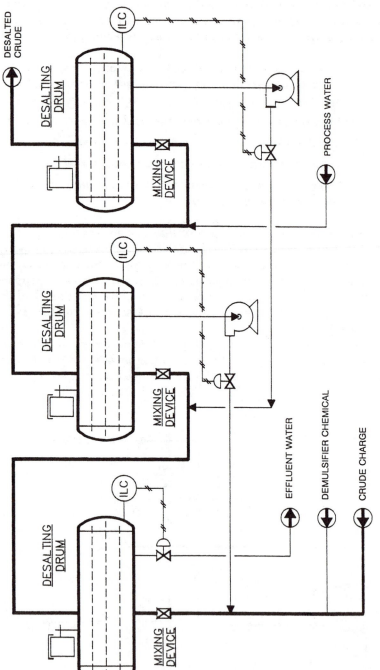

Figure 2. Three-stage electrostatic desalter.

to provide a known crude-oil mixture with a set retention time at a given rate of crude-oil charge to water wash. This retention time is deemed to be adequate to allow emulsion breaking to proceed to a desired end point or oil quality. The preheat exchange system will heat the crude-oil to the desired treating temperature. The desalting temperature is usually determined from previous experience with similar crude oils or from simulation runs using proportionally smaller vessels and flows. Experience with similar crude oils and any simulation runs performed will help to determine the requirements of electrostatic grids; crude-oil distribution systems; desand–desludge equipment; oil-in-water removal equipment; level control equipment; wash-water injection points; mixing equipment; and initial treating chemical injection points, type, and dosage. Experience may dictate that some of this equipment and capabilities may not be necessary with certain crude-oil types. Most modern desalters have all of these capabilities.

Once operational, the process of optimizing the desalting system for maximum performance or cost-effectiveness begins. Through planned manipulation of system parameters concurrent with monitoring of the quality of oil, water, and solids produced, operational data are gained. From these data, optimum set points for temperature, interface level, treating chemicals, addition, wash-water rates and placement, mix-valve pressure differentials, and desand–desludge frequency and rates are determined.

This process may take several months. If initial specifications on product quality cannot be met, then changes may be made to mechanical equipment present, chemicals applied, rate of crude-oil charge, type of crude oil processed, or product quality specifications. The process of optimizing the desalting program then starts again. On particularly difficult crude oils this process may last many years, as is the case in refineries processing heavy crude oil.

Within oil-field production, treating equipment is selected in a similar manner. As the amount of water produced will vary over the life span of an oil field, equipment is often added as needed to an oil-field treating facility. The design of existing facilities will allow the addition of equipment to occur with minimal disruption during periodic maintenance shutdowns (or "turn-arounds") if proper consideration has been given to the potential of changing production fluids.

The equipment variations within oil-field treating equipment are greater than those found in refinery desalting applications. The three basic types of modern vessels are gas separators, free-water knockouts (FWKOs), and treaters. Gas separators allow for early exit of gas from the oil, which helps to reduce the amount of agitation and hence reduces the emulsification tendencies. Treating chemicals may be injected upstream of a gas separator. The FWKO is typically a vessel designed to remove free water before the oil is transferred to a treater, and usually allows additional removal of gas. The FWKO may be heated by fire tubes or exchangers; however, its temperature is usually much lower than that of a treater. This lower temperature allows for removal of water from a treating system without having to heat the water to maximum system temperatures. Substantial fuel-gas savings result.

The treater is the vessel that attempts to deliver pipeline-specification crude oil as its hydrocarbon product. Heater treaters typically aid in resolving emulsions by the addition of heat, the use of electrostatic grids, or the use of mechanical coalescing aids such as hay or baffle systems.

Optimization of oil-field treating performance parameters is accomplished in a similar manner to that described for refinery desalting applications.

Thermal Methods. In both refinery and oil-field emulsion breaking, the addition of heat usually occurs to enhance emulsion breaking. Only rarely does the addition of heat alone provide adequate emulsion resolution. In the oil-field environment, resolution may occur with light oils in which paraffin forms the prime emulsifying agent. An increase in temperature above the paraffin melting point (50–65 °C) may completely destabilize an emulsion. In refinery desalting applications and heavy-oil production applications, the high treating temperatures usually remove paraffin as an active emulsifying agent.

Heat addition in emulsion breaking is usually based on the overall economic picture of a treating facility. Excess heat is not added when it is more cost-effective to add chemical or install electrostatic grids. In refinery desalting, the temperature of the crude oil must far surpass the temperatures present in the desalter as the crude oil approaches the atmospheric fractionation unit. Thus, minimizing heat in the desalter is not a key issue providing water present in the desalter does not boil, temperatures are not high enough to significantly elevate water solubility in a specific crude oil, and high temperatures do not cause significant amounts of asphaltenes to become insoluble in the crude oil and form an interface pad.

Within oil-field emulsion breaking, the economics usually favor minimal heat input because light ends are not lost to the gas phase and fuel-gas consumption is minimized. Other significant effects caused by the addition of heat are an increased tendency toward scale deposition on fire tubes, an increased potential for corrosion in treating vessels, and a tendency to render asphaltenes insoluble (because of loss of light aromatic components), which may produce an interface pad problem.

Electrical Methods. The principle of electrostatic dehydration in emulsion breaking for both refinery desalting and oil-field production is essentially the same. The electric field produced disturbs the surface tension of each droplet, probably by causing polar molecules to reorient themselves. This reorientation weakens the film around each droplet because the polar molecules are no longer concentrated at the droplet's surface. In addition, a mutual attraction of adjacent emulsion particles receives induced and oriented charges from the applied electric field. This mutual attraction places oppositely charged particles in close proximity to each other. As the film is weakened and the droplets are electrically attracted to each other, coalescence occurs.

This process does not usually resolve emulsions completely by itself, although it is an effective and often necessary addition to the use of chemicals or heat. Electric dehydration does have limitations as well. Excess water, typically greater than 6%, will cause shorting in many treater grid systems. Thus, when additional demulsifying power is required during system upsets, the grids may cease to function. This failure occurs with greatest frequency in light-oil demulsification when residence times are very short and the percentage of water within the crude oil may be up to 90%. A high percentage of water minimizes retention time of the oil phase if demulsification is not occurring rapidly enough.

Chemical Methods. The most common method of emulsion resolution in both oil-field and refinery applications is a combination of heat and application of chemicals designed to eliminate or neutralize the effects of emulsifying agents. Addition of suitable chemicals with demulsifying properties specific to the crude oil to be treated will generally provide quick, cost-effective, and flexible resolution of emulsions. Success of chemical demulsifying methods is dependent upon the following:

1. An adequate quantity of a properly selected chemical must enter the emulsion.

2. Thorough mixing of the chemical in the emulsion must occur.

3. Adequate heat may be required to facilitate or fully resolve an emulsion.

4. Sufficient residence time must exist in treating vessels to permit settling of demulsified water droplets.

For the oil producer, the use of chemicals in emulsion breaking is attractive for a variety of reasons. The capital costs of implementing or changing a chemical emulsion-breaking program are relatively small and can be accomplished without a shutdown. This feature is attractive because it means that emulsion-breaking chemical programs can be altered "on the fly" to react to changes in emulsion characteristics or crude-oil slate changes. (Crude-oil slate is a mixture of crude-oil types that the refiner wishes to process.) Chemical emulsion-breaking programs also imply a service commitment from the supplying company, which is an asset. Typically the supplying chemical company monitors the overall cost performance of an emulsion-breaking program on an ongoing basis and reports the findings to the customer. Improvements in chemical application and adjustments to system operating parameters should also be performed on an ongoing basis. In all but the largest installations, the oil-production company or refiner will not have access to a designated expert in this field, or, if an in-house expert is available, the economic justification for designating the labor and resources to cost reductions in emulsion breaking would not exist year after year. Thus, the chemical service company fills a technological need not readily addressed by the oil producer or refiner.

The cost-effectiveness of chemical emulsion-breaking programs is dependent on proper chemical selection and application. Systems that have wide variations in emulsion types and charge rates typically require chemicals that are effective over broad dosage ranges. A hypothetical performance curve of two demulsifiers comparing treated-oil BS&W quantity versus dosage is provided in Figure 3. If applied chemicals do not have a broad treating range, fluctuating overtreat and undertreat conditions will reduce the performance considerably.

The condition of an overtreat (in which excess chemical actually stabilizes or creates new emulsion types) is often very difficult to detect. This situation can be sharply aggravated by inexperienced chemical companies and petroleum company operations staff. Chemical application in emulsion breaking has a large human element involved in its application and interpretation and is therefore subject to misapplication due to human error.

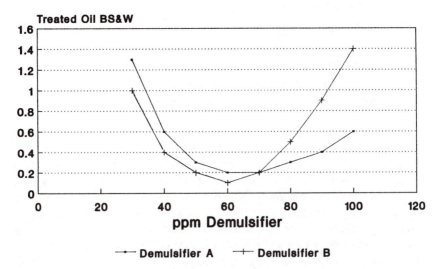

Figure 3. Hypothetical performance curve of two demulsifiers.

Chemical Applications in Commercial Emulsion Breaking

In processing petroleum emulsions, chemical treating compounds may be added to a crude-oil emulsion to produce desirable oil quality and remove water or inorganic solids. The most common types of treating compounds are referred to as emulsion breakers. Various mechanisms are postulated as to how emulsion breakers function, but it is clear that an emulsion breaker must reach the interface of an emulsified droplet and the surrounding liquid. At that point, an emulsion breaker disrupts the interfacial tensions between oil and water and allows the droplets to coalesce and settle by gravity.

The concepts involved in how an emulsion breaker performs this function are as varied as the chemistries that constitute the bulk of commercial emulsion breakers. All concepts may be correct in specific cases.

Emulsion breakers are typically specific for site or crude-oil type. Conventional emulsion breakers are most commonly formulated from the following types of chemistries: polyglycols and polyglycol esters, ethoxylated alcohols and amines, ethoxylated resins, ethoxylated phenol formaldehyde resins, ethoxylated nonylphenols, polyhydric alcohols, and sulfonic acid salts. Commercial emulsion breakers may contain but one type of active ingredient or intermediate or a variety of intermediate types.

Wide chemical variation is possible within the intermediate types as well. The ethoxylated resin group demonstrates variable molecular weight on its resin base with different amounts and placement of ethoxylated groups. These structural variations provide a complete range of solubilities, charge neutralization tendencies, solids-wetting characteristics, and costs.

To affix any definitive quality to any one type of intermediate is unrealistic, with the possible exception of the sulfonic acid salts, which exhibit fast water drop (rapid dehydration) and good solids wetting. However, because of the high dosages required, they are not usually cost-effective enough to be used in any major commercial emulsion-breaking process, with the exception of waste-oil treatment.

Usually, each intermediate has a different effect in each crude oil tested. One intermediate may have a synergistic effect with another intermediate that may far exceed the sum of the two individual intermediates. These intermediate mixtures plus solvent systems (usually aromatic solvents and alcohols) are the ingredients in most emulsion breakers.

Two examples of commercial emulsion-breaking products supplied by one chemical company are provided in Table I. Product 1 in Table I was found to be applicable in some paraffinic crude oils of medium to high API gravity. Product 2 was found to be applicable in some heavy asphaltic crude oils with significant amounts of inorganic solids present.

For most commercial emulsion-breaking applications, chemical compa-

Table I. Composition of Two Commercial Emulsion-Breaking Products

Component	Total Product (wt%)	Chemical Type
Product 1		
Intermediate A	25	Ethoxylated resin (10% ethylene oxide, 40% propylene oxide)
Intermediate B	20	Polyglycol ester (low ethylene oxide content, high propylene oxide ester component, polyglycol of mol. wt. 4000)
Intermediate C	5	Modified polyglycol (polyglycol of mol. wt. 6000 cross-linked with epon resin)
Carrier solvent A	45	Heavy aromatic naphtha (to maintain intermediate solubility)
Carrier solvent B	5	2-Methylpropyl alcohol (to provide product with low pour point)
Product 2		
Intermediate A	10	Polyglycol oxide component, $\rightarrow$ 40% ethylene oxide, 60% propylene oxide
Intermediate B	5	Ethylene oxide resin, 45% ethylene oxide
Intermediate C	25	Modified polyglycol ester, polyglycol of mol. wt. 8000 cross-linked with epon resin
Intermediate D	2	Ethoxylated nonylphenol resin, 20% ethylene oxide, 10% propylene oxide
Carrier Solvent A	53	Heavy aromatic naphtha
Carrier Solvent B	5	2-Propyl alcohol

nies are able to provide emulsion breakers that provide pipeline-specification crude oil or acceptable water contents in desalted crude oils. These compounds are most commonly used at dosages of 1–200 ppm in oil-field production and 5–20 ppm in refinery desalting. Light oils are generally treated in the lower dosage ranges, and heavy crude oils require the higher dosages. Exceptions to these dosage requirements can be found.

In many commercial emulsion-breaking applications, conventional emulsion-breaking chemistries will also achieve desired oil-in-water contents and acceptable interface quality. However, these results are not always accomplished. Reverse emulsions are not usually resolved by conventional emulsion-breaking chemistries. The addition of a specific reverse-emulsion breaker, either to the crude-oil stream or to the water-handling system, may be required to produce desired water quality parameters.

Reverse emulsion-breaking chemicals are usually solution or latex polymers ranging in molecular weight from 10,000 to 30 million, and may be classified as either coagulants or flocculants. Although the classification of coagulant or flocculant can be assumed from molecular weight (flocculants having the higher molecular weights), the most accurate categorization is based on the polymer's effect within a system. Many other types of reverse-emulsion breakers such as zinc chloride, water-soluble ethoxylated resins, nonylphenol resins, and alum are available.

Concepts applied to regular emulsions still apply with reverse emulsions. Emulsifying agents or charges must be neutralized to permit coalescence of the oil. Adequate settling time for the oil to rise out of the water phase must be available. Skim tanks and API separators are used for this purpose. A variety of flotation units will accomplish the same goals at an accelerated rate.

Dosages of reverse-emulsion breakers will vary widely with each application. A 10–100-ppm dosage of reverse-emulsion breakers would include most applications but exclude high-molecular-weight polymers. High-molecular-weight polymer dosages are usually less than 10 ppm if injected into a water system and less than 5 ppm if injected into an oil system. In oil systems, high-molecular-weight polymers are usually added to separate hydrocarbons from inorganic solids (primarily sands, clays, and iron sulfides). This separation reduces interface pads and oil-in-water concentrations.

If accumulated oils from water systems are to be recycled into oil-treating systems, it is crucial to ensure that the accumulated oil (or slop) and associated chemicals are compatible with the oil-treating program. This need for compatibility is particularly true of high-molecular-weight polymers when applied to systems that contain large amounts of asphaltene-coated solids. Dosages as low as 2 ppm may be excessive under these conditions and promote a very stable (essentially untreatable) emulsion in oils accumulated from the water-handling system. High-molecular-weight polymers should not be expected to resolve treating problems arising from the presence of asphaltenes and inorganic solids.

In some crude oils, high amounts of insoluble asphaltenes and inorganic solids with high surface charges (chiefly clays) will combine to form a stable "solids" interface pad. This interface problem is usually accompanied by poor water quality and excessive consumption of emulsion breakers. This type of interface pad is typically removed from a treating vessel by desand–desludging operations to form uneconomically treatable slop oils. Disposal costs of this slop may be high for either the oil producer or refiner.

This problem occurs primarily in treating heavy crude oils and light asphaltic crude oils produced by miscible flooding. The addition of an asphaltene dispersant to the crude oil to prevent accumulation of insoluble asphaltenes may resolve this problem.

The selection of asphaltene dispersant chemistry and dosages requires careful consideration. Most asphaltene dispersants are either light aromatic compounds with polar groups (e.g., cresylic acid) or highly water-dispersible or water-soluble surfactants. These materials will prevent insolubility of asphaltenes, but they also tend to adversely affect the dehydration of emulsions and to increase oil-in-water concentrations.

Successful selection of asphaltene dispersant chemistries and dosages will provide reductions in interface pads, emulsion-breaker consumption, and oil-in-water concentrations and provide oil-free inorganic solids.

Emulsions stabilized by paraffin are usually restricted to light crude oils in oil-field production. If paraffin deposition that restricts production is occurring upstream of an oil-treating facility, it may be feasible to apply a paraffin crystal modifier to the crude oil to prevent paraffin deposition and to eliminate paraffin as an emulsifying agent. A paraffin crystal modifier must enter an oil system at a temperature greater than the cloud point of the crude oil and upstream of the problem area.

To select chemical programs for an oil-treating facility, each facility must be examined on an individual basis. The selection of a chemical or group of chemicals for emulsion breaking must be preceded by valid test procedures and a thorough understanding of the treating system and petroleum company objectives.

Demulsifier–Auxiliary Chemical Selection

The selection of chemicals that will provide the refiner or oil producer with a cost-effective emulsion-breaking program that meets or exceeds all performance parameters is usually the function of a chemical service company. The selection process has historically been viewed as a "black art" that produces as many failures as successes. This assessment of the situation has been realistic. However, with an ever-increasing understanding of emulsions and emulsion-breaking chemicals, the development of new test procedures and devices, and a well-organized method of chemical selection, many of the failures can be eliminated.

Characterization of Crude Oils and Containants.   The first step in selection of emulsion breakers is to obtain as complete an understanding as possible about the crude oil or emulsion. Density (or API gravity) and BS&W ranges should be determined. The crude oil should be classified as asphaltic or paraffinic, and the asphaltene and paraffin content should be determined. If treatment will occur at a temperature below the paraffin melting point, the cloud point of the crude oil should be determined. This information will aid in selecting the treating temperature.

The inorganic solids content of the crude oil should be known, and the types of solids present should be identified. If a treating system is operational, the interface should be examined to determine its composition. From this information, many of the key crude-oil emulsifiers will be identified, and a knowledge of which contaminants require additional treatment will be derived.

In refinery desalting operations, the amount and type of salts (ionized or crystalline) present in the crude oil should be determined. In all treating applications, the chemistry of the water should be known so that the impact of changing treating temperatures and pressures on scaling and corrosion potential can be considered.

Past operating experiences with the same or similar crude oils should be reviewed. If possible, these should be compared with the results obtained from any previous emulsion-breaker selection tests. This comparison will give an indication of the validity of previous test work, eliminate many treating compounds from the selection process, and provide information as to which chemistries may be required to complement a conventional emulsion breaker. This information will also allow changes in emulsion characteristics to be noted very quickly and predict whether an existing emulsion-breaking program should be reviewed for possible improvements.

System Mechanical Parameters and Operating Data.   Before the selection of emulsion-breaking chemicals, a complete understanding of the equipment and operating procedures at a treatment facility should be obtained. Key components will include the following:

- rates of production
- treating vessel retention times
- treating vessel capabilities (i.e., electric ability, temperature limitations, etc.)
- recent treating temperatures and pressures
- existing chemical addition points, methods, and equipment
- sampling point locations
- operating procedures for desand–desludge cycles and slop oil recycling with appropriate frequencies, rates, and time frames
- type of level controllers
- schedules and operating procedures for well treatments, equipment cleaning, and addition of the chemicals
- water-handling equipment and procedures
- problem crude-oil sources
- overall cleanliness of system internal components
- diluent or wash-water rates and quality

Emulsion-breaking performance data should also be gathered. Areas of prime concern are the following:

- oil, water, salt, and solids content of the fluids before and after each treating vessel or process. Test procedures should also be known.

- composition and quality of interface fluids

- all operating and chemical costs associated with emulsion breaking at that facility

- amount of waste oil, solids, and water generated

- downstream effects of the emulsion-breaking process

From this information the economics of the present emulsion-breaking program can be calculated as a standard from which to evaluate changes in the program. A determination of whether chemical, mechanical, or operational changes will provide the greatest improvement to emulsion breaking can be made. If a change in chemicals is required, the information can be used to design a chemical-testing procedure that is most representative of the oil-treating installation.

Prioritization of Performance and Cost Issues. All personnel involved in attempting to improve an emulsion-breaking program must have a clear understanding of which areas require the greatest improvement and how much improvement is required for the exercise to be successful. The improvements to an emulsion-breaking program may occur by allowing compliance to a product or waste-stream quality specification, increasing throughput, decreasing overall treating and operation costs, reducing environmental hazards, improving system reliability, or providing operating information. Any one or all of these areas may require improvement. If requirements for successful problem resolution are not mutually agreed to by all parties involved before beginning to attempt to improve emulsion breaking, it is unlikely that the end goals will be achieved.

Emulsion-Breaker Selection. If a chemical change must be made or investigated to provide improvements in the emulsion-breaking program, a method of selecting new chemicals must be available. In all cases, the test method must simulate the conditions of a treating system as closely as possible. Candidate chemicals must then be compared with chemicals of known operating performance within the treating system.

If chemical treatment was not previously used, a chemical that performs well in a similar crude oil and system should be selected for use as a standard. The relative improvement in performance over the standard must be the criterion for selection of new candidate chemicals. In larger installations that are experiencing severe emulsion problems, many (possibly hundreds) of commercialized emulsion-breaking products or experimental blends of emulsion-breaker intermediates may be tested. The selection process for emulsion-breaking chemicals can be confirmed only by application with a treating system. As a result, selection may be a slow process.

In the selection process, various general testing procedures are available to determine appropriate chemicals. All test procedures will have limitations. In all cases, these limitations can be minimized by using only fresh fluid samples that are representative of the treating system or slip streams of system fluids, by using consistent and appropriate analytical procedures, by ensuring that test fluids contain only the emulsion-breaking chemicals to be tested, and by using test procedures that subject the emulsion to conditions as close as possible to those found in the testing system of study.

Historically, the chemical selection process has been performed on a bench-top scale. Bottle tests (including ratio, elimination, and confirmation test), jar tests, and portable electric desalter tests fall into this test category. No effort will be made to describe these tests or the associated analytical procedures in detail, as they are described in Chapters 3 and 10. Furthermore, significant variation in testing procedures will exist between various chemical companies, oil producers, and refiners. Each test procedure is also tailored to each treating facility.

Bench-top testing will allow variation in chemical type and dosage, temperature, pressure, agitation, treatment time, electrical input (portable electric desalters only), and wash-water or diluent addition. Variations in temperature and pressure will not allow simulation of high pressures and temperatures. The bench-scale tests imply that a batch treatment of the emulsion is used to determine treating chemicals for a dynamic continuous treating system. Thus, results will have limitations even if the parameters of the test procedure are as accurate as possible.

Quantitative information on the BS&W and salt content of oil as a whole or at selected levels in the oil column can be collected from the bottle and portable electric desalter tests. This information is usually accurate when applied within a system, although the chemical dosages necessary to achieve the results noted in the tests is usually significantly lower.

Quantitative information on the nature of the oil–water interface can be obtained from all bench-top tests. Chemicals that produce a poor interface in bench-top tests will almost certainly produce a poor interface in a treating facility. Chemicals that produce a good interface in bench-top tests will not always produce a good interface in a treating facility. The bench-top tests have difficulty producing accumulations of emulsions, solids, and insoluble hydrocarbons that usually produce poor interfaces. This difficulty results because they are small-scale batch tests, not dynamic tests that allow the accumulated effects of materials present in small quantities to be noted.

Oil-in-water content can be measured only on a qualitative basis in all bench-top testing. Water quality is often directly associated with the quality of the interface within a system. If the quality of the interface is not fully predictable, then neither is water quality. Water quality is also greatly affected by a treating system's flow dynamics. The ability to accurately predict trace quantities of oil in water in a system from a small-scale batch bench test does not exist.

In an effort to eliminate the limitations of bench-top testing, various groups and companies are attempting to develop dynamic simulators that are actually scaled-down treating facilities. These treating simulation units provide dynamic continuous flow of emulsions on a smaller scale than the actual treating facility. The charge of emulsion may be transported to the simulator for test runs. This process has been used for many years to size treaters and desalters and select start-up chemicals for a treating facility. The same variations possible in a bench-top test are used; however, the range of variation is much greater. Manufacturers of treaters and desalters as well as large, well-funded petroleum research groups use this method of testing to provide more accurate selection of mechanical equipment and chemicals. This system has a much greater ability to accurately predict treated-oil quality, water quality, and interface quality. A limitation in this method is that the crude oil or emulsion suffers from aging as it is transported to the simulator.

Portable simulators can be transported to the field and receive their crude-oil charge directly from the treating system by diverting a slip-steam through the unit. This step eliminates the aging effect on the test fluids while maintaining the same capabilities noted in stationary oil-treating simulators. A flow diagram of a portable oil-treating simulator developed by Nalco Canada Inc. is shown in Figure 4.

Regardless of the test method used to select chemicals, the true capabilities of the treating chemicals must be determined by commercial application in the treating system of concern. All test work in chemical selection is merely a method of reducing the risk of failure an oil treater is subjected to when changing emulsion-breaking chemicals or testing a new facility.

Once new emulsion-breaking chemicals are entered into a treating system, they are evaluated by the methods noted in the system mechanical and operating data section. Confirmation that the goals of the oil-production company or refiner are being met is then obtained. If the goals are not being met, the areas requiring improvement are determined, and an appropriate course of investigation and manipulation of system parameters is conducted to achieve these goals.

Summary

In the petroleum industry, mixtures of oil and water will occur as emulsions in both production and refining segments. The types of emulsions will vary widely, although all emulsions will be the result of normally immiscible oil

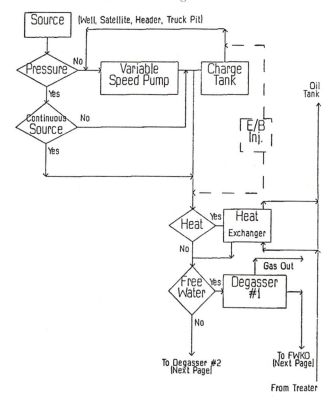

Figure 4. Oil-treating simulator fluid path.

and water subjected to agitation and stabilized by a wide variety of emulsifying agents.

To achieve desired product quality, meet environmental concerns, and improve equipment reliability, it is often necessary for oil producers or refiners to resolve emulsions and eliminate contaminants. These goals can be accomplished with many methods to reach acceptable standards in dehydration of oil, removal of solids and contaminants, oil-in-water interface control, and energy input.

Most commonly, a combination of electrical, thermal, chemical, and time factors is applied to an emulsion in a treating facility designed specifically for that emulsion and for that facility. The economics of emulsion breaking determines which methods and to what degree each method is used to achieve the end goals at that facility.

Chemical programs applied in commercial emulsion breaking are selected from a wide variety of emulsion-breaking chemistries and auxiliary chemicals that control very specific agents within the emulsion. These chemicals are selected on the basis of a variety of tests, both bench-top and

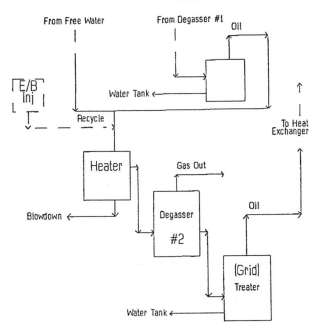

Figure 4. Continued.

simulator, that will provide a measure of the performance of treating chemicals with a specific crude-oil and treating system. The results must be confirmed by use of the chemical at a treating facility.

The factors that influence emulsion formation and emulsion breaking show wide variation from site to site. As a result, no universal rules exist for applying emulsion-breaking technology. Each emulsion-breaking facility must be viewed as an individual or unique case. If this approach is taken, the theories of demulsification may then be applied to a specific situation in a carefully organized, documented, and directed attempt to provide the most cost-effective methods of achieving the goals in emulsion breaking of the producer and refinery.

Bibliography

Jones, T. J.; Neustrader, E. L.; Whittingham, K. P. "Water in Crude Oil Emulsion Stability and Emulsion Destabilization by Chemical Demulsifiers", *J. Can. Pet. Technol.* **1978,** 100–107.

Kronenberger, D. L.; Pattison, D. A. *Mater. Perform.* **1986,** 7, 9–17.

Lissant, K. J. *Demulsification;* Dekker: New York, 1983.

Mackay, D. "Formation and Stability of Water-in-Oil Emulsions", Report EE93; Environment Canada: Ottawa, Ontario, Canada, 1987.

Impurities in Petroleum; Petrolite Corporation: St. Louis, MO, 1981.

Speight, J. G. *The Chemistry and Technology of Petroleum;* Dekker: New York, 1981.

Speight, J. G. *The Structure of Petroleum Asphaltenes: Current Concepts;* Information Series no. 51, Alberta Research Council: Alberta, Canada, 1981.

Strassner, J. E. *J. Pet. Technol.* **1968,** *243,* 303.

Treating Oilfield Emulsions, 3rd ed.; American Petroleum Institute: Austin, TX, 1974.

RECEIVED for review December 18, 1990. ACCEPTED revised manuscript July 9, 1991.

Breaking Produced-Fluid and Process-Stream Emulsions

Gerhard Leopold

Department of Energy, Mines, and Resources, CANMET, Fuel Processing Laboratory, P.O. Bag 1280, Devon, Alberta, Canada, T0C 1E0

This chapter will briefly review the nature and the consequential sources of oil-field emulsions encountered in the handling of produced fluids recovered at a wellhead and subsequently processed (i.e., "broken") at central treatment facilities. The principal factors and agents commonly employed in the separation of both the oil and the water phases found in these produced-fluid streams will be discussed. Subsequently, this chapter will describe sampling and testing techniques that assist in characterizing a process stream's composition and thus in evaluating the effectiveness of a particular separation process. Finally, the major components of a typical oil-field emulsion-treatment facility will be described. Selection and design criteria of appropriate separation equipment will also be presented.

Source and Nature of Process-Stream Emulsions

An emulsion is a system consisting of a liquid dispersed as droplets in a second immiscible liquid, often stabilized by an emulsifying agent. In the oil field, the two basic types of emulsions are water-in-oil and oil-in-water; oil-in-water emulsions are often termed reverse emulsions. More than 95% of the crude-oil emulsions formed in the oil field are of the water-in-oil type. Nonetheless, oil-in-water emulsions are receiving growing interest in pollution abatement as they are readily miscible with water.

Petroleum emulsions vary from one oil field to another simply because crude oil differs according to its geological age, chemical composition, and associated impurities. Furthermore, the produced water's chemical and physical properties, which also are specific to individual reservoirs, will affect emulsion characteristics as well.

0065–2393/92/0231–0341 $012.00/0

Group Gathering Systems.   The gathering system in petroleum production operations consists primarily of the pipes, valves, and fittings necessary to connect the production wellhead to the separation equipment. The configuration of the gathering system typically entails several flow lines connecting the individual wells to a group header or test header system as dictated by the size and layout of the particular oil field. Consequently, the nature of the composite emulsion sent to the treatment facilities is determined by the combination of produced fluids from individual wells. The combination of several wells' production in a particular oil field may generate an average emulsion that can be more manageably treated than production from an individual well. For example, excessive water production from a single well is more effectively treated in a group heater treater than if special equipment for water removal were required.

Trucked versus Pipelined Emulsions.   Over the years transportation of produced liquids has evolved from the use of wooden barrels filled at the wellhead to a system of pipelines and trucks. Inasmuch as it is usually much cheaper to use pipelines than trucks, the selection of either transportation mode is generally not dictated by economics alone. Some of the major advantages of using pipelines are

1. economy

2. reliability (e.g., in the face of weather and breakdowns)

3. control (i.e., variety of flow rates)

4. continuity

Because of their high viscosities and densities (i.e., low API gravities[1]), heavy oils and their emulsions are currently transported mainly by trucks. The major problems in pipelining heavy crude oils are associated with pour point (i.e., wax crystallization problems) and viscosity (i.e., flow problems).

Several techniques are employed by the industry to overcome these problems; the two most commonly used methods are the application of heat and dilution with a solvent (i.e., low-viscosity hydrocarbons such as condensate, natural gasoline, and, most often, naphtha). Both methods serve to reduce the transported crude oil's viscosity; dilution will also reduce the heavy-oil mixture's pour point.

An innovative technique involving the preparation of a lower-viscosity unstable slurry-emulsion system by mixing water with the oil as a means of conveying crude oil has yet to be proven on a commercial scale. The technical problems and issues limiting the application of this technique (analogous to crude-oil emulsion pipelining) are to sustain the two immiscible liquids in a stable emulsion during transport and to destabilize the emulsion

[1]API gravity is defined in the Glossary.

once it arrives at the delivery point (i.e., to separate water from the oil). The selection of an effective pair of emulsifying and demulsifying agents is crucial to the method's success.

Trucking is a more economic and expedient method of transporting produced emulsions to production treatment facilities, particularly for heavy crude oils that are difficult to pump. However, over extended distances (i.e., traveling time in excess of 1 h), free-water settling can be quite pronounced within a tanker truck. Agitation due to travel over rough roads may also promote some degree of coalescence.

Oil Recovery Scheme. The type of oil recovery process employed influences the nature of the emulsion eventually produced from a specific subterranean zone or reservoir. For oil-bearing formations under primary recovery, the produced fluids are withdrawn from a specific zone or reservoir and lifted to the surface by either natural (reservoir pressure) or artificial (bottomhole pump or gas) means. The resulting emulsion is simply a combination of the petroleum and any associated water and gas in the reservoir. Naturally occurring emulsifying agents are usually present in sufficient quantities to stabilize the emulsion. Furthermore, the agitation arising from the turbulent flow of the oil–water mixture through the well casing, tubing, downhole pump (if required), and surface equipment is usually ample to promote emulsification. (The terms "downhole" and "bottomhole" refer to the lowest depth, normally the producing zone, of an oil well.)

To improve ultimate oil recovery from some reservoirs, either water or (natural) gas injection is used to displace the oil to the producing well bore. Such processes are categorized as secondary recovery. The introduction of an injected fluid to the reservoir adds a new constituent to the produced emulsion. Consequently, the source and nature of the selected injection fluid must be closely scrutinized, not only for its effect on the recovery process, but also for its impact on the nature and stability of the produced emulsion. (For example, the compatibility of the injected water with the connate water present in the reservoir must be ensured to prevent precipitation that would result in formation plugging.)

Tertiary or enhanced oil recovery (EOR) incorporates a variety of techniques involving more elaborate injection schemes than employed in secondary recovery. The treatment of EOR-produced emulsions must be approached independently from any primary or secondary production from the same field or reservoir. Standard demulsifiers and treatment methods used during primary and secondary recovery operations may not handle EOR-produced emulsions.

Specifically, EOR activity can be classified into three major types: (1) chemical, (2) miscible displacement, and (3) thermal. The first category includes the injection of surfactants (micellar), polymers, alkaline (caustic), and carbon dioxide. In miscible displacement, a gas or liquid hydrocarbon is injected into the reservoir where it becomes miscible with the hydrocarbons indigenous to the reservoir. The thermal category includes all processes that rely on the addition of heat to the reservoir to lower the viscosity of high-density crude oils. Among the more popular thermal recovery processes are steam injection, hot-water flooding, and in situ combustion.

The introduction of heat to the reservoir affects the properties of the reservoir fluids (e.g., lowering oil viscosity). In the extreme case of fire flooding, where combustion temperatures of 480 °C (900 °F) can be attained, thermal cracking and generation of combustion products can introduce new constituents to the produced emulsion. Under certain conditions of steam injection, milder forms of thermal cracking of in situ heavy crude oils are also possible. Thus the emulsions produced from a particular reservoir under primary and secondary recovery may be physically and chemically distinct from thermally recovered fluids. Changes in the nature of not only the oil and water but also the emulsifying agents (i.e., naturally occurring surfactants) result during the thermal recovery process.

Co-mingled Production. For economic reasons, specific well bores are sometimes completed to produce from more than one oil-bearing zone simultaneously. This system is termed co-mingled production.

The compatibility of the co-mingled produced fluids must be determined before initiating such an operation. Chemical compatibility of the produced waters must be confirmed. Otherwise, the formation of solids (precipitates) could result in operational problems such as plugging, as well as treating difficulties.

Agents Employed in Emulsion Treatment

Chemicals. Chemical treatment of emulsions requires the dispersion of a chemical demulsifier, or emulsion breaker. Demulsifiers are surface-active agents comprising relatively high-molecular-weight polymers. When added to an emulsion, they migrate to the oil–water interface and rupture the film, or at least weaken it sufficiently for the emulsifying agent to be dispersed back into the oil and for droplets of the dispersed phase to attract, collide, and coalesce. The remaining step is to bring these larger droplets into contact without excessive agitation, which might redisperse the droplets.

Several factors affect demulsifier performance: temperature, pH, and the nature of the aqueous-phase salt. In most cases, an increase in temperature results in a decrease in emulsion stability. Consequently, for a particular emulsion, less demulsifier is required at higher treating temperatures to effect the same degree of treatment. Studies (1) on the effect of pH on the instability of crude-oil–water emulsions have shown that a pH of 10.5 produced the least stable emulsions. Furthermore, basic pH produced oil-in-water emulsions and acidic pH generated water-in-oil emulsions.

Selection of a suitable chemical emulsion breaker and dosage is crucial. A particular demulsifier may be effective and efficient for one emulsion yet entirely unsatisfactory for another. Contemporary demulsifiers are formulated with polymeric chains of ethylene and propylene oxides of alcohol, alkyl phenols, amino compounds, and resinous materials that have hydroxy acceptor groups. Each of these polymers is carefully formulated to yield a molecule with a particular affinity for water. Demulsifier dosage is also important; excessive demulsifier addition can inhibit the efficiency of emulsion breakdown.

Demulsifiers are very similar to emulsifiers because both are surfactant in nature. Consequently, the action of the demulsifier in emulsion breaking is to "unlock" the effect of the emulsifying agent(s) present. This unlocking is accomplished in three fundamental steps (2): flocculation, coalescence, and solids wetting.

Flocculation. The first action of the demulsifier on an emulsion involves a joining together or flocculation of the small water droplets. When magnified, the flocks take on the appearance of bunches of fish eggs. If the emulsifier film surrounding the water droplet is very weak, it will break under this flocculation force and coalescence will take place without further chemical action. Bright oil[2] is an indicator of good flocculation. In most cases, however, the film remains intact, and therefore, additional treatment is required.

Coalescence. The rupturing of the emulsifier film and the uniting of water droplets is defined as coalescence. Once coalescence begins, the water droplets grow large enough to settle out. Good coalescence is characterized by a distinct water phase.

Solids Wetting. In most crude oil, solids such as iron sulfide, silt, clay, drilling mud solids, and paraffin complicate the demulsification process. Often such solids are the primary stabilizing material, and their removal is all that is necessary to achieve satisfactory treatment. To remove solids from the interface, they can either be dispersed in the oil or water-wetted and removed with the water.

Agitation. The effectiveness of any demulsifier added to a treatment system is directly dependent upon its making optimum contact with the emulsion. Therefore, the emulsion must be sufficiently agitated after the chemical demulsifier has been added. Increased mild agitation, such as in flow lines and in settling tanks, is beneficial in promoting coalescence. Re-

---

[2]The term "bright oil" refers to the shiny color that is characteristic of treated oil (2). Further discussion on the subject of oil brightness is presented in the subsection "Oil Color" in the section "Interpreting Results".

emulsification may occur if an emulsion is agitated severely once it has broken into water and oil. This sort of agitation may occur in gas separators, pumps, or other locations in the system.

Gravity. Settling is the simplest and one of the oldest techniques to separate two immiscible phases. Consequently, settling is a basic component in all treatment procedures. The treatment vessel should be designed to ensure sufficient time for quiet settling of all associated water from the oil phase.

Energy. *Thermal.* The addition of heat promotes the treatment process. First, it reduces the viscosity of the oil. This effect is especially instrumental in the handling and treatment of heavy crude oils or bitumen emulsions. Second, it weakens or ruptures the film between the oil and water droplets by expanding the water present. Last, heat increases the difference in densities of the fluids and thereby tends to reduce the settling time.

Heat, in effect, accelerates the treating process. Consequently, its use can help to reduce the required size of the treatment vessel. There is, however, an upper limit to how much heat can be added to a system because, at higher temperatures, light ends (that is, the more volatile hydrocarbon fractions) in the oil may vaporize. Unless these lighter ends are conserved, both the treated oil's volume and its API gravity will be reduced (i.e., its density will be increased). The economic trade-offs of longer treating time, heating costs, and ultimate oil-product sales price cannot be ignored.

Mechanical. Various mechanical devices are employed to supplement the performance of the treatment vessel(s) in the breaking of oil-field-produced emulsions. These include, but are not limited to, free-water knockouts (FWKO), gas separators, settling tanks, and gun barrels. Free-water knockout systems are generally used in connection with high water-to-oil ratio production; separation of gas may also occur in the upper section (Figure 1). Gas separators, either horizontal or vertical in configuration, provide tremendous agitation potential, principally through the turbulent evolution of the associated gas from the produced liquids. Without the efficient removal of gas in these vessels, unwanted (and uncontrollable) agitation in the downstream treatment vessels may result.

Settling tanks are fundamentally simple in their principle of operation. The rate of water dropout is not as crucial here as with treatment vessels because the injected chemical(s) may continue to operate over an extensive time period (days versus hours). The water–oil interface is found closer to the bottom of the tank.

Gun barrels are similar to settling tanks in that they commonly operate at atmospheric pressure (Figure 2). The speed of water dropout is generally

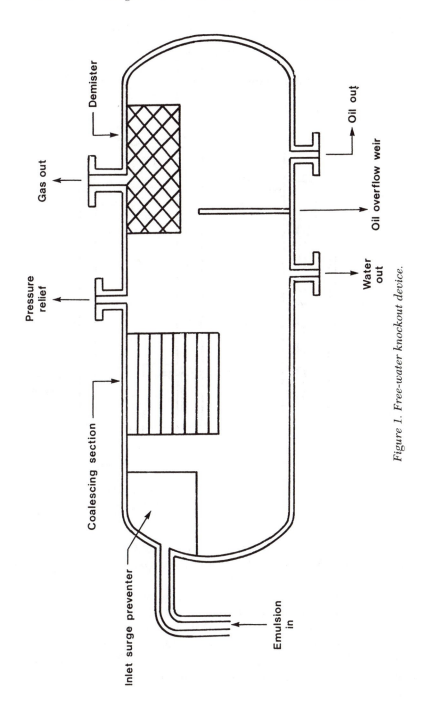

Figure 1. Free-water knockout device.

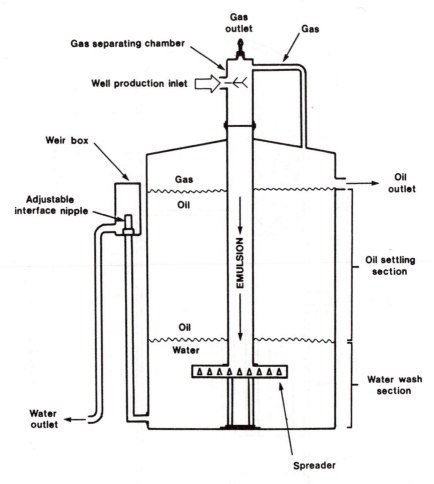

Figure 2. Typical gun barrel settling tank with internal flume. (Reproduced with permission from reference 3. Copyright 1986 Gulf Publishing Company.)

not too important as gun barrels usually have a high volume-to-throughput ratio. As with settling tanks, the chemical may continue acting over a long time, and the interface need not be clean.

Electrical. Electricity is frequently used to supplement (and sometimes replace) heat as an aid to the treating process. Furthermore, where electricity is much cheaper than most other energy sources, electrically induced coalescence techniques are gaining prominence. They are particularly valuable where space is of prime importance because the use of electricity accelerates the settling process even more than heat and allows the use of a smaller vessel.

Testing Process-Stream Emulsions

Sampling Process Streams.   The most critical step in testing process-stream emulsions is to procure a representative sample from the total produced fluids. Test results will otherwise be meaningless. A good sample therefore must be (2):

- representative of the system
- composite
- consistent with the produced fluids to be treated
- free of injected chemicals
- free of contaminants
- stable
- 40–50% water, if possible
- consistently obtainable

Generally, a good place to collect a sample is off the main flow line into the system, before chemical is added. If there is no readily available sampling point, the chemical injection line can be disconnected and a sample can be taken directly from the injection point. However, caution must be exercised, because it is extremely difficult to obtain uncontaminated samples from chemical injection points in spite of purging. Consequently, it is advisable to avoid injection points as sampling points.

An ideal sample for bottle testing (discussed in the next section) is a stable composite emulsion containing 30% water. When a composite sample is not suitable, as evidenced by a lack of stability, chemical contamination, or low water cut (content), a composite of several wellhead samples must be used. ["Cut" refers to the process of determining the BS&W (basic sediment and water) content of the oil phase in a sample of produced fluid. Cuts can be taken from the top, middle, or interface of the oil phase.]

Testing Procedures.   *Bottle Test.*   The most widely adopted procedure for testing emulsions is the so-called bottle test method. A detailed description of the bottle test procedure is provided in Appendix A (2). Also, Chapter 3 gives information on this topic. The purpose of the bottle test is to provide information about the effectiveness of treatment chemicals for a given emulsion.

The bottle test is excellent for screening prospective demulsifier chemicals for plant use. However, it is less reliable when determining the quantity (dosage) of chemical required to treat a specific emulsion under plant conditions. (A rule of thumb that has been proposed (4) correlates 6–8 h in a bottle test with 24 h in the plant). For example, the bottle test may indicate that 0.010 mL of a specific demulsifier is required to treat 100 mL of an emulsion (i.e., 100 ppm); yet, under plant operations only 50 mL is required to treat 1 m^3 of emulsion (i.e., 50 ppm). Several factors may be influencing this divergence (5):

- The chemical, injected continuously in-line during plant operations, may be better dispersed than in the bottle test where a specific volume is added.

- The sample taken for bottle test purposes from a valve or stopcock may be more emulsified than the emulsion in the flow line because of splitting action of the sampling orifice.

- The bottle test sample does not always represent the full stream. Wells (headers) producing little or no water may have been omitted from the test.

- The bottle test sample is typical of the average emulsion only at the time of sampling. A variation in the characteristics of the emulsion flowing to the treater is likely simply because of combinations of varied speeds of pump strokes for wells producing different water-content levels.

- In the bottle test, only one demulsification process occurs for a sample of emulsion treated with a fixed amount of chemical. In the plant, any residual demulsifier in the treated oil or at the interface can act again on fresh emulsion. Experiments have shown (5) that the treated oil from a bottle test can be used again to treat half of its volume of fresh emulsion. Similarly, other experiments have shown that the water recovered from a bottle test can also treat about half its volume of fresh emulsion.

As a rule of thumb, a good estimate of plant demulsifier dosage requirement is half the dosage indicated by a bottle test.

Plant Test. The ultimate proof of a demulsifier's performance is the plant test. The most encouraging bottle test results are only an indication of a given chemical's expected performance under plant conditions. The duration of such a test may be as short as 1 day or as long as 2 weeks, depending upon the particular treatment circumstances (i.e., emulsion characteristics and treatment facilities' nature and size).

Before any chemical change is made in a system that is operating unsatisfactorily, the performance of the existing system conditions should be monitored fully and recorded. The following data should be included (6):

- identification of the system
- date and time
- all production data (oil, water, gas, etc.)
- all chemicals, their consumption (daily), and pump speed
- temperatures of vessels (heater treater, gun barrel, etc.)
- amount of interface in treatment vessels
- water content at various points in the plant:
  1. inlet stream(s)
  2. samples of the oil phase in FWKO taken from the high and low petcocks
  3. all interfaces in all treatment vessels
  4. oil streams out of all treaters and/or tanks
- quality of recovered water (parts per million of suspended oil and insolubles)
- amount of recycled oil
- condition of the chemical pump(s)
- BS&W of sales oil
- any other peculiarities in the system

All of this information is required to establish a base line and determine retention time throughout the plant before introducing the new chemical(s).

The plant test starts when the new chemical is first injected into the system. The performance of the new chemical is well monitored until the system is cleared of the old chemical. Replacing the old chemical entirely has the advantage of demonstrating the quickest results; however, there is also a greater risk of a plant upset. Most chemicals are compatible, but in some circumstances a short period of either poor or exceptionally good treatment during the changeover results. Generally, the dosage of the new chemical is gradually increased until the old material is entirely displaced.

The frequency of data collection should be linked to the plant's retention time. For example, in a system with a 12-h retention time, the cuts on the system should be made every 2 h; a system that has a retention time of 15 min must be monitored continuously.

Interpreting Results.   Objectivity is paramount in interpreting both bottle test and plant test results. Items to consider include the following.

Centrifuge Cut. The centrifuge cut is generally considered to be the most important attribute of bottle testing. (The term "centrifuge cut" refers to the method of analysis rather than the point of sampling. Thus, a centrifuge cut can be taken from the top, middle, or interface of the oil phase.) Although there are a few instances where emulsions can be evaluated without this final step, omitting this step is generally poor practice because small quantities of basic sediment or water may be overlooked. Bottle testing must yield at least pipeline specifications that currently amount to a maximum allowable BS&W of 0.5%. Minimum basic sediment (i.e., the entire portion of BS&W is composed of simply water) is highly favorable in that it will reduce the buildup and subsequent handling of tank bottoms.

Oil Color. Emulsions are characteristically hazy in appearance, in contrast to the brightness of treated oil. The hazy color is due to a high concentration of fine water droplets dispersed in the oil sample. Essentially, oil brightness is a measure of the water droplet size and the amount of dispersed water in the oil phase. When these water droplets become invisible to the naked eye, the oil phase takes on a "cloudy" appearance (6). Alternatively, as an emulsion coalesces into larger droplets, the oil phase becomes brighter. Consequently, the oil brightness can be increased in two ways: either through a reduction in the number of droplets or through an increase in water droplet size. A reduced number of particles implies that the crude oil is becoming drier; this is not necessarily the case for increased droplet size. Consequently, oil brightness can be deceptive as a sole determinant in selecting an appropriate chemical. It follows that color is no guarantee of a successful demulsifier, but haziness is assurance that the chemical is ineffective.

Interface. The desired interface, referred to as a mirror interface, has a shiny oil at its top surface. Solids present in the crude oil frequently prevent the occurrence of a mirror interface even in the presence of selected chemicals. If the crude oil has no known paraffin problems or solids and has a medium to low density, a smooth interface should be expected. However, the opposite circumstances should not be expected to produce a smooth interface. However, if any of the aforementioned problems is present but a smooth interface is still observed, then emulsion is quite likely held up in the oil phase.

Water Quality. This characteristic (i.e., turbidity) is usually difficult to interpret in the bottle test and consequently to correlate with plant behavior. Clear water is definitely the desired result. Complications, however, arise when solids are present in the crude oil. Therefore, it is necessary to remove the solids with the water, along with any oil adsorbed on these solids. Removing the oil will discolor the water somewhat; therefore, a clear product may be unlikely. Wetting agents applied with demulsifiers in treating plants have improved the dehydration and deoiled the solids to a remarkable extent (2). However, this result has not been observed in bottle testing where the water is frequently more turbid when the wetting agents are included. Deoiling of the solids is a concentration effect, and high concentrations of surfactants with partial water solubility make the water phase turbid.

Sludge.   Sludge can form or accumulate at the middle, at the interface, or at the surface of the oil phase. In certain systems, noncoalesced water droplets will result in a loose agglomeration that separates into water and oil without any problem. This separation is a form of "middle" sludge that will not accumulate. However, "interface" sludge may or may not accumulate, depending upon both the system and the sludge stability. This sludge can be stabilized by finely divided solids and other contaminants to form pads. Loose interface sludge can be detected swirling about the axis of the test bottle. Finally, "surface" sludge, the most difficult to describe, is water in the form of droplets that remain floating on top of the oil phase. Surface sludge can be detrimental in systems in which residence time is critical, in that the sludge can be carried over.

Sludging characteristics can appear as overtreatment. The difference between the two is that sludge can be present at any given ratio, low or high, depending upon the crude oil; overtreatment traits are notably associated with excessive chemical. Sludging characteristics of a demulsifier should be identified during the course of the bottle test work. This identification is easily done by using an excess amount of the chemical (i.e., 4 mL of 10% solution) in a bottle and observing the condition of the oil after agitation. If top, middle, or interface sludge is noted or if the oil is re-emulsified, the chemical should be used with caution.

Selecting Appropriate Separation Equipment

This section briefly discusses the more frequently employed equipment types that comprise a typical emulsion-treatment plant configuration (Figure 3). Appendix B describes CANMET's pilot-scaled emulsion-treatment facilities located at the Coal Research Laboratories near Devon, Alberta, Canada.

Free-Water Knockout.   By definition, free water is any water associated with the crude-oil emulsion that settles out within 5 min while the produced fluids are stationary in a settling space within a vessel. Free-water knockouts (FWKO) are simply three-phase separation vessels that separate

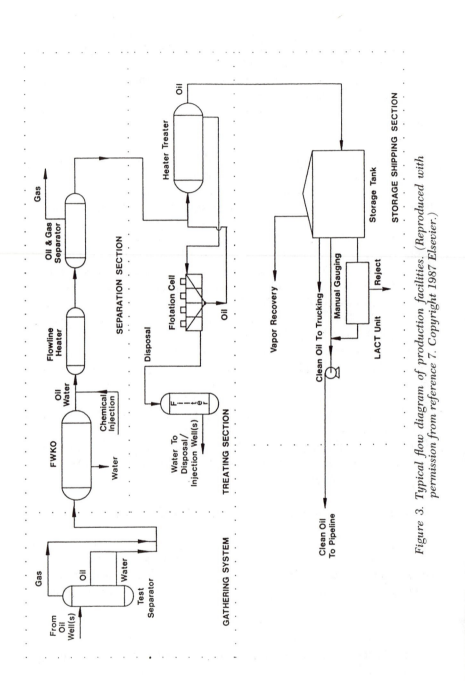

Figure 3. *Typical flow diagram of production facilities. (Reproduced with permission from reference 7. Copyright 1987 Elsevier.)*

large amounts of free water from the crude-oil emulsion. Any produced gases associated with these produced liquids are also separated and removed overhead (Figure 1).

Although FWKOs are not considered to be treatment equipment per se, their discussion here is appropriate because they are used extensively in conjunction with treatment equipment. Specific applications for FWKO vary with each situation. In some cases, the amount of water removed by the FWKO may not be crucial provided the treatment facilities downstream are performing effectively. When excessive amounts of water are produced with the crude-oil emulsion, as in secondary recovery (i.e., water-flood) projects, it may be necessary to specify requirements of 20% water to ensure proper treater performance.

Three-Phase Separators.   In addition to FWKOs, this category includes a broad array of treating vessels designed, as their name suggests, to separate produced fluids into three distinct products: gas and two immiscible liquids of different densities (i.e., oil and water). Generally, such vessels are employed where separation or measurement of all three phases is required (i.e., production testing of individual wells or streams).

Separators employ as many as three different fundamental separation mechanisms: (1) gravitational separation, (2) impingement and coalescence, and (3) centrifugal separation (Figure 4).

Gravitational Separation.   This process is both the simplest and most universally employed in all types of separators. It is based upon the fact that aqueous components of an inlet stream have a greater density than the associated petroleum fractions. Thus, aqueous components are subject to greater downward gravitational force. The liquid (aqueous) droplets settle out of a lighter (petroleum) phase if the gravitational force acting on the droplet is greater than the drag force of the oil flowing around the droplet. Furthermore, for conventional crude-oil emulsions, the water present in the liquid phase is usually heavier than the oil phase, and it subsequently settles below the oil. In heavy-oil emulsions, in which the two liquids are of similar densities, separation by means of gravity is extremely inefficient. Therefore, agents such as chemicals, heat, and electricity are employed to increase the rate of separation.

Impingement and Coalescence.   Impingement separation relies upon the difference in momentum, either between a gas particle and a liquid droplet, or between two liquid droplets. It occurs when liquid-laden gas approaches a coalescing device or target (e.g., wire mesh pad, vane element, or filter cartridge). This coalescing device causes the gas to follow a tortuous path, while the liquid droplets continue in a straighter path as a result of

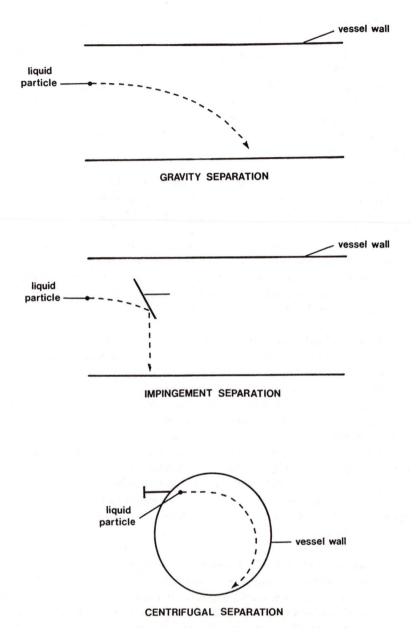

Figure 4. Schematic of separation processes. (Reproduced with permission from reference 6. Copyright 1990 Petroleum Industry Training Service.)

their greater momentum. These liquid droplets consequently impinge the targets or collide with each other and thereby coalesce. As these droplets increase in size, the effects of gravity become significant, and the droplets fall toward the liquid collection section of the vessel.

Centrifugal Separation.   The final separation process employed in the design of three-phase separators relies on the fact that fluid phases with different densities have different momentum. Centrifugal separation takes place when the inlet stream is forced to rotate at high velocities inside a vessel through the use of a tangential entry or deflector. The change in direction forces the liquid droplets to the vessel wall because of their greater momentum. There they coalesce and eventually drop to the vessel's liquid section.

Heater Treaters.   The heater treater is the ultimate processing step in the emulsion-treatment plant schematic. This vessel effects the actual breaking of the emulsion into treated oil and produced oily water streams. Oil-field emulsion treaters are either horizontal or vertical in orientation. Horizontal heater treaters normally have a high throughput requiring rapid demulsifier action. The large interface area (and consequently shallow fluid depth) in a horizontal heater treater requires that the interface be very clean (i.e., sharp, unobstructed). Consequently, this treater can tolerate very little interface buildup. The higher the throughput, the lower the tolerance for interface buildup.

Vertical treaters have a much lower volume-to-throughput ratio than gun barrels. As a result, more complete treatment is necessary in a shorter time. Solids control is as important as interface control, just as with the horizontal treater. Figure 5 illustrates a schematic of a vertical heater treater that also employs a dual-polarity electrostatic grid to effect more efficient coalescence and thereby better oil–water separation.

Induced Gas Flotation.   Mechanically induced gas flotation (IGF) is employed extensively to remove suspended solids, oil, and other organic matter from oil-field and refinery wastewaters. Consequently, these IGF units are particularly suited to the treatment of oil-in-water or reverse emulsions. Such units generally follow gravity oil–water separation units such as FWKOs, gun barrels, and skim tanks in oil-field-produced water-treatment schemes, and also handle the oily water streams generated from all treaters in a specific produced-fluid treatment plant.

Induced gas flotation can clean large quantities of wastewater containing 200–5000 ppm of suspended oil, depending upon the nature of the oil and its emulsion with the produced water. In most cases, the oil content of the IGF effluent water is less than 10 ppm after a 4-min cleaning cycle.

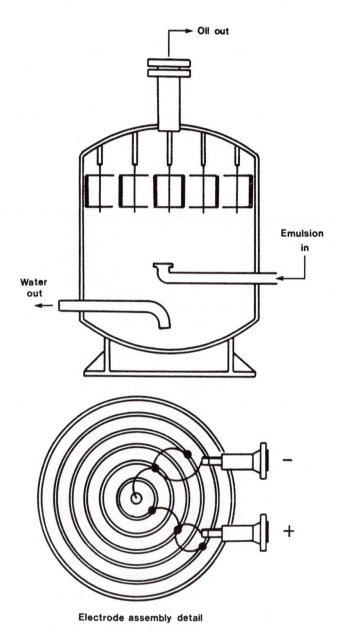

Oil out

Emulsion
in

Water
out

Electrode assembly detail

Figure 5. Dual-polarity electrostatic treater, simplified internal structure.

These units generally consist of four flotation cells (Figure 6), each of which is equipped with a motor-driven self-aerating rotor mechanism. The following describes the principle of operation (8):

> *As the rotor spins, it acts as a pump, forcing water through a disperser and creating a vacuum in the standpipe. The vacuum pulls gas into the standpipe and thoroughly mixes it with the water. As the gas–water mixture travels through the disperser at high velocity, a mixing force is created, causing the gas to form minute bubbles. Oil particles and suspended solids attach to the gas bubbles as they rise to the surface. The oil and suspended solids gather in a dense froth at the surface, are removed from the cell by skimmer paddles, and collected in internal launders.*

The oil and solids collected from the launders are then passed along to the oil-handling system; the recovered produced water, providing it meets the necessary requirements, is either disposed of or reutilized in the oil recovery process. If the water is to be reinjected, cleaning by the IGF unit prevents formation plugging and reduced pump efficiency. If the water is to be used for steam generation, the IGF is used before the traditional boiler pretreatment equipment.

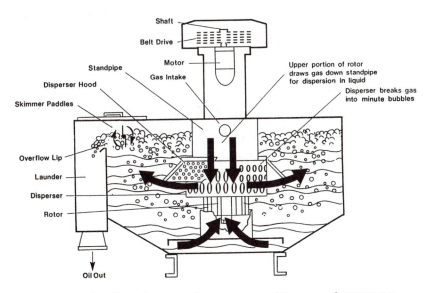

Figure 6. Induced gas flotation cell, section view. (Courtesy of WEMCO Process Equipment Company.)

Process Plant Design Considerations

When determining the size of a treater for a specific service, three significant but independent variables must be defined: (1) the vessel's diameter, (2) the length-to-height ratio of the coalescing section, and (3) the average treating temperature. In the absence of a unique solution, underlying assumptions and engineering judgement must be employed in developing an appropriate treater design. Briefly, this iterative process is as follows (3):

1. A treating temperature is chosen.

2. The oil viscosity at the treating temperature is determined.

3. The diameter of the water droplet that must be removed from the oil at the treating temperature is determined.

4. The treater geometry necessary to satisfy the settling criteria is determined.

5. The geometry is checked to ensure that it provides sufficient retention time.

6. The procedure is repeated for different assumed treating temperatures.

This procedure, although it does not yield the overall dimensions of the treater (including the inlet gas separation and FWKO sections), does provide a methodology for specifying the heating requirements and a minimum size for the coalescing section (where the treatment actually occurs). It is also invaluable when evaluating vendor proposals and when limited laboratory data are available.

Vessel Capacity Determination. *Vessel Diameter.* For conventional oil-treatment systems, the specific gravity[3] difference between the dispersed water droplets and the oil should result in the water "sinking" to the bottom of the treater vessel. The downward velocity of the water droplet must be sufficient to overcome the upward velocity of the oil phase throughout the treater. A general sizing equation is derived by setting these two velocities equal to each other (3). For horizontal vessels,

$$d = 0.2418 \left[\frac{Q_o \mu}{L_{eff} \Delta SG d_m^2} \right] \qquad (1)$$

For vertical vessels,

$$d = 0.1691 \left[\frac{Q_o \mu}{\Delta SG d_m^2} \right]^{1/2} \qquad (2)$$

[3]Specific (i.e., "relative") gravity is defined as the ratio of the weight of a given volume of a liquid at 15.5 °C (60 °F) to the weight of the same volume of water at 15.5 °C (60 °F).

where d is vessel diameter (m), Q_o is the oil flow rate (m³/h), μ is oil viscosity (mPa · s), L_{eff} is the length of the coalescing section (m), ΔSG is the difference in specific gravity between the oil and water (relative to the water), and d_m is the diameter of a water droplet (μm).

In a horizontal vessel, the cross-sectional area of the flow for the upward velocity of the oil is a function of the vessel diameter and the length of the coalescing section. For a vertical treater, the height of the coalescing section does not enter the equation. For larger diameter, vertical-flow treaters (i.e., such as gun barrels) a correction factor for short-circuiting effects must be included in the equation.

Retention Time. To effectively demulsify an oil-in-water emulsion, it must be held at a suitable treating temperature for a specific time period. In the absence of experimental data, 20–30 min is usually a realistic estimate for retention time for conventional oil projects; for heavy-oil recovery operations, retention times could be several hours. Nonetheless, the vessel geometry and specifications required for a specific retention time may not necessarily be the same as those dictated by the settling requirements. The solution is to select the larger geometry and dimensions determined by either of the two criteria. The retention time is determined as follows (3). For horizontal vessels,

$$t_r = \frac{d^2 L_{eff}}{32 Q_o} \tag{3}$$

For vertical vessels,

$$t_r = \frac{d^2 h}{3369 Q_o} \tag{4}$$

where t_r is retention time (min), d is vessel diameter (m), Q_o is oil flow rate (m³/h), L_{eff} is the length of the coalescing section (m), and h is the height of the coalescing section (m).

Water Droplet Size. To find a solution to the settling equation (i.e., for either equation 1 or 2), the water droplet size, d_m, must be known. Qualitatively, the water droplet size is expected to increase with an increase in retention time in the coalescing section and with heat input. Conversely, it should decrease with increase in the oil-phase viscosity. Furthermore, viscosity will have a greater effect on coalescence than temperature. Practical experience in the design of treaters has resulted in a reliable correlation of water droplet size to oil-phase viscosity (3):

$$d_m = 500 \mu^{-0.675} \tag{5}$$

where d_m is the diameter of a water droplet (μm), and μ is the oil viscosity (mPa · s).

Nonetheless, there is no substitute for actual experimental data on droplet coalescence. The universality of equation 5 has still to be proven.

Product-Stream Quality Requirements. In addition to the pipeline-marketing target of 0.5% BS&W that the treated-oil product must satisfy, there are analogous constraints for the various intermediate process streams based upon the downstream equipment's operating design range. For example, if a heavy-oil emulsion contains roughly 60% water, and only half of that quantity can be removed by the FWKO (i.e., resulting in 30% BS&W still present in its effluent stream), a heavy-oil evaporation (HOE) dehydration unit requiring an inlet stream containing no more than 10% water cannot simply be used to reduce the BS&W level to the pipeline specifications of 0.5%. An intermediary stage, preferably a heater treater, must be employed to supplement the treatment effort. The majority of the remaining BS&W preferably should be removed in the first (primary) treatment stage. Thus any unexpected or infrequent excessive treatment demands can be transferred on to the secondary treatment stage. In the foregoing example, it would be better to treat the emulsion in the first stage to an effluent level of 5% (if attainable) and have added treating capacity remaining in the secondary treater, than to merely target for the inlet conditions of the secondary treater. Similar arguments can be made for the addition of a downstream filtering system to the gas flotation unit in the oily water treatment stage.

Material Balance Requirements. Regular production testing of individual oil wells is a mandatory requirement for proper production accounting. A good practice is to place a well on a production test 1 day of every month, if feasible, to determine its individual oil-, water-, and gas-producing rates. The results can then be used to allocate that specific well's contribution to the overall producing rates on a monthly basis based upon a fieldwide proration factor. These allocations are updated with each new production test for that specific well. In essence, the total volume of fluids produced at the well sites must be accounted for at the final effluent streams of the treatment facilities (excluding accumulation within the individual vessels).

To effectively monitor the separation efficiency of the particular treatment equipment, two specific methods are employed: centrifugation (discussed briefly under "Testing Procedures") and the Dean–Stark analysis. The Dean–Stark analysis determines the fractional composition of oil–hydrocarbon, water, and solids of an emulsion stream by using a distillation process. Its results for heavy-oil emulsions are generally more reliable than those obtained by centrifugation; however, the results of centrifugation are available within minutes of sampling, compared to a 1–2-day turnaround for Dean–Stark analyses. These results, when coupled with process-stream flow measurements, provide comprehensive information for conducting component mass balances across the individual treating units and the entire plant.

Summary

This chapter has dealt with the important considerations involved in commercial demulsification equipment. Just as in Chapter 9, which dealt with choosing the right demulsifier, here too, numerous treatment options are available. No unique set of conditions can successfully break all oil-field-produced emulsions. Instead, the resolution of any specific emulsion-treating program involves the selection of a particular set of conditions that, acting together, yield a maximum treating performance. Whether an emulsion comes from a production wellhead or from an upgrader or refinery, the same choices have to be made among chemicals to be used, driving forces to be applied, optimum economics, and necessary product qualities to be achieved.

Testing process-stream emulsions is a necessity not only for characterizing the emulsion itself, but also for establishing performance of the selected treating process. Nevertheless, sampling a process-stream emulsion, the most critical step in the testing procedure, is more of an art than a science. Extensive work has been done to develop standard methods of testing process-stream emulsions. Testing procedures, such as the ubiquitous bottle test, can be used as a first estimate for equipment sizing. In the final analysis, pilot-plant testing of the chosen demulsifying scheme, with actual field or plant samples, will likely be needed to provide realistic data for scale-up to production levels.

The bottle test and the plant test, which is the ultimate determinant of a successful set of treating conditions, have become the oil industry's standards. However, further work should be undertaken to develop and refine sampling techniques (i.e., both procedures and hardware) to ensure that samples taken are truly representative, composite, and consistently obtainable.

Appendix A: Bottle Test Procedure

This procedure is described further in reference 2.

   Equipment.   The equipment needed is the following:

   1. sampling jug
   2. pouring device

3. test bottles or tubes: 175-mL prescription bottles, 100-mL tapered centrifuge tubes, and any test bottles of adequate volume that will allow for suitable agitation and have a reliable 100-mL mark

4. API centrifuge tubes

5. pipettes: 0.2, 1.0, 2.0, and 10.0 mL

6. syringes: 0.25, 1.0, 5.0, and 25.0 mL

7. solution bottles: 15 and 30 mL

8. solvent: aromatic or xylene, isopropyl alcohol, and a 75:25 mixture of xylene and isopropyl alcohol

9. demulsifier samples

10. knockout drops (i.e., Petrolite F-46, F-17, or RN-3003: Champion DN-71)

11. water bath

12. thermometer (0–100 °C)

13. marking pen, labels, and test book

14. centrifuge

15. optional equipment: shaking machine, reading lamp, centrifuge tube rack, and clock timer or stopwatch

Test Solutions. The selection of the most appropriate solution for a specific test requires a general understanding of the volume of chemical that will be required. Emulsions of high-density, high-viscosity crude oils may require 200–500 ppm (1.0–2.5 mL of 2% solution). Certain slop oil tests require 1000–2000 ppm (1.0–2.5 mL of 10% solution). Alternatively, very light oil emulsions may be easily broken with 10–20 ppm (0.05–0.10 mL of 2% solution).

The method used to prepare 2% solutions is as follows:

1. Measure 24.5 mL of solvent (i.e., heavy aromatic naphtha) into a 30-mL bottle.

2. Fill a 0.5-mL syringe past the 0.5-mL mark with the desired chemical demulsifier.

3. Expel all air bubbles by inverting the filled syringe and gradually pushing the piston up to the 0.5-mL mark.

4. Empty the contents of the filled syringe into the bottle containing the solvent, and shake to dissolve.

5. Mark the solution bottle with the name or number of the demulsifier used.

6. Rinse the syringe with solvent before it is used again or put away. Be sure to rinse the pipette or syringe used to place solutions into test bottles before placing it in a different solution. Also be certain to ensure the absence of air bubbles when using syringes with either undiluted chemical or solutions.

Execution of Bottle Test.

1. Prepare solutions (2 or 10%) or mixtures of solutions of the chemicals to be tested.

2. Obtain a chemical-free sample of the crude oil to be tested.

3. Test the sample as soon as possible.

4. Remove free water, if any, from the sample; measure and save. Record the amount of free water in the sample (i.e., 5, 10, or 20%, etc.). Some chemicals are more water-soluble or oil-soluble than others. The presence of free water can alter the test results, so it should be added to the oil when testing.

5. Fill two 12-mL centrifuge tubes with 50% (6 mL) of xylene or gasoline. To one tube add two drops of a 20% knockout drops solution. Agitate the sample well, and fill both tubes to 100% with crude oil to be tested. Mix well. Spin in centrifuge for 5 min. Record on test sheet the amount of water and basic sediment in the tube without chemical and the amount of water in the tube with chemical.

6. Add 100 mL of crude oil to test bottles. If free water was present, add this to the test bottle first, and then fill to the 100-mL mark with crude oil (e.g., if sample contained 10% free water, add 10 mL of this water).

7. Place the bottles in a water bath and allow 30 min to reach the temperature at which the chemical is injected.

8. Add 2% solutions of the various chemicals to the bottles.

   a. From the daily production and amount of chemical used, determine the parts per million used in the system.

      b. On the first test, run ratio tests at 0.5, 0.75, 1, 1.5, and 2 times the rate used in the plant.

      c. After establishing the ratio that will produce acceptable oil, various compounds can be checked at this ratio. Remember that oil must be treated in the system and so must be treated in the bottle.

9. Agitate the bottles a given number of times and place them in a water bath that is at the temperature of the treating vessel. Thirty minutes after being placing in the water bath at treating vessel temperature, agitate the bottles a second time and return to the water bath. Notes:

      a. This is the important step; many tests go astray here.

      b. Normal agitation is 200 times at chemical injection temperature, 10 to 100 times at treating vessel temperature. This level is a good starting place but should not be taken as a standard.

      c. Correct agitation is whatever type of agitation is required to reproduce the system.

      d. If results similar to the plant results can be obtained in the bottle test at the same concentration of chemical, the type of agitation used should duplicate the plant results.

      e. On the first test of an emulsion, chemicals known to have been used in the system should be tried with various types of agitation. A test method should be selected from these tests. If none of the agitation variations produce clean oil at twice the plant concentration or less, different types of agitation should be tried.

10. Record the amount of water down (i.e., settled at the bottom)

      a. immediately before agitation at treating vessel temperature and 1 h after.

      b. at the conclusion of the test.

      c. at other times as desired or that are critical to the system.

11. The length of test time is calculated by dividing the daily production into the capacity of the treating vessel.

12. Take a deep grind on all bottles having 75% or more of the total water down. (A deep grind is a sample of oil taken 10–15 mL above the water–oil interface.)

    a. Place this sample in a centrifuge tube containing 50% solvent. Mix well and centrifuge for 5 min.

    b. Total the sum of basic sediment and water. Record the total on the test sheet.

    c. Add two drops of 20% knockout drops solution, mix well, and centrifuge for another 5 min.

    d. Record the amount of water with chemical on the test sheet. Because only 50% oil is used, readings should be doubled.

    e. For crude oils of density 986 kg/m^3 or more (i.e., API gravity 12 or less), 9 mL of solvent should be added to centrifuge tubes before filling them to 100% with oil.

    f. Clean the sampling device well after each use.

13. Take a mixed grind on any sample containing 1% water or less after having added the knockout drops. (A mixed grind involves removing the entire portion, or as much as possible, or the water that has separated during the settling period.)

    a. Remove all water from the test bottle, being careful not to remove any of the oil at the interface.

    b. Agitate the bottle 10 times, pour the contents into a centrifuge tube containing 50% solvent, and mix well.

    c. Centrifuge and record the total amount of water and basic sediment.

    d. Add two drops of 20% knockout drops solution, mix well, and record the total amount of water with chemical.

14. The best chemical is one that produces saleable oil at the lowest concentration on the deep grind and that has very little or no basic sediment in the mixed grind. If a mixed grind has been extracted properly, up to 1% of the water may be due to free water remaining in the test bottle. Any amount over this is probably due to water being held up in the crude oil or to a poor interface.

15. Miscellaneous notes.

    a. Include the compounds being used in the system and that you are trying to beat in each test.

    b. As testing progresses, reduce the dosage in order to select the best compound. Only the first test should be at a rate where the competitive product will barely treat. Tests thereafter should be at about a 20% lower ratio.

    c. Low-density crude oils should be agitated frequently to ensure uniform samples in all bottles.

Observations.   The following are the more important characteristics and observations of effective demulsification bottle tests.

Regarding the speed of water dropout, in a sample with high water volume, a chemical or demulsifier with a fast water dropout is required. Where FWKOs are involved, the speed of water dropout may become the most important factor.

In samples with low-water volume, or those with more than normal residence time, the speed of water dropout may be of lesser significance in selecting the best demulsifier. Nonetheless, in all cases the speed of water dropout and volume should be recorded.

The speed of water break is important and should be evaluated. It is sometimes misleading in that a fast-water-dropout chemical will sometimes quit treating before oil that meets pipeline specifications is obtained. A good rule of thumb is to never use any faster water break than is needed.

Among the other principal items to observe, which have been discussed in the principal text (*see* "Interpreting Results"), are centrifuge cut, oil color, interface, water quality, and sludge.

Possible Errors.   Errors such as the following can occur.

1. Sampling errors. A bad sample is one that is not representative, is aged, or is contaminated with chemical or recycled oil.

2. Pouring into sample bottles can cause errors if unequal amounts of emulsion sample are poured into each bottle or if the sample characteristics vary among individual bottles (usually because of the presence of free water).

3. Errors in adding the chemical solutions are adding the wrong mixture to a bottle or adding an inappropriate amount of solution to a bottle.

4. Misplacing or confusing sample bottles.

5. Centrifuging the sample insufficiently. Look for emulsion in the oil phase and water adhering to the sides of the centrifuge tube. Check for emulsion or water droplets on the sides of the tube after centrifuging by turning the sample upside down.

6. Drawing the interface material off along with the free water in the sample bottle.

Appendix B: Pilot-Scaled Plant for Heavy-Oil Emulsions Treatment

The Canada Centre for Mineral and Energy Technology (CANMET) is the main research and technology development arm of Energy, Mines, and Resources Canada. As one of CANMET's five laboratory groups, Coal Research Laboratories (CRL) performs and sponsors research to enhance Canadian industry's competitive position. With laboratories in Devon, Alberta, and Sydney (Nova Scotia), CRL is well located to serve both coal and other industry clients, notably those in the recovery and processing of oil sands and heavy oil.

CANMET has a pilot-scaled emulsion-treatment plant (Figure B.1) available to industry for pilot-scaled investigation of heavy-oil–bitumen separation from oil-field-produced waters. This facility is designed to process emulsions at a throughput between 130 L/h (20 barrels per day) and 460 L/h (70 barrels per day) for raw bitumen–oil of API gravity between 8 and 15 (i.e., density between 1014 and 966 kg/m³, respectively).

The unit operations in the miniplant employ proven emulsion-treatment principles: free-water knockout, dual-polarity electrostatic treatment (DPET), heavy-oil evaporation (HOE) dehydration, and induced gas flotation (IGF). The overall process configuration provides maximum flexibility and allows for performance evaluation of units on either an individual basis or in various combinations.

CANMET researchers, working in close cooperation with industry, will seek to achieve the following objectives:

- determine the parameters that govern oil–water separation and that can be extrapolated to larger industrial plants
- evaluate the performance of conventional emulsion-treatment and separation equipment in processing unique and difficult-to-separate emulsions
- test and evaluate new and innovative on-line process monitoring and control equipment
- perform component and material balances associated with various process streams
- identify an integrated strategy incorporating equipment selection, plant layout, and operating strategy (e.g., the usage of demulsifiers or diluents) that would optimize emulsion-breaking operations for specific troublesome emulsions

Laboratory facilities and bench-scale equipment are also available in support of the work conducted in the miniplant. Laboratory facilities are gener-

NOTE: Hanna F. Sieben was a contributor to this appendix.

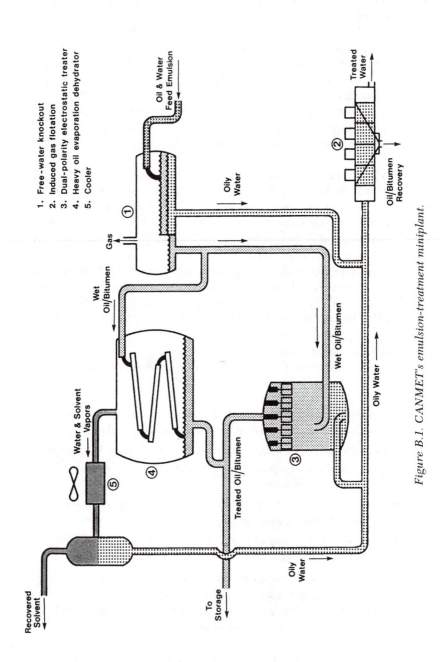

Figure B.1. CANMET's emulsion-treatment miniplant.

ally employed in conducting precursory studies aimed at defining prospective operating conditions for the pilot-scaled operations.

Description of Major Equipment.   The principal kinds of equipment comprising the emulsion-treatment miniplant are briefly described in Table B.1, and their interrelationship is presented schematically in Figures B.2 through B.7.

Feed Tanks (T-1, T-1A).   Figure B.2 depicts schematically the process flow around both emulsion feed storage tanks. The emulsion feed tank (T-1) is designed to hold 15,900 L (100 bbl). This horizontal tank is equipped with two identical mixers and a recirculation loop to ensure that emulsion feed is well mixed and consistent throughout a typical run. The tank is also equipped with a steam-plate coil-tank heater capable of raising the feed temperature to 70 °C (160 °F) with a heat load of 18.3 kW (62,500 Btu/h) during heat-up. A feed pump (P-1) is employed either to recirculate the tank's contents or to transfer it to the plant for processing. This pump, a progressive cavity type, is designed to handle liquids with a viscosity range of 500 to 10,000 mPa·s at a discharge pressure of 1034 kPa (150 psi) and a maximum flow rate of 680 L/h (3.0 U.S. gallons per minute or USGPM).

The auxiliary feed tank (T-1A) holds 2840 L (18 bbl). This tank is equipped with a mixer and a heating coil that is rated at 48 kW (163,800 Btu/h) and is capable of raising the feed temperature to 150 °C (300 °F) in 2 h; normal operating temperature is expected to be approximately 80 °C (176 °F). A sensor in the tank is used as a low-level alarm to shut off the heater. The auxiliary feed pump (P-1A) is capable of generating a discharge pressure of 517 kPa (75 psi) and a maximum flow rate of 1590 L/h (7.0 USGPM). It may also be used to deliver feed to the other process units or to recirculate the tank's contents. Instrumentation is in place downstream of both feed tanks to control the feed flow rate and to measure the temperature and pH of the feed. Connections are also available for the addition of demulsifier or diluent.

Free-Water Knockout (FWKO, V-1).   Figure B.3 is a process flow schematic of the free-water knockout unit's operation. The FWKO vessel is a three-phase horizontal separator designed to hold 940 L (5.9 bbl) and to operate safely at pressures up to 517 kPa (75 psi) at 177 °C (350 °F). Emulsion feed containing greater than 15% (v/v) water enters the FWKO, first passing through a baffled "coalescing zone", and then into a quiescent separation area where free water collects in the bottom of the vessel and any evolved gas leaves from the top of the vessel through a 304 SS (stainless steel) wire-mesh mist eliminator. The remaining emulsion flows over a weir into the oil recovery side of the vessel.

Instrumentation on the unit includes pressure and temperature mea-

Table B.1. Major Equipment List

Tag	Name	Description
Feed Tanks (Figure B.2)		
HT-1	Feed tank heater	Double-embossed plate coil; 25-kW heating duty with steam
HT-1A	Auxiliary feed tank heater	48 kW; 460 VAC; three-phase; 60 Hz
MX-1e	Feed tank mixer (east end)	1.5 hp; 575 VAC; three-phase; 60 Hz; 1720 rpm; dual impeller at 350 rpm
MX-1w	Feed tank mixer (west end)	1.5 hp; 575 VAC; three-phase; 60 Hz; 1720 rpm; dual impeller at 350 rpm
MX-1A	Auxiliary feed tank mixer	1.5 hp; 460 VAC; three-phase; 60 Hz; 1720 rpm; dual impeller at 350 rpm
P-1	Feed pump	eccentric screw; 680 L/h at 1034 kPa (250 rpm); 1.5 hp; 460 VAC; three-phase; 60 Hz; 1720 rpm
P-1A	Auxiliary feed pump	Rotary vane; 1560 L/h at 517 kPa (40 rpm); 2.0 hp; 460 VAC; three-phase; 60 Hz; 1140 rpm
T-1	Feed tank	Horizontal; 2.4 m o.d. × 3.7 m long; 15,900-L capacity
T-1A	Auxiliary feed tank	Horizontal; 1.2 m o.d. × 2.4 m long; 2840-L capacity
Free-Water Knockout (Figure B.3)		
V-1	FWKO vessel	Horizontal; three-phase separator; 0.8 m o.d. × 2.4 m long; 940-L capacity; MAWP 517 kPa at 177 °C c/w inlet flow baffles, coalescing plate section, oil overflow wire, 304 SS gas outlet mist eliminator, liquid outlet vortex breakers
Dual-Polarity Electrostatic Treater (Figure B.4)		
HT-3	DPET inlet heaters	3–20 kW; 480 VAC; three-phase; 60 Hz
MX-2	DPET inlet mixer	38-mm helical coil in-line mixer
P-7	DPET feed pump	Progressing cavity; 570 L/h at 517 kPa (590 rpm); 0.5 hp; 480 VAC; three-phase; 60 Hz; 1140 rpm
V-2	DPET vessel	Vertical; 1.2 m o.d. × 1.8 m high; 1720-L capacity; MAWP 517 kPa at 150 °C; c/w 2 kVA; 480 VAC; single-phase; 60 Hz primary & 9.0, 12.5, 16.5, and 23.0 kVDC selectable secondary transformer
Heavy-Oil Evaporation Dehydrator (Figure B.5)		
HT-4	HOE inlet preheater	Horizontal; shell–tube heat exchanger; 56-kW process duty with superheated steam at 690 kPa; MAWP: shell (steam) 1034 kPa at 176 °C, tube (emulsion) 1034 kPa at 149 °C

Tag	Name	Description
HT-5	Treated oil cooler–vapor condenser	Fin-fan–tube heat exchanger; cooling duty 20 kW; condensing duty 44 kW; MAWP: 1034 kPa at 177 °C (oil), 1034 kPa at 149 °C (vapors)
P-6	HOE oil discharge pump	Gear; 684 L/h at 345 kPa (590 rpm); 1.0 hp; 460 VAC; three-phase; 60 Hz; 1725 rpm
P-8	Skimmer feed pump	Gear; 342 L/h at 345 kPa (590 rpm); 0.25 hp; 460 VAC; three-phase; 60 Hz; 1725 rpm
V-3	HOE vessel	Horizontal; 1.4 m i.d. × 2.1 m long; 3150-L capacity; MAWP 345 kPa at 177 °C
V-5	Liquid accumulator	Vertical; surge vessel; 0.2 m o.d. × 0.8 m high; 15-L capacity; MAWP 517 kPa at 65 °C
V-6	Skimmer vessel	Vertical; two-phase separator; 0.2 m o.d. × 1.5 m high; 30-L capacity; MAWP 517 kPa at 65 °C

Treated-Oil–Bitumen Tanks (Figure B.6)

Tag	Name	Description
HT-2	Treated-oil tank heater	Double-embossed plate coil; 1140 kPa; 25 kW heat load
HT-2A	Auxiliary treated-oil tank heater	20 kW; 460 VAC; three-phase; 60 Hz
MX-2	Treated-oil tank mixer	3.0 hp; 575 VAC; three-phase; 60 Hz; 350 rpm; dual impeller
P-2	Treated-oil recirc. pump	Eccentric screw; 680 L/h at 1034 kPa; 1.5 hp; 460 VAC; three-phase; 60 Hz
P-2A	Auxiliary treated-oil recirc. pump	Rotary vane; 1560 L/h at 517 kPa; 2.0 hp; 460 VAC; three-phase; 60 Hz
T-2	Treated-oil tank	Vertical; 2.9 m o.d. × 1.9 m high; 11,130-L capacity
T-2A	Auxiliary treated-oil tank	Horizontal; 1.2 m o.d. × 2.4 m long; 2840-L capacity

Induced Gas Flotation (Figure B.7)

Tag	Name	Description
P-3	Produced water pump	0.25 hp; 230 VAC; one-phase; 60 Hz; 1725 rpm
T-3	Produced water tank	Vertical; 0.9 m o.d. × 1.5 m high; 950-L capacity
V-4	IGF unit	Four-cell flotation unit; 2280 L/h capacity at 25 °C; c/w one froth skimmer and disperser–impeller per cell

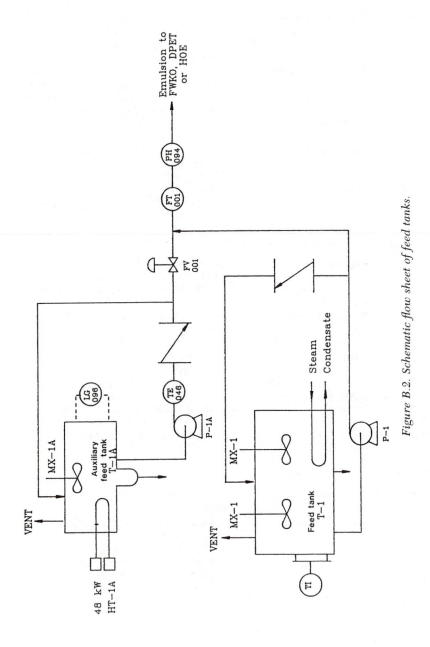

Figure B.2. Schematic flow sheet of feed tanks.

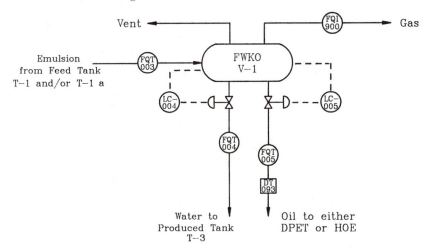

Figure B.3. Schematic flow sheet of free-water knockout.

surement; level control; oil, water, and gas outlet flow-rate measurement; and determination of the outgoing oil and bitumen density. The water leaving the FWKO is sent directly to the induced gas flotation (IGF, V-4) unit, and the exiting emulsion may be sent to either the dual-polarity electrostatic treater (DPET, V-2) or the heavy-oil evaporation (HOE, V-3) dehydrator for further treatment. Demulsifier, diluent, or both may also be added to the exiting emulsion stream.

Dual-Polarity Electrostatic Treater (DPET, V-2).   The DPET (a C-E Natco vertical VFH-CWW model) process flow diagram is presented in Figure B.4. The purpose of this vertical vessel is to provide for both free-water removal and coalescence of entrained water droplets. Within the miniplant's configuration, it principally serves as a primary treating stage followed by the HOE unit as the final treating step. Nonetheless, either treating unit may be operated independently of the other in achieving the ultimate goal, a treated oil–bitumen meeting marketing specifications: a maximum of 0.5% BS&W.

The DPET is designed to hold 1720 L (10.8 bbl) and operates safely at pressures up to 517 kPa (75 psi) and temperatures up to 150 °C (300 °F). The application of a high-voltage, dual-polarity electric potential to electrodes inside the vessel is used to coalesce and remove small droplets of water in the oil emulsion. The oil should be degassed and have a water content less than 15% before entering the vessel; however, the treater does have the capability for free-water knockout. Preheated emulsion is pumped into the bottom portion of the vessel, below the electrodes, where free water generated by heating or chemical treatment may drop out. As more emulsion is

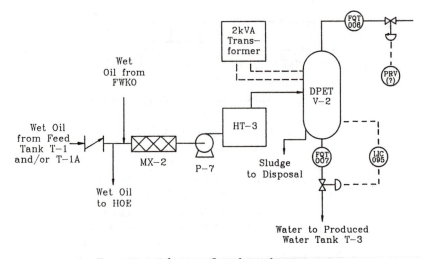

Figure B.4. Schematic flow sheet of DPET treater.

introduced to the vessel, the emulsion rises until it comes into contact with the electrodes.

The DPET's electrodes are in the form of concentric cylindrical plates suspended from the top of the vessel and are connected to a high-voltage transformer such that adjacent plates are given opposite charges. As the oil passes through the electrodes, water droplets are influenced by the field to create a sinusoidal migration between plates of opposite charge. This motion serves two purposes: to restrict the upward flow of the water in relation to the oil; and to enhance the rate of collision of water droplets, which are distorted to form dipoles under the electric field, and thereby increase the rate of coalescence. Clean oil leaves from the top of the vessel, while water is drained out from the bottom. Instrumentation available on this unit permits pressure and temperature measurement of both the feed and vessel conditions and the treated oil and outlet water flow rates.

The treated oil–bitumen, which should contain less than 0.5% water, is then transferred to the treated oil storage tank (T-2). The exiting water-rich stream is pumped to a produced water storage tank (T-3) from which it may be pumped to the IGF.

Heavy-Oil Evaporation Dehydrator (HOE, V-3). Figure B.5 illustrates the process flow schematics for the HOE unit. This horizontal vessel can serve as either an alternative to the DPET (should the oil–bitumen-rich stream exiting the FWKO contain less than 10% water) or as a secondary treater to the DPET. The HOE unit is designed to process an emulsion composed of oil–bitumen (70%), water (10%), and diluent (20%) at a combined rate of 460 L/h (70 barrels per day) for raw oil–bitumen of API gravity between 8 and 15 (i.e., density between 1014 and 0.966 kg/m³, respectively). Chart I summarizes the unit's process design basis.

The emulsion fed to the HOE can be preheated to approximately 125 °C (257 °F) by a shell-and-tube heat exchanger (HT-4). The hot emulsion is then fed into the evaporator vessel (V-3), through a spreader, onto a wide, shallow tray. The spreader ensures that a thin uniform coat of oil is deposited onto the tray, which is sloped downward and is heated from below by steam coils. As the emulsion runs down the tray, water, light ends, and any remaining diluent are evaporated. This process is repeated consecutively on two additional trays.

The treated oil–bitumen is collected in the bottom of the vessel, while the vapors exit at the top. The contact temperature in the vessel is approximately 150 °C (300 °F), and the total area of the three trays is approximately 3.34 m² (36 sq ft). The treated oil, which should contain less than 0.5% (v/v) water, is then pumped from the bottom of the evaporator, fan-cooled and sent to the treated-oil storage tank (T-2). The water–diluent vapors are also fan-cooled, sent to an accumulator, and then transferred to a gravity separator for diluent recovery. Water is sent to the produced-water tank (T-3) before being processed through the IGF unit.

The instrumentation on this unit includes pressure and temperature control of the preheated emulsion; pressure and temperature measurement

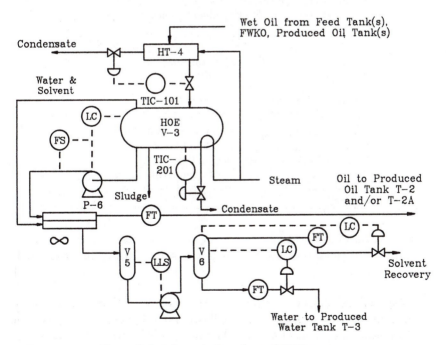

Figure B.5. Schematic flow sheet of HOE treater.

Chart I. Heavy-Oil Evaporation Dehydrator (HOE; V-3) Process Design Basis

Process Flow to Preheater: at 40 °C (104 °F)
  to 127 °C (260 °F)

Oil-bitumen (1,020 kg/m^3): 324 L/h (729 lb/h)
Water (1,000 kg/m^3): 42 L/h (104 lb/h)
Diluent (800 kg/m^3): 93 L/h (163 lb/h)
● pressure: 207 kPag (30 psig)
● water evaporates in the exchanger
● 95% of diluent flashes in the exchanger
● process heating duty: 56 kW (192 000 Btu/h)

Evaporator: contact temperature, 149 °C (300 °F)

● remaining 5% of diluent flashes
● steam is superheated
● 2% (v/v) of raw-oil-bitumen flashes
● tray duty: 7 kW (25 000 Btu/h)

Vapor Condenser: from 108 °C (227 °F)
  at 34 kPag (5 psig) to 38 °C (100 °F)

● condense 46 kg/h (102 lb/h) steam
● condense 74 kg/h (163 lb/h) diluent
● condense 7 kg/h (15 lb/h) produced condensate
● cool all products to 38 °C (100 °F)
● condensing duty: 44 kW (149 000 Btu/h)
● based upon 27 °C (80 °F) ambient air cooling

Oil Cooler: from 149 °C (300 °F) to 38 °C (100 °F)

● cool 318 L/h (714 lb/h) oil--bitumen
● cooling duty: 20 kW (69 300 Btu/h)

Recovered Products: at 38 °C (100 °F)
  and 345 kPag (50 psig)

Oil-bitumen (1,020 kg/m^3): 317 L/h (714 lb/h)
Water (1,000 kg/m^3): 41 L/h (102 lb/h)
Diluent (800 kg/m^3): 102 L/h (178 lb/h)
● based upon 100% mass transfer equations
● treated-oil--bitumen contains less than 0.5% (v/v)
  BS&W

and level control in the evaporator vessel, the accumulator, and the gravity separator; and measurement of the outlet flow rates of the oil, water, and diluent.

Treated-Oil Tanks (T-2 and T-2A). The flow schematics for both treated oil–bitumen storage tanks are illustrated in Figure B.6. The treated oil–bitumen storage tank (T-2) holds 11,130 L (70 bbl). Treated oil–bitumen from either the DPET or the HOE treater can be fed to this tank. The

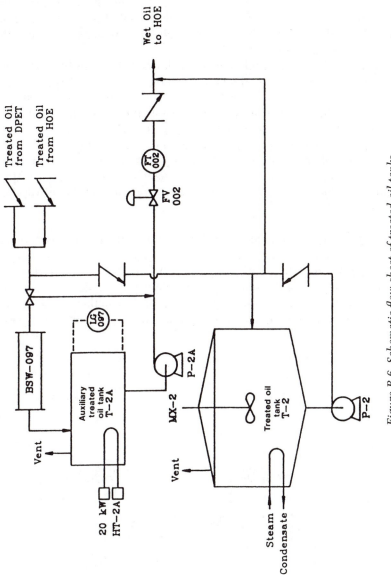

Figure B.6. Schematic flow sheet of treated-oil tanks.

treated-oil pump (P-2), which is identical to the emulsion feed pump (P-1), can also be employed to (1) transfer any off-specification (i.e., "wet") oil–bitumen back to the HOE (with the option of passing it through the auxiliary treated oil tank (T-2A) where blending with diluent can occur) for further processing or (2) recirculate the contents of tank T-2. Tank T-2 is also equipped with an electric mixer and a steam-plate coil-tank heater identical to the unit installed in the feed tank (T-1).

The auxiliary treated-oil tank (T-2A) holds 2840 L (18 bbl). The inlet, which can accept flow from the DPET and the HOE or the larger T-2 tank, is equipped with an in-line BS&W meter to monitor the water content of the treated-oil–bitumen stream. If the water content is not sufficiently low, the product may be recycled back to the HOE treater. This tank also has a heater and a level sensor to act as a low-level alarm to turn off the heater.

Additional instrumentation on the unit allows for measurement of temperature and pressure inside the vessel, flow-rate control of the off-specification oil, and measurement of rheological properties of the off-specification and produced oil.

Induced Gas Flotation (IGF, V-4).   Figure B.7 schematically depicts the process flow for the IGF unit. This unit processes all the water streams generated by the other unit operations (FWKO, DPET, and HOE) in the miniplant by moving the remaining oil from it prior to disposal or further treatment. The oil in the water collects at the surface of these bubbles and is drawn out of the bulk liquid as the bubbles rise to the top. An oily foam

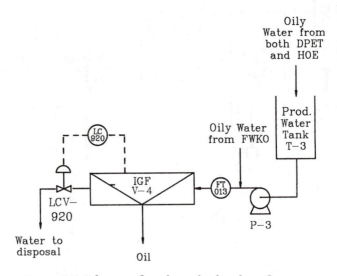

Figure B.7. Schematic flow sheet of induced gas flotation.

forms at the top of the cells and is skimmed off and collected. The oil content of the water may be reduced from as high as 5000 ppm to less than 10 ppm.

Instrumentation on the unit consists of a magnetic flowmeter to measure feed rate and a level controller to maintain proper liquid height in the flotation cells.

Demulsifier and Diluent Injection.   The system available for demulsifier addition consists of a feed pump and flexible hose connections that allow for chemical injection at various process points. The maximum flow rate generated by the demulsifier feed pump is 6.5 L/h (1.7 U.S. gallons per hour or USGPH). The diluent injection system is identical except that the feed pump may deliver up to 76 L/h (20 USGPH).

Computer Monitoring System.   A data acquisition system is used to monitor and log 27 various measurements from the miniplant. The signals from the instruments are fed into an analog-to-digital converter, and the resulting digital signal is converted into an appropriate value based upon a scale factor for its respective instrument.

Data and Results.

Feed Characteristics.   To date, plant test runs have been conducted with a bitumen-in-water emulsion comprising 28.8% (v/v) bitumen of 9 API gravity (i.e., density of 1007 kg/m^3). This emulsion was produced at an in situ oil-sand thermal recovery project in northern Alberta. Approximately 11.1 m^3 (70 bbl) of emulsion was treated. Separation of the bitumen from the water phase by gravity segregation was not a practical solution in this case because of the slight difference in the two phases' densities (1010 and 1002 kg/m^3 for bitumen and water, respectively). Consequently, determination of stream composition through centrifugation (i.e., ASTM D 96, "Standard Test Method for Water and Sediment in Crude Oils") did not generally produce reliable results. Dean–Stark analysis, which was also employed, proved to be more reliable in determining bitumen, water, and solids fractions in the emulsion. However, whereas the results from centrifugation are virtually immediate, Dean–Stark analysis requires roughly 24 h. Table B.2 summarizes the comparative results of both methods for a set of emulsion feed samples.

Plant Operation Performance.   Table B.3 summarizes the targeted flow compositions for the integrated plant operation of all four process units required to treat the aforementioned emulsion feed to acceptable product compositions: The bitumen (diluted with naphtha) must meet the petroleum marketing requirements of no more than 0.5% BS&W; for both economical and environmental reasons, the recovered produced water should contain no more than 10 ppm of bitumen–oil.

Table B.4 represents a composite mass balance for the entire pilot-plant operation. Any differences in the totals are directly attributable to sampling error, experimental error in conducting the Dean–Stark analysis, and the flow measurement error.

Table B.2. Dean–Stark Analysis vs. Centrifuge Comparative Results of Bitumen Cut

Sample Number	Dean–Stark Analysis	Centrifugation Method	% Difference
1	30.37	30.62	0.83
2	28.67	30.02	4.59
3	24.49	17.69	32.26
4	37.40	18.29	68.62
5	26.59	29.91	11.77
6	27.23	27.99	2.77
7	25.33	32.54	24.93
8	26.56	27.19	2.33
Avg.	28.33	26.78	6.71

NOTE: All values are percent of total weight.

Table B.3. Typical Results of Emulsion-Treatment Miniplant Dean–Stark Analysis

Process Stream	Emulsion Feed	FWKO Oil Outlet	DPET Oil Outlet	HOE Oil Outlet	IGF Inlet
Original Sample (g)	450.0	434.0	421.5	452.5	435.0
Bitumen (g)	129.0	126.0	371.5	448.6	5.0
Water (g)	319.1	67.0	48.3	2.2	427.8
Solids (g)	0.4	0.3	1.0	0.1	0.3
Total Recovered (g)	448.5	193.3	420.8	450.9	433.1
Recovery (%)	99.67	44.55	99.83	99.64	99.56
Bitumen Cut (%)	28.76	65.17	88.28	99.49	1.15
Water Cut (%)	71.15	34.65	11.48	0.49	98.78
Solids Cut (%)	0.09	0.18	0.24	0.02	0.07

Table B.4. Typical Example of Emulsion-Treatment Miniplant Mass Balance

Process	Bitumen kg	Bitumen %	Water kg	Water %	Solids kg	Solids %	Total (kg)
Emulsion feed	3,202	28.8	7,893	71.1	9	0.1	11,104
Bitumen outlet	3,176	99.4	16	0.5	4	0.1	3,196
Water outlet	23	0.3	7,910	99.6	6	0.1	7,939
Total outlet	3,199	28.7	7,926	71.2	10	0.1	11,135
Mass balance difference	−3	−0.1	33	0.4	1	11.1	31 (0.3%)

Acknowledgment

Thanks to K. C. McAuley for preparation of the figures.

References

1. Strassner, J. E. *J. Pet. Technol.* **1968,** *3*, 303–312.
2. Bessler, D. U. *Demulsification;* Petrolite Corporation: St. Louis, MO, 1984.
3. Arnold, K.; Stewart, M. *Surface Production Operations: Design of Oil-Handling Systems and Facilities;* Gulf Publishing Company: Houston, TX, 1986; Vol. 1.
4. Bessler, D. U. *Waste Oil Treatment;* Petrolite Corporation: St. Louis, MO, 1981.
5. Claassen, E. J.; Harlan, J. T.; Trial, C. B. *Emulsions;* Champion Chemicals: Houston, TX, 1976.
6. "Conventional Oil Production Facilities: Selection and Design Concepts"; presented by the Petroleum Industry Training Service, Calgary, Alberta, Canada, 1990.
7. Chilingarian, G. V.; Robertson, J. O., Jr.; Kumar, S. *Surface Operations in Petroleum Production;* Developments in Petroleum Science, 19A; Elsevier: Amsterdam, Netherlands, 1987; Vol. I.
8. "WEMCO Depurator 1+1 Flotation Machine"; Technical Bulletin No. F8-B5(8-85-5M), WEMCO: Sacramento, CA, 1985.

Additional Reading

Angelidou, C.; Keshavarz, E.; Richardson, M. J.; Jameson, G. J. *Ind. Eng. Chem. Process Des. Dev.* **1977,** *16(4)*, 436–441.

Arnold, K.; Stewart, M., Jr. *World Oil* **1985,** *2*, 31–36.

Arnold, K.; Stewart, M., Jr. World Oil **1985,** *5*, 91–98.

Arnold, K. E.; Koszela, P. J. *SPE Prod. Eng.* **1990,** *5(1)*, 59–64.

Bessler, D. U. *Demulsification of Enhanced Recovery Produced Fluids;* Petrolite Corporation: St. Louis, MO, 1984.

Bradley, B. W. *Oil Gas J.* **1985,** *12*, 42–45.

Brunsmann, J. J.; Cornelissen, J.; Eilers, H. *J. Water Pollut. Control Fed.* **1962,** *34*, 44–54.

Donaldson, E. C.; Chilingarian, G. V.; Yen, T. F. *Enhanced Oil Recovery;* Developments in Petroleum Science, 19A; Elsevier: Amsterdam, Netherlands, 1985; Vol. I.

Powers, M. L. Presented at the 63rd Annual Technical Conference of the Society of Petroleum Engineers, Houston, TX, October 2–5, 1988, paper SPE 18205; pp 241–252.

Sport, M. C. *J. Pet. Technol.* **1970,** *8*, 918–920.

RECEIVED for review December 18, 1990. ACCEPTED revised manuscript May 23, 1991.

Glossary of Petroleum Emulsion Terms

Laurier L. Schramm

Petroleum Recovery Institute, 3512 33rd Street N.W., Calgary, Alberta, Canada T2L 2A6

This glossary provides brief explanations for nearly 200 important terms in the science and engineering of petroleum emulsions. The field of petroleum emulsions encompasses aspects of so many different disciplines that there exists a voluminous body of terminology. A selection of frequently encountered terms has been made including scientific terms related to the basic principles and properties of emulsions, and petroleum production and processing terms used to describe practical emulsions and their treatment. In addition, cross-references for the more important synonyms and abbreviations are included.

A VAST POTENTIAL LEXICON is associated with the field of petroleum emulsions, partly because of the diversity of occurrences and properties of petroleum emulsions, and partly because of the many scientific disciplines involved in their study and treatment, each discipline bringing elements of its own special language. This glossary presents some of the more common terms used in the science and technology of emulsions. The selections have been chosen to encompass the more important emulsion terms used in the chapters of this book, and the explanations presented are consistent with the usage in those chapters.

No attempt has been made to include every term that may be encountered in dealing with petroleum emulsions. Some basic knowledge of underlying fields such as physical chemistry and chemical engineering is assumed. Many named emulsions and phenomena (such as Pickering emulsions) have been included, but named equations and constants have generally not been included.

0065–2393/92/0231–0385 $06.25/0

Some terms are used in other ways by other researchers, or in other countries, and may have legal definitions different from those given here. The distinctions drawn among light, heavy, extra-heavy, and bituminous crude oils were made on the basis of United Nations Institute for Training and Research (UNITAR)-sponsored discussions aimed at establishing such definitions (1–3). For terms drawn from the area of colloid and interface science, much reliance was placed on the recommendations of the IUPAC Commission on Colloid and Surface Chemistry (4). For important emulsion terms that are frequently used in industrial practice, the aim was to be consistent with the standard petroleum dictionaries such as references 5–7.

Absolute Viscosity   A term used to indicate viscosity measured by a standard method, with the results traceable to fundamental units. Absolute viscosities are distinguished from relative measurements made with instruments that measure viscous drag in a fluid, without known and/or uniform applied shear rates. *See also* Viscosity.

Acid Number   *See* Total Acid Number.

ACN   Alkane Carbon Number, *see* Equivalent Alkane Carbon Number.

Adsorption   The increase in quantity of a component at an interface. In most usage it is positive, but it can be negative. Adsorption may also denote the process of components accumulating at an interface.

Aggregation   The process of forming a group of droplets that are held together in some way. For emulsions, this process is sometimes referred to as coagulation or flocculation.

Aging   The properties of emulsions, and of crude oils, may change with time in storage. Aging in crude oils may refer to changes in composition due to oxidation, precipitation of components, bacterial action, or evaporation of components that have low boiling points. Aging in emulsions may refer to any of aggregation, coalescence, creaming, or chemical changes. Aged emulsions frequently have larger droplet sizes.

Alkane Carbon Number (ACN)   *See* Equivalent Alkane Carbon Number.

Amphoteric Surfactant   A surfactant molecule for which the ionic character of the polar group depends on solution pH. For example, lauramidopropylbetaine, $C_{11}H_{23}CONH(CH_2)_3N^+(CH_3)_2CH_2COO^-$, is positively charged at low pH but is electrically neutral, having both positive and negative charges at intermediate pH. Other combinations are possible, and some amphoteric surfactants are negatively charged at high pH. *See also* Zwitterionic Surfactant.

Anionic Surfactant A surfactant molecule whose polar group is negatively charged. Example: sodium dodecyl sulfate, $CH_3(CH_2)_{11}SO_4^-Na^+$.

Anisokinetic Sampling *See* Isokinetic Sampling.

API Gravity A measure of the relative density (specific gravity) of petroleum liquids. The API gravity, in degrees, is given by

$$API = (141.5/relative\ density) - 131.5$$

where the relative density at temperature T (°C) equals the density at T divided by the density of water at 15.6 °C.

Apparent Viscosity Viscosity determined for a non-Newtonian fluid without reference to a particular shear rate for which it applies. Such viscosities are usually determined by a method strictly applicable to Newtonian fluids only.

Asphalt A naturally occurring hydrocarbon that is a solid at reservoir temperatures. An asphalt residue may also be prepared from heavy (asphaltic) crude oils or bitumen, from which lower boiling fractions have been removed.

Asphaltene A polyaromatic component of some crude oils that has a high molecular mass and also high sulfur, nitrogen, oxygen, and metal contents. In practical work asphaltenes are usually defined operationally by using a standardized separation scheme. One such scheme defines asphaltenes as those components of a crude oil or bitumen that are soluble in toluene but insoluble in *n*-pentane.

Basic Sediment and Water That portion of solids and aqueous solution in an emulsion that separates on standing, or is separated by centrifuging, in a standardized test method. Basic sediment may contain emulsified oil as well. Also referred to as BS&W, BSW, bottom settlings and water, or bottom solids and water.

Batch Treating In oil production or processing, the process in which emulsion is collected in a tank and then broken in a batch. This method is as opposed to continuous, or flow-line, treating of emulsions.

Bicontinuous Microemulsion A possible structure for middle-phase microemulsions is one in which both oil and water phases are continuous throughout the microemulsion phase. An analogy is the structure of porous rock, in which both the mineral phase and the pore–throat channels can be continuous at the same time. *See also* Middle-Phase Microemulsion.

Bitumen   A naturally occurring viscous hydrocarbon having a viscosity greater than 10,000 mPa·s at ambient deposit temperature, and a density greater than 1000 kg/m^3 at 15.6 °C. In addition to hydrocarbons of high molecular mass, bitumen contains appreciable quantities of sulfur, nitrogen, oxygen, and heavy metals.

Bottle Test   An empirical test in which varying amounts of a potential demulsifier are added into a series of tubes or bottles containing subsamples of an emulsion to be broken. After some specified time, the extent of phase separation and appearance of the interface separating the phases are noted. There are many variations of this test. In addition to the demulsifier, a diluent may be added to reduce viscosity. In the centrifuge test, centrifugal force may be added to speed up the phase separation. There are also many variations of the centrifuge test.

Bottom Settlings and Water   *See* Basic Sediment and Water.

Breaking   The process in which an emulsion separates, the formerly dispersed phase becoming a continuous phase, separate from the original continuous phase.

BS&W   *See* Basic Sediment and Water.

Bulk Phase   Usually refers to a dispersion as a whole. For example, in an emulsion the term "bulk phase viscosity" refers to the emulsion viscosity, as opposed to the continuous-phase viscosity. Thus the bulk phase is *not* a separate, single phase at all and may contain dispersed solid and liquid phases.

Capillary Forces   The interfacial forces acting among oil, water, and solid in a porous medium. These determine the pressure difference (capillary pressure) across an oil–water interface in a pore. Capillary forces are largely responsible for oil entrapment under typical reservoir conditions.

Capillary Number (N_c)   A dimensionless ratio of viscous to capillary forces. One form gives N_c as velocity times viscosity divided by interfacial tension. It is used to indicate how strongly trapped residual oil is in a porous medium.

Capillary Pressure   The local pressure difference across the oil–water interface in a pore contained in a porous medium. One of the liquids usually preferentially wets the solid; thus the capillary pressure is normally taken as the pressure in the nonwetting fluid minus that in the wetting fluid.

Cationic Surfactant   A surfactant molecule whose polar group is positively charged. Example: cetyltrimethylammonium bromide, $CH_3(CH_2)_{15}$-$N^+(CH_3)_3Br^-$.

CCC   *See* Critical Coagulation Concentration.

Centrifugal Separator   *See* Separator.

Centrifuge Test   *See* Bottle Test.

Chocolate Mousse Emulsion   A name frequently used to refer to the water-in-oil emulsions having a high water content that are formed when crude oils are spilled on the oceans. The name reflects the color and very viscous consistency of these emulsions. It has also been applied to other petroleum emulsions of similar appearance.

CMC   *See* Critical Micelle Concentration.

Coagulation   *See* Aggregation.

Coalescence   The merging of two or more droplets into a single droplet. In an emulsion coalescence reduces the total number of droplets and also the total interfacial area.

Coefficient of Viscosity   *See* Viscosity

Colloidal   A state of subdivision in which the particles, droplets, or bubbles dispersed in another phase have at least one dimension (e.g., diameter) between ~1 and 1000 nm. This definition includes colloids as a category of dispersions.

Condensate   Any light-hydrocarbon liquid mixture obtained from the condensation of hydrocarbon gases. Condensate typically contains mostly propane, butane, and pentane.

Contact Angle   When two immiscible fluids (e.g., liquid–gas or oil–water) are both in contact with a solid, the angle formed between the solid surface and the surface of the more dense fluid phase is termed the contact angle. By convention, if one of the fluids is water, then the contact angle is the angle measured through the water phase.

Continuous Phase   In an emulsion, a liquid phase in which are dispersed droplets of an immiscible liquid of a different composition. Also called external phase.

Continuous Treating   *See* Flow-Line Treating.

Cosurfactant   Any chemical, whether surface active by itself or not, that may be added to a system to enhance the effectiveness of a surfactant.

Creaming   The process of emulsion droplets floating upwards under gravity or a centrifugal field to form a concentrated emulsion (cream) quite distinct from the underlying dilute emulsion. Creaming is not the same as the breaking of an emulsion. *See also* Sedimentation.

Critical Coagulation Concentration (CCC)   The electrolyte concentration that marks the onset of coagulation. The CCC is very system-specific, although the variation in CCC with electrolyte composition has been empirically generalized. *See also* Schulze–Hardy Rule.

Critical Micelle Concentration (CMC)   The surfactant concentration above which micelles begin to be formed. In practice a narrow range of surfactant concentrations represents the transition from a solution in which only single, unassociated surfactant molecules (monomers) are present to a solution containing micelles.

Crude Oil   A naturally occurring hydrocarbon produced from an underground reservoir. *See also* Asphalt, Bitumen, Extra-Heavy Crude Oil, Heavy Crude Oil, Light Crude Oil, Oil.

Cuff Layer Emulsion   *See* Interface Emulsion.

Darcy's Law   *See* Permeability.

Demulsifier   1. Chemical: Any agent added to an emulsion that causes or enhances the rate of breaking of the emulsion (separation into its constituent liquid phases). Demulsifiers may act by any of a number of different mechanisms.
2. Device: Any device that is used to break emulsions. Such devices may employ chemical, electrical, or mechanical means, or a combination, to break an emulsion and cause separation into its constituent liquid phases.

Desalter   An oil-field or refinery apparatus used to separate water and associated dissolved salts from crude oil.

Detergent *See* Surfactant.

Differential Viscosity The rate of change of shear stress with respect to shear rate, taken at a specific shear rate ($\eta_D = d\tau/d\dot\gamma$).

Diffuse Layer *See* Electric Double Layer.

Dilatant A fluid for which viscosity increases as the shear rate increases. Also termed shear thickening.

Diluent A low-boiling-point petroleum fraction, such as naphtha, that is added to a more viscous high-boiling-point petroleum liquid or oil-continuous emulsion. The diluent is usually added to reduce viscosity.

Dispersed Phase In an emulsion, the droplets that are dispersed or suspended in an immiscible liquid of a different composition. Also called internal phase.

Dispersion 1. In colloids, a system in which finely divided droplets, particles, or bubbles are distributed in another phase. As it is usually used, dispersion implies a distribution without dissolution. An emulsion is an example of a colloidal dispersion (*see also* Colloidal).
2. In oil recovery from a reservoir, the mixing by convection of fluids flowing in a porous medium.

DLVO Theory An acronym for a theory of the stability of colloidal dispersions developed independently by B. Derjaguin and L. D. Landau in one laboratory and by E. J. W. Verwey and J. Th. G. Overbeek in another. The theory was developed to account for the stability against aggregation of electrostatically charged particles in a dispersion.

Double Layer *See* Electric Double Layer.

Drilling Fluid The circulating fluid used when drilling a well. The drilling fluid lubricates the drill bit, forces cuttings out of the wellbore up to the surface, and may also prevent blowouts. Drilling fluids are usually suspensions, but emulsions and foams may also be used. The terms "drilling fluid" and "drilling mud" are used interchangeably.

EACN *See* Equivalent Alkane Carbon Number.

Electric Double Layer An idealized description of the distribution of free charges in the neighborhood of an interface. Typically a particle or droplet surface is viewed as having a fixed charge of one sign (one layer), while oppositely charged ions are distributed diffusely in the adjacent liquid (the second layer). The second layer may be considered to be made up of a relatively less mobile Stern layer in close proximity to the surface, and a relatively more diffuse layer at greater distance.

Electrophoresis   The motion of colloidal species caused by an imposed electric field. The species move with an electrophoretic velocity that depends on their electric charge and the electric field gradient. The electrophoretic mobility is the electrophoretic velocity per unit electric field gradient and is used to characterize specific systems.

Electrophoretic Mobility   *See* Electrophoresis.

Electrostatic Treater   A vessel used to break emulsions by promoting coalescence through the application of an electric field. *See also* Treater.

Emulsifier   Any agent that acts to stabilize an emulsion. The emulsifier may make it easier to form an emulsion, provide stability against aggregation, or provide stability against coalescence. Emulsifiers are frequently but not necessarily surfactants.

Emulsion   A dispersion of droplets of one liquid in another, immiscible liquid, in which the droplets are of colloidal or near-colloidal sizes.

Emulsion Drilling Fluid   *See* Drilling Fluid.

Emulsion Test   In general, emulsion tests range from simple identifications of emulsion presence and volume to detailed component analyses. The term "emulsion test" frequently refers simply to the determination of sediments in an emulsion or oil sample. *See also* Basic Sediment and Water.

Emulsion Treater   *See* Treater.

Enhanced Oil Recovery   The third phase of crude-oil production, in which chemical, miscible fluid, or thermal methods are applied to restore production from a depleted reservoir. Also known as tertiary oil recovery. *See also* Primary Oil Recovery, Secondary Oil Recovery.

Equivalent Alkane Carbon Number (EACN)   Each surfactant or surfactant mixture in a reference series will produce a minimum interfacial tension (IFT) for a different n-alkane. For any crude oil or oil component, a minimum IFT will be observed against one of the reference surfactants. The EACN for the crude oil refers to the n-alkane that would yield minimum

IFT against that reference surfactant. The EACN thus allows predictions to be made about the interfacial tension behavior of a crude oil in the presence of surfactant. *See* references 8 and 9.

External Phase   The continuous phase of an emulsion, in which droplets of a second phase are dispersed.

Extra-Heavy Crude Oil   A naturally occurring hydrocarbon having a viscosity less than 10,000 mPa·s at ambient deposit temperature and a density greater than 1000 kg/m^3 at 15.6 °C.

Flocculation   *See* Aggregation.

Flotation   *See* Sedimentation.

Flow-Line Treating   In oil production or processing, the process in which emulsion is continuously broken and separated into oil and water bulk phases. Also called continuous treating. This method is as opposed to batch treating of emulsions. *See also* Batch Treating, Treater.

Foam   A dispersion of gas bubbles, in a liquid or solid, in which at least one dimension falls within the colloidal size range. Thus a foam typically contains either very small bubble sizes or, more commonly, quite large gas bubbles separated by thin liquid films.

Free Water   The readily separated, nonemulsified, water that is coproduced with oil from a production well.

Free-Water Knockout (FWKO)   A vessel designed to separate the readily separated (nonemulsified or "free") water from oil or an oil-containing emulsion. Further water and solids removal may be accomplished in a treater.

FWKO   *See* Free-Water Knockout.

Gas Emulsion   A term used to describe crude oil that contains a small volume fraction of dispersed gas.

Gravity Separator   *See* Separator.

Gun Barrel   A type of settling vessel used to separate water and oil from an emulsion. Typically, heated emulsion is treated with demulsifier and introduced into the gun barrel, where water settles out and is drawn off. Any produced gas is also drawn off.

Hamaker Constant   In a description of the London–van der Waals attractive energy between two dispersed bodies, such as droplets, the Hamaker constant is a proportionality constant characteristic of the droplet composition. It depends on the internal atomic packing and polarizability of the droplets.

Heater Treater   *See* Treater.

Heavy Crude Oil   A naturally occurring hydrocarbon having a viscosity less than 10,000 mPa·s at ambient deposit temperature and a density between 934 and 1000 kg/m^3 at 15.6 °C.

Heterodisperse   A colloidal dispersion in which the dispersed species (droplets, particles, etc.) do not all have the same size. Subcategories are paucidisperse (few sizes) and polydisperse (many sizes). *See also* Monodisperse.

Hydrophile–Lipophile Balance (HLB) Scale   An empirical scale categorizing surfactants in terms of their tendencies to be mostly oil soluble or water soluble, hence their tendencies to promote W/O or O/W emulsions, respectively.

Hydrophilic   A qualitative term referring to the water-preferring nature of a species (atom, molecule, droplet, particle, etc.). In emulsions, hydrophilic usually means that a species prefers the aqueous phase over the oil phase. In this sense hydrophilic has the same meaning as oleophobic.

Hydrophobic   A qualitative term referring to the water-avoiding nature of a species (atom, molecule, droplet, particle, etc.). In emulsions, hydrophobic usually means that a species prefers the oil phase over the aqueous phase. In this sense hydrophobic has the same meaning as oleophilic.

Impingement Separator   *See* Separator.

Interface   The boundary between two phases, sometimes including a thin layer at the boundary within which the properties of one bulk phase change over to become the properties of the other bulk phase.

Interface Emulsion   An emulsion occurring between oil and water phases in a process separation or treatment apparatus. Such emulsions may have a high solids content and are frequently very viscous. In this case the term interface is used in a macroscopic sense and refers to a bulk phase separated by two other bulk phases of higher and lower density. Other terms are cuff layer, pad layer, or rag layer emulsions.

Interfacial Film   A thin layer of material positioned between two immiscible phases, usually liquids, in which the composition of the layer is different from either of the bulk phases.

Interfacial Tension   *See* Surface Tension

Interfacial Viscosity   The two-dimensional analog of viscosity acting along the interface between two immiscible fluids. It is also called surface viscosity, especially where one fluid is a gas. *See also* Viscosity.

Internal Phase   The phase in an emulsion that is dispersed into droplets; the dispersed phase.

Intrinsic Viscosity   For emulsions, the limit, as the dispersed-phase concentration approaches infinite dilution, of the specific increase in viscosity, at low shear rate, divided by the dispersed phase concentration ($[\eta] = \lim_{c \to 0} \lim_{\dot{\gamma} \to 0} \eta_{SP}/C$).

Inversion   The process by which one type of emulsion is converted to another, as when an O/W emulsion is transformed into a W/O emulsion, and vice versa.

Invert-Oil Mud   An emulsion drilling fluid (mud) of the water-dispersed-in-oil (W/O) type, and having a high water content. *See also* Drilling Fluid, Oil Mud, Oil-Base Mud.

Isoelectric Point   The solution pH for which the electrokinetic, or zeta, potential is zero. *See also* Point of Zero Charge.

Isokinetic Sampling   Collecting samples of a flowing dispersion using a method in which the sampling velocity (in the sampling probe) is equal to the upstream local velocity. If these velocities are not the same (anisokinetic sampling) then fluid streamlines ahead of the probe will be distorted; collection of particles or droplets will be influenced by their inertia, which varies with particle size; and sampling will not be representative.

Kinematic Viscosity   The absolute viscosity of a fluid divided by the density.

Laminar Flow   A condition of flow in which all elements of a fluid passing a certain point follow the same path, or streamline. Also referred to as streamline flow.

Light Crude Oil   A naturally occurring hydrocarbon having a viscosity less than 10,000 mPa·s at ambient deposit temperature and a density less than 934 kg/m^3 at 15.6 °C.

Loose Emulsion   A petroleum industry term for a relatively unstable, easy-to-break emulsion, as opposed to a more stable, difficult-to-treat emulsion. *See also* Tight Emulsion.

Lower-Phase Microemulsion   A microemulsion, with a high water content, that is stable while in contact with a bulk oil phase, and in laboratory tube or bottle tests tends to be situated at the bottom of the tube, underneath the oil phase. *See also* Microemulsion.

Lyophilic   General term referring to the continuous-medium-preferring nature of a species. *See also* Hydrophilic.

Lyophobic   General term referring to the continuous-medium-avoiding nature of a species. *See also* Hydrophobic.

Macroemulsion   In enhanced oil recovery terminology, the term macroemulsion is sometimes employed to identify emulsions having droplet sizes greater than some specified value, and sometimes simply to distinguish an emulsion from the microemulsion or micellar emulsion types. *See also* Emulsion.

Marangoni Elasticity   In an emulsion, a thin film of the continuous phase is created between two droplets closely approaching each other. Any stretching in this film causes a local decrease in the interfacial concentration of adsorbed surfactant. This decreased concentration causes the local interfacial tension to increase, which in turn acts in opposition to the original stretching force. With time the original interfacial concentration of surfactant is restored. The time-dependent restoring force is referred to as Marangoni elasticity, or the Marangoni effect.

Metastable Emulsion   A system in which the droplets do not participate in aggregation, coalescence, or creaming at a significant rate.

Micellar Emulsion   An emulsion that forms spontaneously and has extremely small droplet sizes (<10 nm). Such emulsions are thermodynamically stable and are sometimes referred to as microemulsions (q.v.).

Micelle   An aggregate of surfactant molecules in solution. Such aggregates form spontaneously at sufficiently high surfactant concentration, the critical micelle concentration. The micelles are of colloidal dimensions.

Microemulsion   A special kind of stabilized emulsion in which the dispersed droplets are extremely small (<100 nm) and the emulsion is thermodynamically stable. These emulsions are transparent and may form spontaneously. In some usage a lower size limit of about 10 nm is implied in addition to the upper limit. *See also* Micellar Emulsion.

Middle-Phase Microemulsion A microemulsion, with high oil and water contents, that is stable while in contact with either bulk-oil or bulk-water phases. This stability may be due to a bicontinuous structure in which both oil and water phases are continuous at the same time. In laboratory tube or bottle tests involving samples containing unemulsified oil and water, a middle-phase microemulsion will tend to be situated between the two former phases. *See also* Bicontinuous Microemulsion.

Miniemulsion The term miniemulsion is sometimes used to distinguish an emulsion from the microemulsion or micellar emulsion types. Thus a miniemulsion would contain droplet sizes larger than 100 nm and smaller than 1000 nm, or some other specified upper size limit. *See also* Emulsion.

Monodisperse A colloidal dispersion in which all the dispersed species (droplets, particles, etc.) are the same size. Otherwise, the system is heterodisperse (paucidisperse or polydisperse).

Mousse Emulsion *See* Chocolate Mousse Emulsion.

Multiple Emulsion An emulsion in which the dispersed droplets themselves contain even more finely dispersed droplets of a separate phase. Thus, there may occur oil-dispersed-in-water-dispersed-in-oil (O/W/O) and water-dispersed-in-oil-dispersed-in-water (W/O/W) multiple emulsions. More complicated multiple emulsions are also possible.

Naphtha A petroleum fraction that is operationally defined in terms of the distillation process by which it is separated. A given naphtha is thus defined by a specific range of boiling points of its components. Naphtha is sometimes used as a diluent for W/O emulsions.

Nelson Type Emulsions The different types of phase behavior in microemulsions are denoted as Nelson type II($-$), II($+$), and III. These refer to equilibrium phase behaviors and distinguish, for example, the number of phases that may be in equilibrium and the nature of the continuous phase. *See also* reference 10. Winsor type emulsions are similarly identified, but with different type numbers.

Newtonian Fluid or Emulsion A fluid or emulsion whose rheological behavior is described by Newton's law of viscosity. Here shear stress is set proportional to the shear rate. The proportionality constant is the coefficient of viscosity, or simply, viscosity. The viscosity of a Newtonian fluid is a constant for all shear rates.

Nonionic Surfactant   A surfactant molecule whose polar group is not electrically charged. Example: polyoxyethylene alcohol, C_nH_{2n+1}-$(OCH_2CH_2)_mOH$.

Non-Newtonian Fluid or Emulsion   A fluid whose viscosity varies with applied shear rate (flow rate). *See also* Newtonian Fluid.

Oil   Liquid petroleum (sometimes including dissolved gas) that is produced from a well. In this sense oil is equivalent to crude oil. The term oil is, however, frequently more broadly used and may include, for example, synthetic hydrocarbon liquids, bitumen from oil (tar) sands, and fractions obtained from crude oil.

Oil-Base Mud   An emulsion drilling fluid (mud) of the water-dispersed-in-oil (W/O) type, and having a low water content. *See also* Drilling Fluid, Invert-Oil Mud, Oil Mud.

Oil Color   A qualitative test for the presence of emulsified water in an oil. Emulsified water droplets will tend to impart a hazy appearance to the oil.

Oil-Emulsion Mud   An emulsion drilling fluid (mud) of the oil-dispersed-in-water (O/W) type. *See also* Drilling Fluid, Oil Mud.

Oil Mud   An emulsion drilling fluid (mud) of the water-dispersed-in-oil (W/O) type. A mud having a low water content is referred to as an oil-base mud, and a mud having a high water content is referred to as an invert-oil mud. *See also* Drilling Fluid, Oil-Emulsion Mud.

Oil Sand   A sandstone deposit that contains bitumen. Also referred to as Tar Sand.

Oleophilic   Refers to the oil-preferring nature of a species. *See also* Hydrophobic.

Oleophobic   Refers to the oil-avoiding nature of a species. *See also* Hydrophilic.

Optimum Salinity   In microemulsions, the salinity for which the mixing of oil with a surfactant solution produces a middle-phase microemulsion containing an oil-to-water ratio of 1. In micellar enhanced oil recovery processes, extremely low interfacial tensions result, and oil recovery tends to be maximized when this condition is satisfied.

O/W   Abbreviation for an oil-in-water emulsion, that is, oil droplets dispersed in water.

O/W/O   Abbreviation for an oil-dispersed-in-water-dispersed-in-oil multiple emulsion. Here the water droplets have oil droplets dispersed within them, and the water droplets themselves are dispersed in oil forming the continuous phase.

Pad Layer Emulsion   *See* Interface Emulsion.

Paucidisperse   A colloidal dispersion in which the dispersed species (droplets, particles, etc.) have a few different sizes. Paucidisperse is a category of heterodisperse systems. *See also* Monodisperse.

Permeability   A measure of the ease with which a fluid can flow (fluid conductivity) through a porous medium. Permeability is defined by Darcy's law. For linear, horizontal, isothermal flow, permeability is the constant of proportionality between flow rate times viscosity and the product of cross-sectional area of the medium and pressure gradient along the medium.

Petroleum   A general term that may refer to any hydrocarbons or hydrocarbon mixtures, usually liquid, but sometimes solid or gaseous.

Phase Inversion Temperature (PIT)   Temperature at which the hydrophilic and oleophilic natures of a surfactant are in balance. As temperature is increased through the PIT, a surfactant will change from promoting one kind of emulsion, such as O/W, to another, such as W/O.

Pickering Emulsion   An emulsion stabilized by fine particles. The particles form a close-packed structure at the oil–water interface, with significant mechanical strength, which provides a barrier to coalescence.

PIT   *See* Phase Inversion Temperature.

Point of Zero Charge   The solution pH at which an interface is electrically neutral. This pH is not always the same as the isoelectric point, which refers to zero charge at the shear plane that exists a small distance away from the interface.

Polydisperse   A colloidal dispersion in which the dispersed species (droplets, particles, etc.) have a wide range of sizes. Polydisperse is a category of heterodisperse systems. *See also* Monodisperse.

Porosity   The ratio of the volume of all void spaces to total volume in a porous medium.

Porous Medium   A solid containing voids, or pore spaces. Normally such pores are quite small compared to the size of the solid and are well distributed throughout the solid.

Pour Point   The lowest temperature at which an emulsion, or other petroleum material, will flow under a standardized set of test conditions.

Power Law Fluid or Emulsion   A fluid or emulsion whose rheological behavior is reasonably well-described by the power law equation. Here shear stress is set proportional to the shear rate raised to an exponent n, where n is the power law index. The fluid is pseudoplastic for $n < 1$, Newtonian for $n = 1$, and dilatant for $n > 1$.

Primary Oil Recovery   The first phase of crude-oil production, in which oil flows naturally to the well bore. *See also* Enhanced Oil Recovery, Secondary Oil Recovery.

Protection   The process in which a material adsorbs onto droplet surfaces and thereby makes an emulsion less sensitive to aggregation and/or coalescence, by any of a number of mechanisms. *See also* Sensitization.

Pseudoplastic   A fluid whose viscosity decreases as the applied shear rate increases. Also termed shear thinning.

Rag Layer Emulsion   *See* Interface Emulsion.

Reduced Viscosity   For emulsions, the specific increase in viscosity divided by the dispersed-phase concentration ($\eta_{Red} = \eta_{SP}/C$).

Relative Viscosity   In emulsions, the viscosity of the emulsion divided by the viscosity of the continuous phase ($\eta_{Rel} = \eta/\eta_0$).

Replica   A metal film duplicate of a sample used in scanning electron microscopy. For example, an emulsion sample may be fast-frozen in a cryogen, fractured to reveal interior structure, then coated with a metal film to preserve the structure.

Reverse Emulsion   A petroleum industry term used to denote an oil-in-water emulsion (most wellhead emulsions are W/O).

Rheology   Strictly, the science of deformation and flow of matter. Rheological descriptions usually refer to the property of viscosity and departures from Newton's law of viscosity.

Rheomalaxis A special case of time-dependent rheological behavior in which shear rate changes cause irreversible changes in viscosity. The change can be negative, as when structural linkages are broken, or positive, as when structural elements become entangled (like work-hardening).

Rheopexy Dilatant flow that is time dependent. At constant applied shear rate, viscosity increases; in a flow curve, hysteresis occurs (but opposite to the thixotropic case).

Salinity Requirement *See* Optimum Salinity.

Schulze–Hardy Rule An empirical rule summarizing the general tendency of the critical coagulation concentration (CCC) of an emulsion or other dispersion to vary inversely with the sixth power of the counterion charge number of added electrolyte. *See also* Critical Coagulation Concentration.

Secondary Oil Recovery The second phase of crude-oil production, in which water or an immiscible gas are injected to restore production from a depleted reservoir. *See also* Enhanced Oil Recovery, Primary Oil Recovery.

Sedimentation The settling of suspended particles or droplets due to gravity or an applied centrifugal field. Negative sedimentation, when droplets rise upwards, is also called flotation and is a part of the creaming process.

Sensitization The process whereby small amounts of added colloidal material make an emulsion more sensitive to coagulation by electrolyte. Additions of larger amounts of the same material usually make the emulsion less sensitive to coagulation, and this feature is termed protection.

Separator A vessel designed to separate the oil phase in a petroleum fluid from some or all of the other three constituent phases: gas, solids, and water. Free-water knockouts fall under this category, but so do separators capable of breaking and removing water and solids from emulsions. Separators range from gravity to impingement (coalescence) to centrifugal types.

Shear Plane Any species undergoing electrophoretic motion moves with a certain immobile part of the electric double layer that is commonly assumed to be distinguished from the mobile part by a sharp plane, the shear plane. The zeta potential is the potential at the shear plane.

Shear Rate The rate of deformation of a liquid when subjected to a mechanical stress (shear stress), shear rate is a measure of the relative motion between fluid particles. In straight parallel flow, the shear rate is the velocity gradient perpendicular to the direction of flow.

Shear Stress A certain applied force per unit area is needed to produce deformation in a fluid. For a plane area around some point in the fluid, and in the limit of decreasing area, the component of deforming force per unit area that acts parallel to the plane is the shear stress.

Shear Thickening *See* Dilatant.

Shear Thinning *See* Pseudoplastic.

Soap *See* Surfactant.

Specific Increase in Viscosity The relative viscosity minus unity ($\eta_{SP} = \eta_{Rel} - 1$).

Stability Stability refers to the fact that a system may be in one of a number of states of thermodynamic potential energy. The state with the lowest potential energy is referred to as the stable state. A system may also be in a metastable state. A colloidally metastable emulsion is a system in which the droplets do not participate in aggregation, coalescence, or creaming at a significant rate.

Stern Layer A part of the electric double layer that lies between the surface and a hypothetical boundary—the Stern surface. Ions within the Stern layer are considered to be adsorbed. *See also* Electric Double Layer.

Streamline Flow *See* Laminar Flow.

Surface *See* Interface.

Surface Area The area of a surface or interface, especially that between a dispersed and a continuous phase. The specific surface area is the total surface area divided by the mass of the appropriate phase.

Surface Tension The contracting force per unit length around the perimeter of a surface. This force is usually referred to as surface tension if the surface separates gas from liquid or solid phases, and interfacial tension if the surface separates two nongaseous phases. Although not strictly defined the same way, surface tension can be expressed in units of energy per unit surface area. For practical purposes surface tension is frequently taken to reflect the change in surface free energy per unit increase in surface area.

Surfactant   Any substance that lowers the surface or interfacial tension of the medium in which it is dissolved. Soaps (fatty acid salts) are surfactants. Detergents are surfactants or surfactant mixtures whose solutions have cleaning properties.

Suspension   A system containing solid particles dispersed in a liquid.

TAN   *See* Total Acid Number.

Tar Sand   *See* Oil Sand.

Tertiary Oil Recovery   *See* Enhanced Oil Recovery.

Thixotropic   Pseudoplastic flow that is time dependent. At constant applied shear rate, viscosity gradually decreases; in a flow curve, hysteresis occurs. That is, after a given shear rate is applied and then reduced, it takes some time for the original dispersed species' alignments to be restored.

Three-Phase Separator   *See* Separator.

Tight Emulsion   A petroleum industry term for a practically stable emulsion, in contrast to a less stable, or "loose" emulsion (q.v.).

Total Acid Number (TAN)   A property of crude oil. The acid number expresses the amount of base (potassium hydroxide) that will react with a given amount of crude oil in a standardized titration procedure. A large acid number indicates a high concentration of acids in the oil, usually including natural surfactant precursors.

Treater   A vessel used for the breaking of emulsions and the consequent removal of solids and water (BS&W). Emulsion breaking may be accomplished through some combination of thermal, electrical, chemical, or mechanical methods. A treater might be applied to break an emulsion and separate solids and water that could not be removed in a separator.

Turbidity   A coefficient related to the scattering of light when it passes through a dispersion (including emulsions). Turbidity is a function of the size and concentration of droplets or particles.

Turbulent Flow   A condition of flow in which all components of a fluid passing a certain point do not follow the same path. Turbulent flow refers to flow that is not laminar.

Ultrasound Vibration Potential (UVP)   The electrokinetic potential of colloidal species detected by the electric field generated when the species are made to move by an imposed ultrasonic field. This method can be applied to W/O emulsions that do not transmit light. For low potentials, the UVP can be quantitatively related to the electrophoretic mobility.

Upper-Phase Microemulsion   A microemulsion, with a high oil content, that is stable while in contact with a bulk-water phase, and in laboratory tube or bottle tests tends to be situated at the top of the tube, above the water phase. *See also* Microemulsion.

UVP   *See* Ultrasound Vibration Potential.

Viscoelastic   A liquid (or solid) with both viscous and elastic properties. A viscoelastic liquid will deform and flow under the influence of an applied shear stress, but when the stress is removed the liquid will slowly recover from some of the deformation.

Viscosity   A measure of the resistance of a liquid to flow. It is properly the coefficient of viscosity, and expresses the proportionality between shear stress and shear rate in Newton's law of viscosity.

Wet Oil   An oil containing free water and/or emulsified water.

Wettability   A qualitative term referring to the water- or oil-preferring nature of surfaces, such as the mineral surfaces in a porous medium (rock) of an oil-bearing reservoir. The flow of emulsions in porous media is influenced by the wetting state of the walls of pores and throats through which the emulsion must travel. *See also* Contact Angle.

Winsor Type Emulsions   The three types of microemulsions denoted as Winsor types I, II, and III refer to equilibrium phase behaviors and distinguish, for example, the number of phases that may be in equilibrium and the nature of the continuous phase. *See* reference 11. Nelson type emulsions (q.v.) are similarly identified, but with different type numbers.

W/O   Abbreviation for a water-in-oil emulsion, that is, water droplets dispersed in oil.

W/O/W   Abbreviation for a water-dispersed-in-oil-dispersed-in-water multiple emulsion. Here the oil droplets have water droplets dispersed within them, and the oil droplets themselves are dispersed in water forming the continuous phase.

Yield Stress   For some fluids, the shear rate (flow) remains at zero until a threshold shear stress, termed the yield stress, is reached. Beyond the yield stress flow begins.

Zero Point of Charge   *See* Point of Zero Charge.

Zeta Potential   Properly called the electrokinetic potential, the zeta potential refers to the potential drop across the mobile part of the electric double layer. Any species undergoing electrophoretic motion moves with a certain immobile part of the electric double layer that is assumed to be distinguished from the mobile part by a sharp plane, the shear plane. The zeta potential is the potential at that plane.

Zwitterionic Surfactant   A surfactant molecule whose polar group contains both negatively and positively charged groups. Example: lauramidopropylbetaine, $C_{11}H_{23}CONH(CH_2)_3N^+(CH_3)_2CH_2COO^-$ at neutral and alkaline solution pH. *See also* Amphoteric surfactant.

References

1. Martinez, A. R. In *The Future of Heavy Crude and Tar Sands;* Meyer, R. F.; Wynn, J. C.; Olson, J. C., Eds.; UNITAR: New York, 1982; pp ixvii–ixviii.
2. Danyluk, M.; Galbraith, B.; Omana, R. In *The Future of Heavy Crude and Tar Sands;* Meyer, R. F.; Wynn, J. C.; Olson, J. C., Eds.; UNITAR: New York, 1982; pp 3–6.
3. Khayan, M. In *The Future of Heavy Crude and Tar Sands;* Meyer, R. F.; Wynn, J. C.; Olson, J. C., Eds.; UNITAR: New York, 1982; pp 7–11.
4. *Manual of Symbols and Terminology for Physicochemical Quantities and Units, Appendix II;* Prepared by IUPAC Commission on Colloid and Surface Chemistry; Butterworths: London, 1972.
5. Williams, H. R.; Meyers, C. J. *Oil and Gas Terms*, 6th ed.; Matthew Bender: New York, 1984.
6. *A Dictionary of Petroleum Terms*, 2nd ed.; Petroleum Extension Service, The University of Texas at Austin: Austin, TX, 1979.
7. *The Illustrated Petroleum Reference Dictionary*, 2nd ed.; Langenkamp, R. D.; PennWell Books: Tulsa, OK, 1982.
8. Cash, L.; Cayias, J. L.; Fournier, G.; Macallister, D.; Schares, T.; Schechter, R. S.; Wade, W. H. *J. Colloid Interface Sci.* **1977,** 59, 39–44.
9. Cayias, J. L.; Schechter, R. S.; Wade, W. H. *Soc. Pet. Eng. J.* **1976,** *December,* 351–357.
10. Nelson, R. C. *Chem. Eng. Prog.* **1989,** *March,* 50–57.
11. Winsor, P. A. *Solvent Properties of Amphiphilic Compounds;* Butterworths: London, 1954.

AUTHOR INDEX

AFFILIATION INDEX

SUBJECT INDEX

Copy editing and indexing: Janet S. Dodd
Production: Peggy D. Smith
Acquisition: Cheryl Shanks

Typeset by AGT/Unicorn Graphics, Washington, DC
Books printed and bound by Maple Press, York, PA